AF788247

Computational Fluid Dynamics Simulations

Edited by Guozhao Ji and Jiujiang Zhu

Published in London, United Kingdom

IntechOpen

Supporting open minds since 2005

Computational Fluid Dynamics Simulations
http://dx.doi.org/10.5772/intechopen.83278
Edited by Guozhao Ji and Jiujiang Zhu

Contributors
Like Li, David Korba, José Luis Velázquez Ortega, Alejandro Rincón-Casado, Magdalena Hajdukiewicz, F. Sánchez de la Flor, Enrique Angel Rodriguez Jara, Masayuki Ochiai, Fuma Sakai, Hiromu Hashimoto, Syamsul Rizal Abd Shukor, Siti Nor Azreen Ahmad Termizi, Edwin Lenin Chica Arrieta, Ainhoa Rubio Clemente, Valeriu Vilag, Jeni Vilag, Razvan Carlanescu, Andreea Mangra, Florin Gabriel Florean, Raúl Armando Olvera-Corral, Elva Marcela Coria-Quiñones, Manuel Alberto Flores-Hidalgo, Miriam López-Guzmán, Carlos Omar Omar Ríos-Orozco, Sandra Iliana Torres-Herrera, Abel Hurtado-Macías, Adolfo Ruiz-Soto, Diana Barraza-Jiménez, Pablo D. Druetta, Olayinka Omowunmi Omowunmi Adewumi, Mubashir O. Quadri, Matthew N. Ottah, Ayowole A. Oyediran, Adil Loya, Shoaib Arif, Muhammad Arsalan, Siraj Anis, Arsalan Khan, Muhammad Saad Saeed, Abdul Hameed Siddiqui

© The Editor(s) and the Author(s) 2020
The rights of the editor(s) and the author(s) have been asserted in accordance with the Copyright, Designs and Patents Act 1988. All rights to the book as a whole are reserved by INTECHOPEN LIMITED. The book as a whole (compilation) cannot be reproduced, distributed or used for commercial or non-commercial purposes without INTECHOPEN LIMITED's written permission. Enquiries concerning the use of the book should be directed to INTECHOPEN LIMITED rights and permissions department (permissions@intechopen.com).
Violations are liable to prosecution under the governing Copyright Law.

CC BY

Individual chapters of this publication are distributed under the terms of the Creative Commons Attribution 3.0 Unported License which permits commercial use, distribution and reproduction of the individual chapters, provided the original author(s) and source publication are appropriately acknowledged. If so indicated, certain images may not be included under the Creative Commons license. In such cases users will need to obtain permission from the license holder to reproduce the material. More details and guidelines concerning content reuse and adaptation can be found at http://www.intechopen.com/copyright-policy.html.

Notice
Statements and opinions expressed in the chapters are these of the individual contributors and not necessarily those of the editors or publisher. No responsibility is accepted for the accuracy of information contained in the published chapters. The publisher assumes no responsibility for any damage or injury to persons or property arising out of the use of any materials, instructions, methods or ideas contained in the book.

First published in London, United Kingdom, 2020 by IntechOpen
IntechOpen is the global imprint of INTECHOPEN LIMITED, registered in England and Wales, registration number: 11086078, 7th floor, 10 Lower Thames Street, London, EC3R 6AF, United Kingdom
Printed in Croatia

British Library Cataloguing-in-Publication Data
A catalogue record for this book is available from the British Library

Additional hard and PDF copies can be obtained from orders@intechopen.com

Computational Fluid Dynamics Simulations
Edited by Guozhao Ji and Jiujiang Zhu
p. cm.
Print ISBN 978-1-83880-749-8
Online ISBN 978-1-83880-750-4
eBook (PDF) ISBN 978-1-83880-751-1

For EU product safety concerns:
IN TECH d.o.o., Prolaz Marije Krucifikse Kozulić 3, 51000 Rijeka, Croatia, info@intechopen.com or visit our website at intechopen.com.

We are IntechOpen, the world's leading publisher of Open Access books

Built by scientists, for scientists

5,000+
Open access books available

125,000+
International authors and editors

140M+
Downloads

151
Countries delivered to

Our authors are among the
Top 1%
most cited scientists

12.2%
Contributors from top 500 universities

WEB OF SCIENCE™

Selection of our books indexed in the Book Citation Index in Web of Science™ Core Collection (BKCI)

Interested in publishing with us?
Contact book.department@intechopen.com

Numbers displayed above are based on latest data collected.
For more information visit www.intechopen.com

Meet the editors

Dr. Guozhao Ji majored in chemical engineering and received his PhD degree in 2014 at the University of Queensland, Australia; his MS and BS in mechanical engineering at Northeastern University, People's Republic of China, in 2010 and 2008, respectively. He currently works in the School of Environmental Science and Technology at Dalian University of Technology in China as an Associate Professor. His research interests are computational fluid dynamic application in chemical engineering processes, modelling of gas transport in microporous material, gas separation by inorganic membranes, high temperature CO_2 capture, solid waste gasification, and kinetic modelling of thermochemical conversions. He has authored over 50 refereed journal publications and one book chapter.

Prof. Zhu is an experienced mathematician with a bachground in thermodynamics and continuum mechanics. He has 20 years of teaching experience in international universities and more than 30 years of research experience in Singapore, the UK, the US, and China. He has worked successfully in several different research areas, such as solid and fluid mechanics, materials science, image analysis, computational biology, and computational fluid dynamics. He also possesses particular skills for mathematical modeling and numerical simulation and was awarded "One Hundred Excellent Young Scientists" by the Chinese Academy of Sciences in 1997. He is Editor-in-Chief of the journal "Global Journal of Engineering Sciences."

Contents

Preface

Computational fluid dynamic (CFD) simulation is a trans-disciplinary technique across fluid mechanics, mathematical algorithms, and computer science. This technique was based on finite difference methods (FDM) and finite element methods (FEM) by iteratively solving the partial differential governing equations (mass equation, momentum equation, and energy equation), and providing the numerical results. The temporal and spatial solutions from the CFD simulation could accurately reproduce the real flow phenomenon, and all the key information of a flow can be captured for further analysis. This alleviates the trouble of measuring the flow information by experimental tests, and solves numerous practical problems in industry with higher precision and much lower cost. CFD simulation is a computation-intensive process. With the rapid development of high-performance computers (HPC) in recent decades, CFD application has made substantial progress in different fields such as mechanical engineering, chemical engineering, environmental engineering, and thermal engineering.

This book, "Computational Fluid Dynamic Simulations", collects the recent work of leading researchers, and the contents covers a variety of theoretical studies as well as experimental validation. Despite the interdisciplinary nature of the different applications involved, there is a common need for identifying the distribution of fluid dynamic parameters and detecting its effects. The advances described by the participating authors have significantly helped accomplish this point. For instance, Rincón-Casado et al. employed ANSYS-CFX to obtain the air temperature distribution in a room with an air conditioning unit mounted on an internal wall, which facilitates further analysis of comfort level and energy demand. Adewumi et al. presents an essential study of scale analysis and double diffusive free convection boundary layer laminar flow of low Prandtl fluids over an inclined wall, and investigated the velocity, concentration, and thermal boundary layer thicknesses in a series of geometrical conditions. Velázquez Ortega applied the Lattice Boltzmann Method to study the flow of a non-Newtonian fluid between two plates. Li et al. presented a review study of the interface schemes within the scope of the Lattice Boltzmann Method for conjugate transport between multi-phases or different materials. Flores-Hidalgo et al. conducted a case study of the photocatalytic degradation of water pollutants, and analyzed the effects of pressure and velocity distribution in the photocatalytic reactors with the assistance of CFD. Termizi et al. investigated the effect of inlet velocity toward mixing intensity over two different microchannel configurations. The CFD profile showed inlet velocity has significance effects on the mixing performance and provided information on the mixing length requirement to achieve complete mixing. Druetta studied the enhanced oil recovery process by solving a set of momentum and mass conservation equations in a reservoir simulator. Loya et al. developed a flight dynamic model for aircraft using CFD, which was then used to optimize the aerodynamic performance. This study successfully demonstrated how CFD is a great tool for designing a flight dynamic model of an unknown aircraft. To improve their efficiencies and understand the performance of hydrokinetic turbines, Chica et al. used CFD to analyze the fluid dynamic parameters and power generation efficiency of turbines with complex geometries. Ochiai et al. presented a multi-phase CFD model for oil lubricated high-speed journal bearings. This CFD analysis of the two-phase flow of VOF with vapor pressure and surface tension allowed for the calculation of the gaseous-phase area and temperature of the journal bearing under flooded and starved lubrication conditions. Vilag et al. performed a simulation of a complex process taking place in the combustion chamber of a gas turbine. This work not only showed good agreement between CFD simulations and experiments, but also provided detailed information from inside the gas turbine such as temperature field, component fraction, and velocity which allowed for further analysis and deeper understanding of the combustion process.

During the preparation of this book, all the participating authors spent significant efforts in composing the chapters with their extraordinary knowledge and high motivation, and performed serious revision where needed. Without their ongoing support the publication of this book would not have been possible. The time that they have taken away from their busy schedules to contribute to this book was valuable and greatly appreciated. Also, my appreciation is especially dedicated to Ms. Rebekah Pribetic who helped me in every editing step throughout the entire publishing process.

Guozhao Ji
Dalian University of Technology,
Dalian, People's Republic of China

Jiujiang Zhu
University of Wuyi,
China

Section 1

Basics of Computational Fluid Dynamics Simulation

Chapter 1

Calibration Methodology for CFD Models of Rooms and Buildings with Mechanical Ventilation from Experimental Results

Alejandro Rincón Casado, Magdalena Hajdukiewicz, F. Sánchez de la Flor and Enrique Rodríguez Jara

Abstract

This chapter describes a methodology for the development and calibration of computational fluid dynamics (CFD) models of three-dimensional enclosures for buildings with combined forced and natural convection from experimental result. The models were validated with physical test measurements of room air temperature. The developed CFD models included a model of an internal wall-mounted air conditioning (HVAC) split unit. The methodology proposed here aims at selecting the correct grid size and the appropriate boundary conditions from experimental data. The experimental campaign took place in an empty office room within an educational building. A set of experiments was performed with varying boundary conditions of two main variables, the fan speed of the HVAC unit and the surface wall temperature of the opposite wall to the HVAC unit. The developed CFD models used the standard k-ε turbulence model and the SIMPLE algorithm. The variable of interest was the room air temperature and its distribution within the internal environment. The application of the methodology has shown satisfactory results, finding a maximum error of 9% between the CFD model and the experimental result. This methodology can be used by other researchers to calibrate CFD models in existing rooms and then carry out detailed studies of temperature distribution, comfort and energy demand analysis.

Keywords: room ventilation, forced convection, CFD simulation, indoor environment, mixed-mode ventilation

1. Introduction

Airflow inside internal environments is mainly caused by two main physical phenomena. The first is the temperature gradient in a given volume of air that produces natural buoyancy, and the second cause is the pressure difference created by mechanical fans. Transparent fluids such as the atmospheric air are difficult to study by simple observation. In order to investigate the properties of the indoor airflow, tracer gas techniques or the measurement of variables such as air temperature, surface temperature, air velocity or heat flow through boundary elements is used.

In the scientific literature, we can find works such as those reported by Chen and Srebric [1], where they recommend verifying and validating a CFD code for indoor environment modelling based on the following aspects: basic flow and heat transfer features, turbulence models, auxiliary heat transfer and flow models and numerical methods, assessing CFD predictions and drawing conclusions. Although the format for reporting of CFD analysis does not necessarily have to be the same, the chapter suggests to include all the aspects used in verification and validation for technical readers. This work presents the CFD methodology to follow but does not apply the methodology to a real experimental case. The calibration methodology proposed in our work explains step by step the procedure to be followed for the calibration of the CFD model with the experimental results, also evaluating the error reached and its applicability. Another work published by the mentioned authors [2] describes how to use the verification, validation and reporting manual for the CFD analysis proposed by ASHRAE. The article validates a CFD model with the experimental results in an office with furniture. The conditioning system is composed of a diffuser in the ceiling, and there is an error in speed of 20%. The measurement plane is located in the middle of the office, and the variables obtained are speed, temperature, concentration and turbulence intensity. The measuring points are 6 points in the vertical. However, different points of the plane are not analysed for the stratification phenomenon. Neither the mesh optimization process nor the effect is analysed when the boundary conditions are changed, such as speed and temperature in the walls.

A published overview of the tools used to predict ventilation performance in buildings has shown that the CFD analysis was the most popular among others, contributing to 70% of the reviewed literature [3]. However, the reliability of CFD methods is a big concern. While the CFD analysis can quickly provide extensive information about the indoor temperature and velocity distribution in the form of visually appealing results, the accuracy of CFD predictions must be considered with extreme caution. In order to achieve valid CFD models of indoor environments, comprehensive verification and validation studies must be performed [4, 5]. A particular aspect of the CFD model development is the right choice of the boundary conditions, which is not always straightforward. When simulating the conditions obtained during the experimental setups, it is necessary to calibrate the model in order to achieve agreement between the experimental and CFD results. Although there are good practice guidelines available for the generation, verification and validations of CFD models, like the German Guideline [6], there is lack of methodological procedures for the validation of CFD models focused on internal environments that account for a specific process to adjust input parameters according with experimental measurements [4].

In recent years, the use of experimental studies to perform validation of CFD models has risen. In the study of Stamou and Katsiris [7], an experimental test was performed in an office room with furniture and occupied by people. These conditions were reproduced in a CFD model. The study focused on comparing the results of different turbulence models, including k-ε, RNG k-ε, SST k-ω and the laminar model. Among all the turbulence models studied, the k-ε provided the best results in agreement with the experimental data. However, this reference only takes into account the natural convection mechanism, and there is no mechanical ventilation. In our work standard k-ε model provided better convergence and the best results in agreement with the experimental data. Another study [8] utilised CFD models with coupled convection and radiation to investigate the behaviour of a vertical radiant cooling panel system with condensation installed in an office space. The authors performed validation of the CFD model based on the field measurements. The standard k-ε turbulence model was used, reaching a good accuracy and providing useful information regarding the temperature distribution and the air velocity in the environment. Lin et al. [9] investigated gaseous and particulate contaminant transport, air motion and air temperature profile in a naturally ventilated office room with furniture. The experiment involved the use of smoke tracers and the installation of 17 temperature, air velocity and CO_2 concentration sensors. The measurements obtained during the experiment were used to validate the CFD model of the internal environment. Despite some big discrepancies between the measured and simulated data, in general, the model produced acceptable results with regard to air temperature distribution in the office. Yongson et al. [10] developed a CFD model of an occupied and furnished room, which was mechanically cooled by a refrigeration unit. The aim of the study was to focus on the optimised position of the HVAC unit in relation to the thermal comfort conditions in the room [11]. Thus, the numerical models of the room were developed; however there was a lack of experimental data to validate the model.

Recently correlations have been developed to implement them in thermal simulation programmes of buildings [12]. These correlations are used for convective heat transfer calculations. However, this work does not take into account the phenomena of forced convection, which are very important in mechanical ventilation. More recently, researchers in Ireland have developed a methodology for the validation of CFD models of naturally ventilated indoor environments [4]. The methodology was supported by the field measurements in an office room occupied by people and furniture. The results showed very small air temperature vertical gradient against a more relevant one in comparison with the CFD results. The authors used the response surface method (RSM) to identify the variables with more impact in the results.

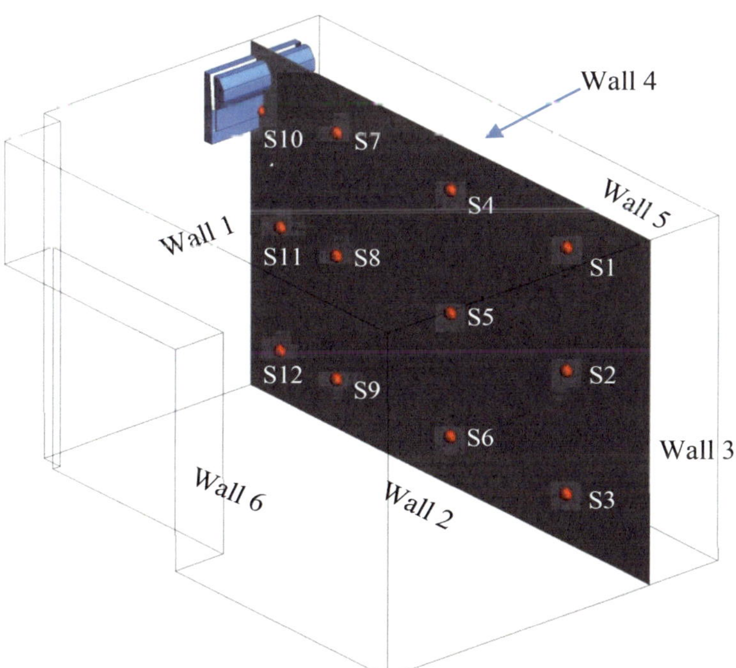

Figure 1.
Location of temperature sensors in the CFD model room.

Finally, the main objective of this research is the development of a methodology for the calibration of CFD models for rooms existing buildings from experimental results. This methodology can be used by other researchers to calibrate CFD models in existing rooms and then carry out detailed studies of temperature distribution, comfort and energy demand analysis. In addition, different conditioning systems, or different boundary conditions, can be tested, and the comfort or energy demand effect can be studied. The methodology is demonstrated by reproducing the experimental results measured in a mechanically cooled test room using CFD model. The calibration analysis is focused on a 2D plane of the room that was perpendicular to the HVAC discharge outlet, where 12 temperature sensors where deployed (**Figure 1**). The variable of interest was the sensor air temperatures, measured at a steady-state regime in order to be compared with the CFD results. The boundary conditions of the CFD model were taken based on the measurements in the test room (i.e. surface temperatures, air velocity and air temperature of the HVAC discharge outlet, etc.).

2. Materials and methods

2.1 Calibration methodology

CFD is today one of the most accurate tools to predict the movement of air within an internal enclosure. CFD simulations require adequate computational power in order to solve the governing equations the fluid flows. It is also of a paramount importance that in order to get reliable results, a validation procedure based on trusted experimental data should be performed. A mesh verification is also necessary to achieve a good agreement between model accuracy and computational cost.

In this work, a validation methodology for CFD models that combine natural and forced convection heat and flow transfer using experimental results is proposed. The validation steps and necessary parameters are described in the workflow shown in **Figure 2**. The diagram is divided into two parts, the left part represents the experimental test and the right part of the workflow represents the CFD model. The proposed method involves using the experimental boundary conditions set up at the room test as CFD model inputs. The variables used to feed the CFD models were (1) HVAC outlet air velocity, (2) HVAC air outlet temperature and (3) surface temperatures. The surface temperatures (3) were taken from the experimental test when steady-state condition was reached and were used as imposed inputs at the internal surfaces of the CFD model.

The validation process starts with the design of the experiment, consisting of room preparation, air temperature sensors and surface temperature sensor placement (see **Figures 1** and **3**) and definition of case studies (see **Table 1**). In parallel, building geometry is introduced in the CFD tool. For every case study, the HAVC temperature and fan velocity are fixed. These values are used as boundary condition for the CFD model. During the experimental campaign, air temperatures and surface temperatures are collected, until the steady-state conditions are reached (see **Figure 4**). This process finalises with surface temperatures to feed the boundary conditions of the CFD model and air temperatures to be compared with the simulation ones. On the CFD side, once all boundary conditions have been introduced, simulations are performed keeping mesh goodness and convergence criteria (see sections 4.3 and 4.4). Previous air temperature measurements are compared with CFD model results. If the differences are larger than the own sensor accuracy error, the input parameters (1) and (2) are adjusted. This last step needs to be repeated

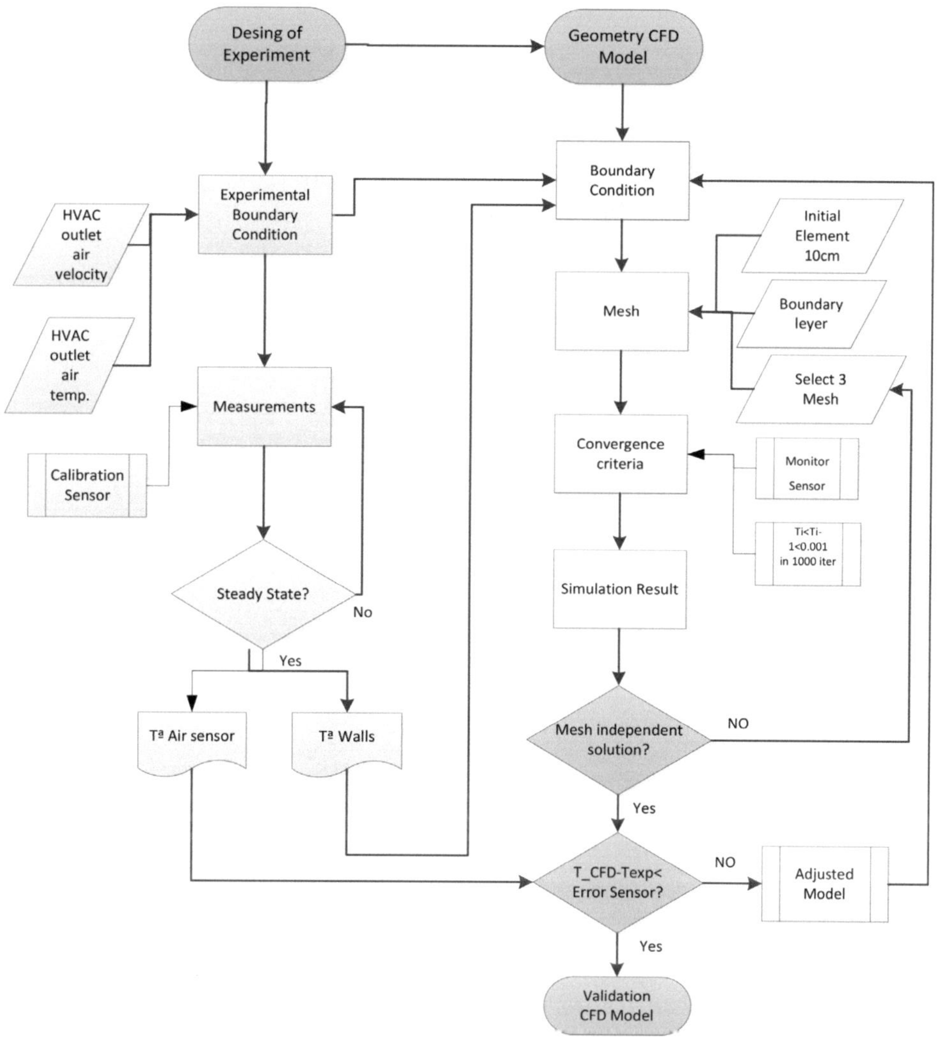

Figure 2.
Workflow for the validation methodology of CFD models using experimental results.

iteratively until the residual error falls within the admittance threshold of the sensor error established.

2.2 Experimental model

2.2.1 Test room description

The building used for the experimental campaign belongs to the *Instituto de Investigacion Tecnologica* within the *Escuela Politecnica Superior de Algeciras* and is located at the Algeciras University Campus of the University of Cadiz (Spain). An external view of the building is shown in **Figure 5**. The building is an educational facility dedicated mainly to work spaces, offices and meeting rooms. The internal spaces in the building are conditioned by a variable refrigerant volume (VRV) cooling system, placed on the top of Wall 1 (see **Figure 5**). The dimensions of the room were W = 2.92 m width, L = 4.22 m length and H = 2.80 m height (**Figure 5b**). Its external wall, which was partially underground, contained an operable window. The ceiling was a concrete slab with suspended ceiling modules. A standard door is

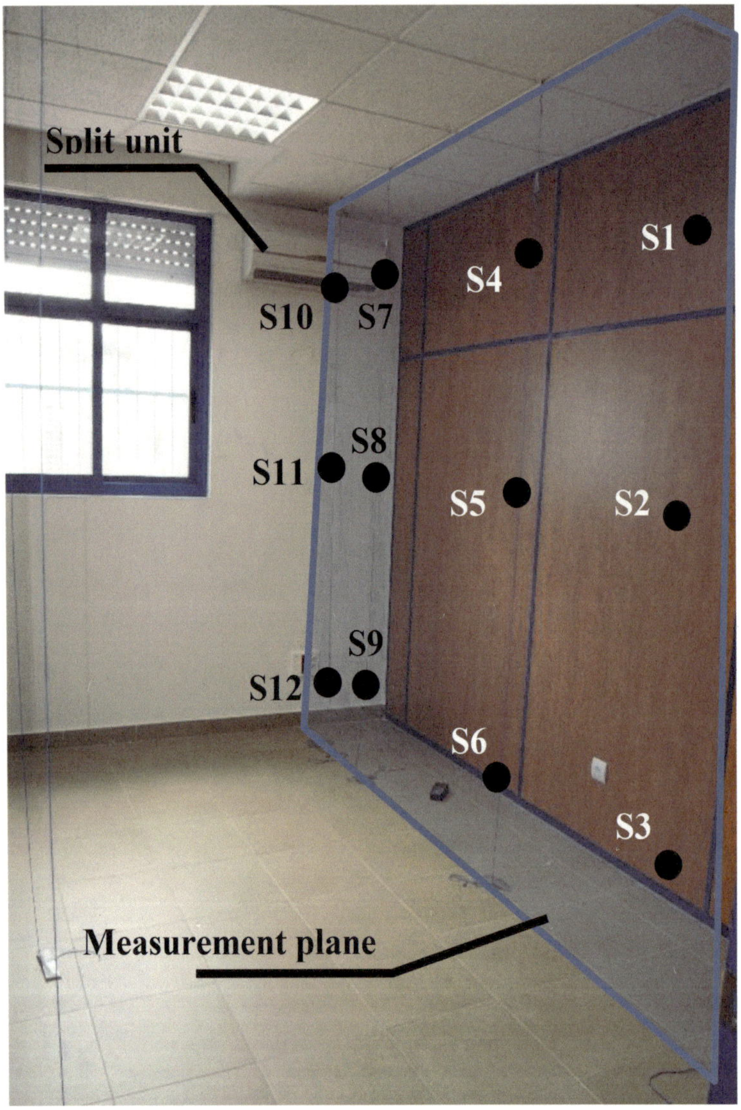

Figure 3.
Location of temperature sensors in the experimental office room.

Experiment number	Date/start time	Date/end time	Boundary conditions			
			Wall 3 temperature		HVAC fan speed	
			Low	High	Low (2.2 m/s)	High (2.7 m/s)
1	09.07.2016/13:30	10.07.2016/10:00	•		•	
2	10.07.2016/12:00	11.07.2016/10:00	•			•
3	11.07.2016/12:00	12.07.2016/10:00		•		•
4	12.07.2016/12:00	13.07.2016/10:00		•	•	

Table 1.
Chronogram of the experiments campaign.

DOI: http://dx.doi.org/10.5772/intechopen.89848

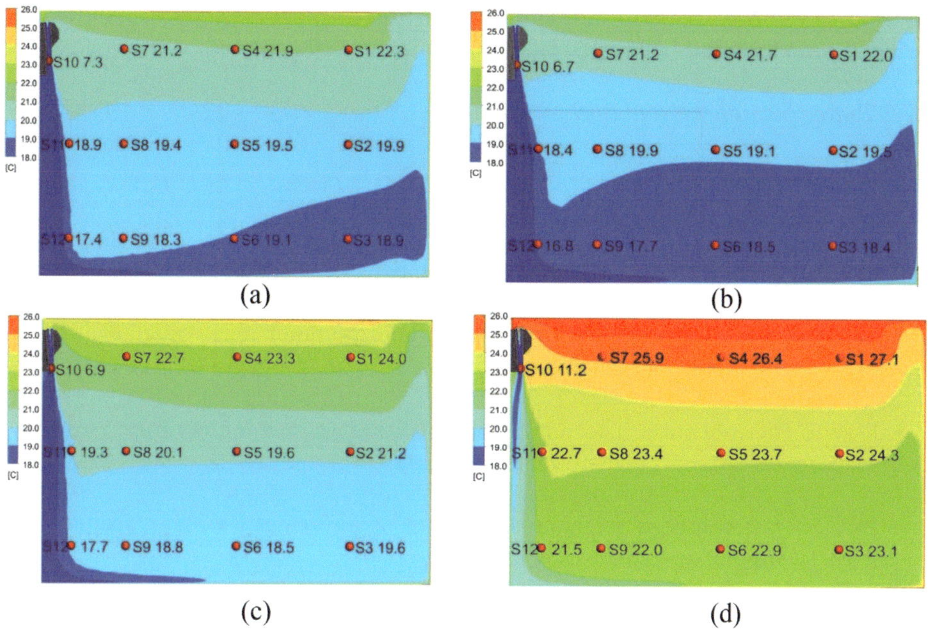

Figure 4.
CFD isotherm contour map and sensor measurements in red points. Experiment 1 (a), 2 (b), 3(c) and 4(d).

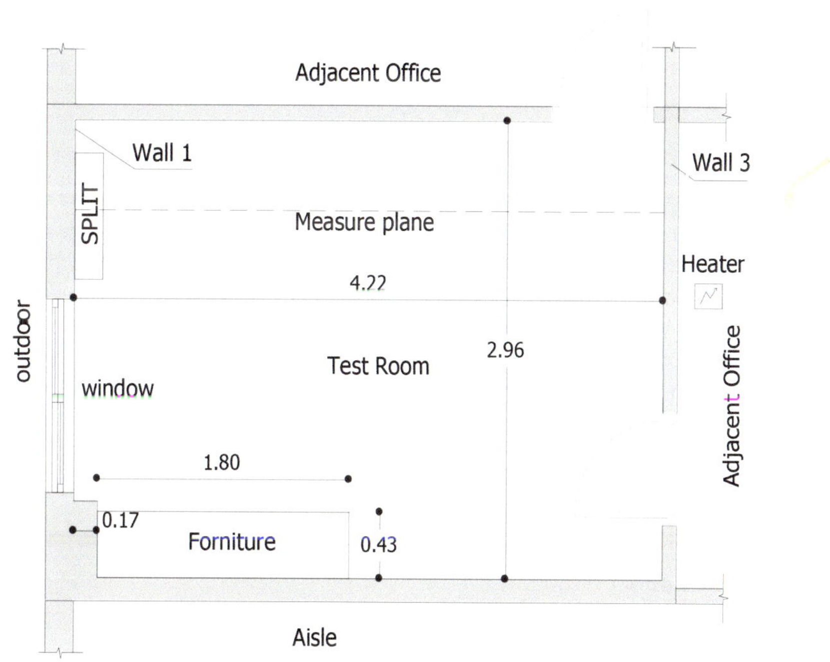

Figure 5.
Floor plan of the investigated office room.

located on the wall opposite the window, contained a H = 2.1 m and W = 0.72 m standard door. The two internal walls separated the room from the adjacent offices and the internal corridor, with similar ventilation characteristics. During the experiment, the room was empty, without any furniture or occupants. **Figure 3** shows the location of the vertical strings with sensors and the internal HAVC split unit. Also in the waiting room heater is installed to increase the temperature of Wall 3 and see its influence on the indoor air temperature.

2.2.2 Testing instruments and calibration

The measurement equipment included:

- Data Logger Testo 174 measuring air temperature with an accuracy of ±0.5°C in a range of −20°C to +40°C (www.testo.es)
- Maxthermo-Gitta ref.: YC-7XXUD series Thermometer measuring air temperature with an accuracy of ±0.1°C (www.maxthermo.com.tw)
- K-type Thermocouple Thermometer measuring surface temperature with an accuracy of ±0.5°C (www.hannainst.com/)
- PKT-5060 hot-sphere anemometer measuring air velocity with an accuracy of ±3% (www.pce-instruments.com)

To calibrate the temperature sensors, the more precise YC-7XXUD (±0.1°C error) temperature metres were used. All the temperature sensors used were calibrated introducing the sensor in an adiabatic isolated chamber to obtain the bias error of each temperature sensor against the readings of the precision temperature metre. The temperature of each sensor was tuned according to their specific bias error registered using this method. Similarly, the surface temperature metres were also calibrated.

2.2.3 Case studies

The investigated office room was conditioned with an internal split unit, which was connected to a central VRV system for the general conditioning of the offices. The test room remained unoccupied during the whole period of the experiment, with the HVAC unit functioning continuously. The external blind was closed during the experiment with the aim of blocking all incident solar radiation to the room. Similarly, the access door remained closed during the duration of the test, to minimise air infiltration from adjacent rooms. These rooms remained nonconditioned and unoccupied during the experiment. Different setup configurations were tested, with different boundary conditions, in order to evaluate the impact of:

1. HVAC outlet air velocity
2. HVAC air outlet temperature
3. Surface temperature of the interior wall (Wall 3), opposite to the façade (Wall 1)

In order to assess the influence of the fan speed on the indoor conditions, the fan was operated at two levels: high speed and low speed. The air speed at the discharge outlet of the HVAC unit was measured for each speed level. For the high-speed setting, the air velocity was 2.7 m/s, while for the low-speed position, the air propelled by the unit reached 2.2 m/s. Similarly, the surface temperature of Wall 3 was tested according to two settings: low temperature and high temperature of the wall surface (see **Table 1**). For the low-temperature setting, the adjacent room to

Wall 3 remained nonconditioned, while for the high-temperature setting, the adjacent room was heated with a heater. It is also important to notice that the air direction in the unit was fixed in a vertical position with the intention of minimising the air turbulence and favouring the temperature stratification of the room air. Eventually, four different configurations were chosen to perform the experiments, which are summarised in **Table 1**, alongside the test chronogram. The ultimate intention of these four experiments was to achieve a high temperature difference in the air of the test room.

Under the test conditions described previously, and for each experiment, a set of air temperature and surface temperature were taken. These values were taken using 12 temperature sensors distributed in a square grid in the measurement plane, as shown in **Figures 1** and **6**. This plane was placed orthogonal to Wall 1 at the middle of the HVAC unit. **Figure 6** shows the exact locations of the sensors. One of these sensors (sensor 10) was purposely placed at the exit of the HVAC outlet to measure the air temperature at that point. The 2D measurement plane includes the walls and the ceiling, where 21 surface temperature sensors were installed uniformly (**Figure 6**). The measurement results are used as boundary conditions of the CFD computational model. An additional temperature sensor (sensor 13) was located in the adjacent room in order to measure the air temperature when the heater was operating (experiments 3 and 4). These temperatures were taken at the specified time at the end of the experiments using surface temperature metres. As previously mentioned, the purpose of heating up the adjacent room was to heat Wall 3.

The experiments were carried out for 20 hours, as seen in **Table 1**, in order to achieve steady-state conditions inside the room. Air temperatures were measured every minute during each experiment, while the surface temperatures only were measured at the end of the experiments, once a steady-state condition was reached. The measurement of the air speed at the HVAC discharge outlet was also taken at the end of each experiment (the fan's setpoint air speed was constant during the experiments).

2.3 Computational model

The computational domain is a three-dimensional enclosure, and the used mesh type was a nonstructured mesh formed with tetrahedral cells. To develop the CFD simulation, the commercial software ANSYS CFX v.17 [1] was used. The model

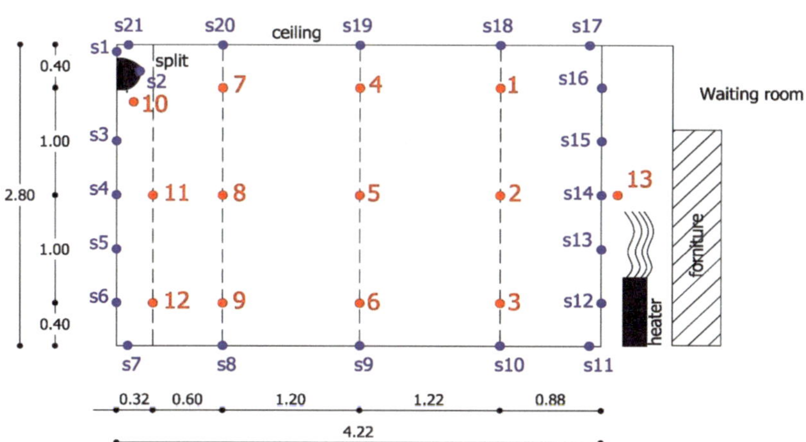

Figure 6.
Vertical view of the measurement plane containing the superficial sensor (blue) and air sensors (red) location.

developed reflected the geometry and boundary conditions of the experimentally investigated room, for the purpose of model validation. In the computational model studied, steady state, 3D geometry, and Newtonian fluid are considered. All of the fluid properties remain constant except for the density, which depends on the temperature difference. The studied phenomenon is forced and natural convection; thus, buoyancy effects are studied due to the gravity effect. The CFD results are obtained by solving the Navier-Stokes equations and the energy equation via finite volumes using the commercial software ANSYS CFX v.17 [13]. The numerical algorithm used is SIMPLE (semi-implicit method for pressure linked equations), which was developed by Patankar and Spalding (1972) and recently Kengni Jotsa, A. C. and Pennati, V. A. (2015) using in a cost-effective FE method 3D Navier-Stokes equations. One of the discretization schemes is the QUICK scheme which has been used for convective flux in incompressible flow on unstructured grids, and the validation was developed by Hua, Xing, Chu and Gu. (2009). In the equations solution, the Boussinesq approximation was considered for buoyancy. Although the problem to be solved is a steady-state problem, due to the computational complexity of the problem, it is necessary to solve the problem as a transient problem until a steady-state solution is reached.

In the CFD simulations, a crucial factor is the choice of the convergence criteria. The convergence of the simulation depends on a number of factors. Convergence is reached when a stable solution is found that does not change significantly with more iterations. The convergence criteria for residuals of the mass, energy, momentum and k and ε equations were under 10^{-7}, and variables of interest show stable behaviour. **Figure** 7 shows the monitored air temperature values (Y-axis), for air temperature sensors 2, 3 and 4, as a function of the number of iterations of the CFD simulation. Convergence of the monitored variables was reached approximately at 6000 iterations remaining constant during 2000 iterations. However, there are cases with high speeds where the steady state is not reached. In these cases the calculation mode must be transient state, and the time step must be calculated. To determine the time step size, the criteria $\Delta t = (L/\beta g \Delta T)^{1/2}$ for difference temperature of wall and inlet recommended by Ansys was used. In order to obtain accurate and meaningful numerical solution, meshing the computational domain is the crucial first step. This importance is more pronounced especially in fast-moving

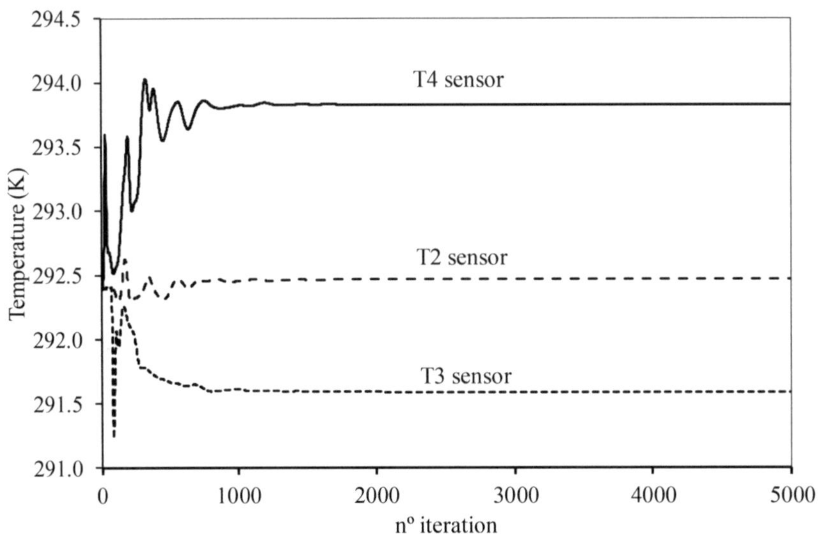

Figure 7.
Convergence of the monitored variables (T2, T3 and T4) over n. of iterations (medium mesh).

flows due to steep gradients occurring within the boundary layers. The simulations for the mesh optimised according to Section 4.4 have the order of 500,000 elements and the mean simulation time of 22 hours.

2.3.1 Geometry description

Model geometry was represented by a 3D enclosure (**Figure 8**). It is worth to mention the importance of a good detailed model of the split unit to fully reproduce the details of the air enclosure boundaries. The only HVAC zonal equipment was a wall-mounted split unit located in the higher part of Wall 1. This unit supplied cool air to the room procuring a high-temperature gradient between the air temperature sensors. The most complex element to model with the CFD tool was the HVAC unit. This equipment contains in its interior a coil where the refrigerant circulates and a fan that forces air to pass through the coil and exchange heat through them. The equipment air inlet is located in the top part and takes the air from the room, while the air outlet, located at the bottom part, discharges the cooled air to the room. To model this unit behaviour in the CFD simulation, the unit was defined as a closed volume with an air passage through the volume. The HAVC computational model has as boundary condition the temperature and velocity of internal walls, this behaviour is like an internal duct (**Figure 8b**), and this values are fixed according with the experimental measurements.

2.3.2 Model setup and boundary conditions

The steady-state conditions were used in the CFD analysis of the single-phase airflow inside the room. The full buoyancy model was considered, where the fluid density was a function of temperature or pressure, and was applied. The air was modelled as ideal gas with the reference buoyancy density of 1.185 kg/m3 (an approximate value of the domain air density). The solution scheme is a pressure-velocity coupled with a pressure-based solver. The standard k-ε turbulence model was chosen for good results' accuracy with the robustness of the solution [4]. The wall function considered was scalable wall function. The mesh should be

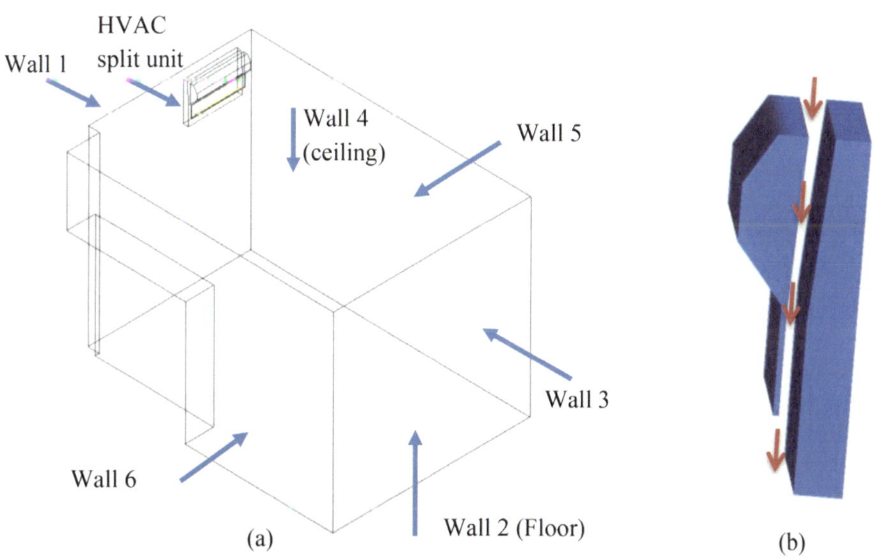

Figure 8.
3D model view and HVAC split unit of the computational modelled room.

sufficiently fine to accurately model convective heat transfer and fluid flow near the walls; for this reason the parameter y + must have a value of approximately to 11.25. The turbulence parameters were defined by testing three values of turbulence intensity 1% (low), 5% (medium) and 10% (high) as a variable in ANSYS CFX setup. The temperatures of S10, S12, S5 and S2 sensors were monitored until reaching the steady state. The results obtained for the S10 sensor were 6.69°C, 6.70° C and 6.75°C for the low, medium and high intensity, respectively. For the S12 sensor, the results were 18.21°C, 18.18°C and 18.22°C; for the S5 sensor, 19.13°C, 19.12°C and 19.13°C; and for the S2 sensor, 19.22°C, 19.21°C and 19.22°C. Therefore, the maximum deviations are 0.2°C; this value is lower than the error of the Data Logger Testo 174 used (0.5°C).

Table 2 summarises the energy (surface temperature) and momentum (air velocity) boundary conditions used in the CFD models, each wall and experiment. Experimental measurements show a linear relation between each surface temperature and the width (floor and ceiling) or height (vertical walls). As described before, the surface temperature measurements were carried out using K-type Thermocouple. These readings were done manually at the end of each experiment, when the room reached a steady-state condition. As a matter of example of the surface temperature gradient, **Figure 9** shows the surface temperatures plotted against the sensor location height for Wall 3. The graph shows also a linear regression function linking both variables.

2.3.3 Mesh verification

An important aspect when developing CFD models is the selection of an appropriate mesh. The number of cells and their shape and size should guarantee a mesh-independent solution while achieving a good trade-off between the result accuracy and computational cost. The used mesh type was a nonstructured mesh formed with tetrahedral cells. The tetrahedral cell offered less degrees of freedom per cell and fixed better the desired geometry. The mesh was denser at the proximities of the wall surfaces and at the split unit discharge outlet, being zones where the temperature and velocity gradients are more pronounced. A first approximation of the minimum numbers of cells of the domain was calculated using the formula recommended by the German Guideline [6] shown in Eq. (1).

$$N = 44.4 \cdot 10^3 \cdot V^{0.38} \tag{1}$$

where:

N = number of finite elements of the volume.

V = volume of the studied space.

In the case of the experiment of this chapter, the volume accounts for 34.47 m3 (12.4 m^2 x 2.8 m), and the number of elements according to the above formula is of 170,924 cells. On the other hand, a common recommendation [6] for CFD cell size when applied to internal environment in buildings is around the 10 cm size for rooms of less than 5 m length. This size should be smaller on zones where significant temperature or velocity gradients were to be expected [14].

In order to capture the temperature and velocity gradients inside of the velocity boundary layer and thermal boundary layer, it is necessary to analyse at least 10 nodes that fall inside these boundary layers. Therefore, this effect can be considered relevant when the sensors are located in a near-wall position or when calculating local convective heat transfer coefficients. To correctly capture gradients inside the boundary layer, the parameter that controls the correct solution of the viscous sub-layer is y+. This dimensionless parameter depends on the turbulence model. Thus, for standard k-ε and the scalable wall function, the parameter y + must have a value

Experiment	Limits	Energy	Momentum
Experiment 1	HVAC	Tin = 8 [°C]	Vin = 2.2 [m/s]
Wall 1 to 6	Wall 1	T1(Y) = 20 [°C]	No slip wall
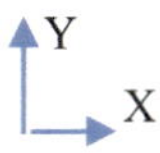	Floor	T2(X) = 1.315·X + 19.97 [°C]	No slip wall
	Wall 3	T3(Y) = 0.9857·Y + 23.12 [°C]	No slip wall
	Ceiling	T4(X) = 0.9788·X + 26.54 [°C]	No slip wall
	Wall 5	T5(Y) = 0.9857·Y + 22.52 [°C]	No slip wall
	Wall 6	T6(Y) = 0.9857·Y + 22.52 [°C]	No slip wall
Experiment 2	HVAC unit	Tin = 8 [°C]	Vin = 2.7 [m/s]
	Wall 1	T1(Y) = 20 [°C]	No slip wall
	Floor	T2(X) = 1.315·X + 19.97 [°C]	No slip wall
	Wall 3	T3(Y) = 0.9857·Y + 23.12 [°C]	No slip wall
	Ceiling	T4(X) = 0.9788·X + 26.54 [°C]	No slip wall
	Wall 5	T5(Y) = 0.9857·Y + 22.52 [°C]	No slip wall
	Wall 6	T6(Y) = 0.9857·Y + 22.52 [°C]	No slip wall
Experiment 3	HVAC	Tin = 8 [°C]	Vin = 2.7 [m/s]
	Wall 1	T1(Y) = 20 [°C]	No slip wall
	Floor	T2(X) = 1.315·X + 19.97 [°C]	No slip wall
	Wall 3	T3(Y) = −1.467·Y + 34.18 [°C]	No slip wall
	Ceiling	T4(X) = 0.9788·X + 26.54 [°C]	No slip wall
	Wall 5	T5(Y) = 0.9857·Y + 22.52 [°C]	No slip wall
	Wall 6	T6(Y) = 0.9857·Y + 22.52 [°C]	No slip wall
Experiment 4	HVAC	Tin = 8 [°C]	Vin = 2.2 [m/s]
	Wall 1	T1(Y) = 20 [°C]	No slip wall
	Floor	T2(X) = 1.315·X + 19.97 [°C]	No slip wall
	Wall 3	T3(Y) = −1.467·Y + 34.18 [°C]	No slip wall
	Ceiling	T4(X) = 0.9788·X + 26.54 [°C]	No slip wall
	Wall 5	T5(Y) = 0.9857·Y + 22.52 [°C]	No slip wall
	Wall 6	T6(Y) = 0.9857·Y + 22.52 [°C]	No slip wall

Table 2.
Boundary conditions for CFD model.

of approximately to 11.25. To obtain a mesh configuration that offers a good trade-off between accuracy and computing costs, it is necessary to establish a mesh refinement process. The method chosen was the one developed by Celik et al. [15]. This process consists in selecting three different grids with different coarseness definition: a coarse grid, a basic grid and a fine grid. The CFD results of the variables of interests are compared with one another to justify the best compromise between accuracy and computational cost.

The first step is to estimate the coarse grid features. This is calculated according to the before mentioned criteria. The number of cells was 170,924 and the average cell size (h) was of 10 cm. This last parameter can also be estimated using Eq. (2) [15], where h is the cell size, ΔV is the volume of each i element, and N is the number of grid cells:

$$h = \left[\frac{1}{N}\sum_{i=1}^{N}(\Delta V_i)\right]^{1/3} \tag{2}$$

The second step entails calculating the refinement degree of the grid "r" using the relationship between the number of the elements of the studied mesh and the number of elements of the refined mesh. According to the cited methodology by Celik [15], the recommended value for this refinement factor needs to be lower than 1.3. With this criteria, the number of elements of the finer and the basic mesh can be calculated having the number of elements of the coarse mesh using Eq. (3) and Eq. (4), where r_{ij} is the refinement degree, h_i is the cell size, and N is the number of grid cells [16]. Finally, the selected meshes are shown in **Table 3**.

$$r_{21} = \frac{h_2}{h_1} = \left(\frac{N_1}{N_2}\right)^{1/3} \tag{3}$$

$$r_{32} = \frac{h_3}{h_2} = \left(\frac{N_2}{N_3}\right)^{1/3} \tag{4}$$

The grid quantitative verification was completed using the grid convergence index (GCI) and based on the Richardson extrapolation formula [17]. These methods are helpful to estimate the grid convergence error. The formula is developed as follows:

$$GCI^{21} = \frac{1.25 \cdot e_{21}}{r_{21}^{p} - 1} \tag{5}$$

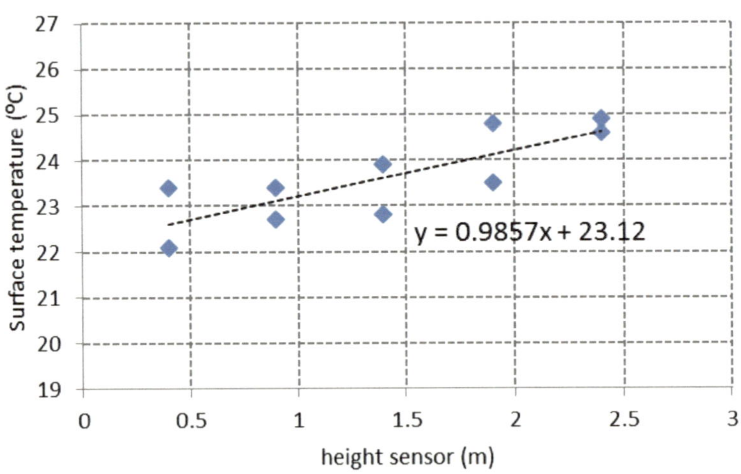

Figure 9.
Trend line of wall surface temperature vs. room height. Wall 3 in Experiment 1 (blue) and 2 (red).

Grid	Number of elements	Refining degree
Grid 1 (fine)	1,151,812	1.37
Grid 2 (medium)	442,939	—
Grid 3 (coarse)	181,938	1.34

Table 3.
Selected grids for the grid refinement study.

$$p = \frac{1}{\ln(r_{21})} \left| \ln|\varepsilon_{32}/\varepsilon_{21}| + q(p) \right| \tag{6}$$

$$\varepsilon_{32} = \varnothing_3 - \varnothing_2; \varepsilon_{21} = \varnothing_2 - \varnothing_1 \tag{7}$$

$$q(p) = \ln\left(\frac{r_{21}^p - s}{r_{32}^p - s}\right) \tag{8}$$

$$s = sgn\left(\varepsilon_{32}/\varepsilon_{21}\right) \tag{9}$$

where $\varnothing_i$ is the variable of interest (sensor temperature), ε_{ij} is the error between i and j mesh, e_{ij} is the error between i and j mesh (%), and $p, q(p)$ and s are the

Sensor	$\varnothing_1$ (°C)	$\varnothing_2$(°C)	$\varnothing_3$(°C)	ε_{21} (%)	ε_{32} (%)	GCI_{12} (%)	GCI_{23} (%)	Measurement bias error (%)
1	18.21	18.19	17.93	0.08%	1.46%	0.01%	0.21%	2.27%
2	17.02	17.03	16.76	0.02%	1.61%	0.00%	0.03%	2.57%
3	16.44	16.47	16.20	0.16%	1.65%	0.02%	0.23%	2.71%
4	18.10	18.11	17.97	0.05%	0.80%	0.01%	0.11%	2.31%
5	17.07	17.07	16.70	0.04%	2.19%	0.01%	0.31%	2.62%
6	16.75	16.77	16.40	0.09%	2.22%	0.01%	0.32%	2.70%
7	17.98	17.92	17.70	0.31%	1.27%	0.04%	0.18%	2.36%
8	17.15	17.12	16.86	0.16%	1.52%	0.02%	0.22%	2.65%
9	16.89	16.87	16.52	0.10%	2.10%	0.01%	0.30%	2.82%
10	11.07	11.15	11.08	0.70%	0.60%	0.08%	0.09%	7.41%
11	17.22	16.94	16.53	1.63%	2.46%	0.19%	0.35%	2.72%
12	16.64	16.45	16.07	1.18%	2.35%	0.14%	0.33%	2.97%

Table 4.
Necessary parameters used for the calculation of the GCI during the grid refinement process (example for experiment n. 2).

Figure 10.
3D cell grid used for medium mesh. Cross-section at the measurement plane.

auxiliary parameters. In **Table 4**, all the necessary variables for the calculation of the GCI for the 12 temperature sensors are shown. The table shows the values of the GCI_{12}, the GCI_{23} indexes and the temperature sensor measurement bias error (0.5 error sensor, divided by measurement temperature sensor in percent). The small values for GCI confirm that the solution is grid independent. On the other hand, high GCI values confirm a larger relative error close to the sensor error. However, in this case the values of the GCI_{12} are low, and therefore the relative errors are far away from the values of the sensor error. It was concluded that mesh number 2 is the option that provides an optimum equilibrium between accuracy and computational cost, and therefore it was the author's choice for the CFD model developed. **Figure 10** shows a cross-section at the measurement plane of the 3D grid generated.

3. Results and discussion

The experiment performed is aimed at measuring the temperature distribution of the air in the internal environment of a 3D test room equipped with an air conditioning device. The measurements were taken with 12 temperature sensors, 1 thermocouple sensor and 1 anemometer. The test room was emptied for the experiment, so no people or furniture was considered in order to simplify the airflow trajectory and to ease the CFD modelling efforts. Regarding the boundary conditions, the intention of the authors was to create big temperature gradients that minimise relative errors. The available sensors were installed in a two-dimensional grid contained in an orthogonal plane positioned perpendicularly to the A/C equipment outlet direction; this arrangement was chosen to better capture the fluid stratification. **Figure 6** shows the location of the surface temperature measurement points. The temperature sensors readings were gathered for 24-hour periods for each test. An example of the sensor measurements gathered in test n. 2 is shown in the time series of **Figure 11**. It can be noticed that sensor n.10 register periodical fluctuations. This sensor is located at the A/C outlet, being approximately sinusoidal fluctuations of 4-minute period. This phenomenon was found to be caused by the specific behaviour of the VRV inverter A/C unit which varies the refrigerant flow

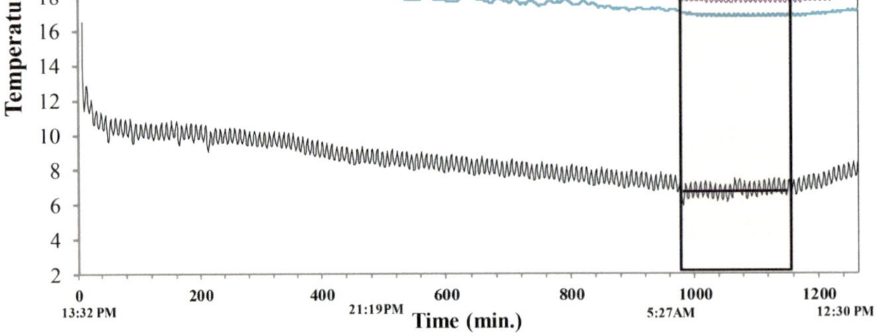

Figure 11.
Air temperature measurements of Experiment 2 during 24 h.

and causes the air temperature variation. This behaviour was present in all the tests and was more pronounced during the daytime, where more thermal load is experienced. Due to the short period of this fluctuation, only the mean temperature of the refrigeration cycles was considered. The steady-state conditions were reached at the end of the measurement campaign, where the sensor temperature was stable and maintained during a certain period of time.

The CFD simulation results are represented in **Figure 4**; this set of 2D graphs show a vertical view of the measurement plane. The variable displayed is the contour air temperature and is plotted according to a colour scale; the different colour areas are delimited by isotherm lines. The red point is the sensor position and the measured values in the experiment. It is worth to notice that, due to the precision error of the temperature sensors, the temperatures measured during the experiments could vary within ±0.5°C. In **Figure 4a** and **b**, it can be appreciated that in both experiments, there is a clear temperature stratification. The difference between experiments 1 and 2 relies mainly in the A/C fan speed. The fan speed at the Experiment 1 was set to "low", while the fan was set to "high" speed at Experiment 2. In the latest case, the temperatures reached within the room were lower due to shortened cooling cycles of the A/C unit. On the other hand, the differences between experiments 3 and 4 rely on the surface temperature of Wall 3, opposite to the A/C unit (Wall 1). In these experiments, the adjacent rooms are heated to warm the cited Wall 3 and analyse the effect on the room air temperature of the test room. The isotherm contour plots corresponding to Experiments 3 and 4 are shown in **Figure 4c** and **d**, respectively. In **Figure 4**, a stratification of the room air is also clearly noticeable. Likewise in Experiments 1 and 2 (**Figure 4a** and **b**), the fan speed of the A/C unit is a very important factor in the average temperature of the internal air. In Experiment 3, the fan speed is set to "high" (2.7 m/s), and the average air temperature reached is 19.8°C, while in Experiment 4 (**Figure 4d**) the fan speed was set to "low" (2.2 m/s), and the average room air temperature was 22.8°C. It can be concluded that the differences in fan speed and consequently the changes in cooling cycles of the A/C unit can result on average thermal differences of 3°C.

Summarising, there is a general good agreement between the experimental results and the CFD models. The stratification phenomenon caused by the fluid natural buoyancy is also clearly reflected in the results, with cool air near to the room floor surface and the hot air at the room upper zones. In the model analysed, the natural convection is also enhanced by the position of the HVAC unit. This device takes the room air through its inlet located in the top part, cools it passing it through the coil and discharges it through the outlet pointing downwards direction, hence working in favour of the natural buoyancy flow and causing and increased temperature gradient. It can be concluded that the complexity of modelling the HVAC unit plus the uncertainty of the surface temperature measurements can be considered the two main causes of discrepancy between CFD model results and experimental results.

In order to facilitate the comparison between the CFD and the experimental results, **Table 5** shows the value of (1) measured sensor temperature once the steady-state condition is reached including the mentioned and (2) temperature simulated produced by the CFD models of the position. In Experiment 1, **Table 5** has shown a good fit of the CFD results, being the larger differences in sensors S1, S4 and S7, near to the room ceiling, and in sensors S12 and S9, near the floor surface closed to the HVAC equipment. However, in the central space, the results of the CFD match closely the experimental measurements. Experiment 2 differs from Experiment 1 in the fan speed of the HVAC unit. The shortened cooling cycle effect explained previously makes the air to circulate at a higher rate around the room, thus producing a sustained cooling effect. In Experiments 3 and 4, Wall 3 (opposite

	Experiment 1			Experiment 2			Experiment 3			Experiment 4		
Sensor	Texp (°C)	Tsim (°C)	ΔT (°C)	Texp (°C)	Tsim (°C)	ΔT (°C)	Texp (°C)	Tsim (°C)	ΔT (°C)	Texp (°C)	Tsim (°C)	ΔT (°C)
S1	22.3	21.0	1.3	22.0	20.7	1.3	24.0	22.5	1.6	27.1	25.3	1.9
S2	19.9	19.6	0.3	19.5	19.3	0.2	21.2	20.3	0.9	24.3	23.2	1.0
S3	19.0	18.8	0.2	18.4	18.4	0.0	19.6	19.2	0.4	23.1	22.2	0.8
S4	21.9	21.1	0.8	21.7	20.9	0.7	23.3	22.8	0.5	26.4	25.4	1.0
S5	19.5	19.6	−0.1	19.1	19.1	−0.1	20.5	20.2	0.3	23.7	23.2	0.5
S6	19.1	19.0	0.1	18.5	18.5	0.0	19.6	19.3	0.3	22.9	22.4	0.5
S7	21.2	20.7	0.5	21.2	20.5	0.7	22.7	22.4	0.3	25.9	25.3	0.6
S8	19.4	19.6	−0.2	18.9	19.2	−0.3	20.1	20.1	0.0	23.4	23.2	0.2
S9	18.3	19.2	−0.9	17.7	18.8	−1.1	18.8	19.6	−0.8	22.0	22.6	−0.5
S10	7.3	7.9	−0.6	6.7	7.5	−0.8	6.9	7.0	−0.1	11.2	10.9	0.3
S11	18.9	19.7	−0.7	18.4	18.7	−0.3	19.3	19.8	−0.5	22.7	22.8	−0.1
S12	17.4	18.7	−1.3	16.8	18.2	−1.4	17.7	19.2	−1.5	21.5	22.3	−0.8

Table 5.
Air temperature comparison of CFD results and experimental measurements.

to the HVAC unit) was maintained at a warmer temperature by heating the adjacent room using heaters. The warm surface temperature of this wall directly influences the average temperature of the internal environment of the test room. Again, the maximum divergences are in the near-ceiling and near-floor locations and closed to the HVAC unit. In this experiment, sensor S2 indicates a larger deviation than in previous experiments due fundamentally to the warm surface effect that produced a larger temperature gradient between the wall and the room air. The difference between Experiments 3 and 4 is the higher fan speed of Experiment 3 versus the one of Experiment 4. This variation makes the cooling cycles of the HVAC units in Experiment 4 fewer than in Experiment 3. Therefore, the average room temperature reached in this example is higher than in the previous experiment. The stratification phenomenon is similar as in Experiment 3, although the temperatures registered are higher due to the lower HVAC fan speed. The error in the temperature prediction of line 4 (Sensors S1, S2 and S3) is quite higher than in Experiment 3.

In general terms, the results show that the CFD model and the test results agree at predicting the stratification effect and the temperature trend distribution inside the room air. In the central space of the room, the temperature is similar across the room air. Temperature increases steadily when approaching the ceiling and diminishes when moving towards the floor surface. For most of the measurement points, the CFD results fell inside the error threshold of the sensor measurements, except for some of the sensors located near the ceiling, floor and Wall 3 surfaces, which registered larger differences. Therefore, it can be stated that the CFD is less accurate at predicting air temperatures in the zones with larger temperature gradients, in contrast with the central spaces, where temperature gradients are of smaller extent and the CFD predictions were more accurate.

4. Conclusions

A methodology has been developed for the calibration of CFD models of rooms and buildings from experimental results. The application of the methodology has shown satisfactory results, finding a maximum error of 9% between the CFD model and the experimental model. In this work it has been shown that the CFD model calibrated can be used to predict the air temperature distribution at any point of the room. Validated 3D models can be a useful tool to assess multiple changes in boundary conditions that would be otherwise very difficult to reproduce in experimental test due to limitations in the number of sensor available and uncertainty and the complexity of changing boundary conditions in a real physical facility.

The biggest difficulty encountered in the CFD model is the modelling of the HVAC split unit, where the inner conduit shape showed satisfactory results with respect to the experimental results. It is worth to highlight the difficulties experienced at collecting reliable surface temperature measurements in the experimental tests. The surface temperatures collected were taken when the steady state was reached at the end of the experiments. As previous authors have [4], inaccurate results at some specific points of the model were to be expected. In the experiments performed, these differences were more remarkable in zones where the temperature gradient was higher, like in the areas closer to the walls, floor and ceiling surfaces and also in the zones near the HVAC equipment.

Summarising the methodology it is necessary to first consider the geometry of the computational domain, where it is advisable to eliminate obstacles and elements to simplify the calibration of the CFD model. Subsequently the meshing, which must be optimised by the GCI method, and find the mesh with a balance between precision and computational cost. Another important aspect is the turbulence model and the wall function chosen, presenting the k-ε model with scalable wall function and y + 11.25 satisfactory results. Regarding the solver of ANSYS CFX, in the method based on pressure and using air as the ideal gas, good results are obtained. Another important aspect is to monitor the variables of interest to be studied, such as the temperature of certain points within a 2D plane. The boundary conditions must be measured when the experimental test reaches the steady state, and in case of stratification, the variable temperature conditions with the height present better results. Finally, in the case of bad convergence, the transient model can be used with a small time step until reaching the steady state.

The highlight of this work is the methodology carried out to calibrate the CFD model with experimental results. The methodology is useful for other researchers to calibrate the CFD model of building rooms. In addition, the calibrated CFD model can be used to study the effect of different boundary conditions on comfort or energy demand. CFD analysis reveals as a powerful technique to overcome the limitations of physical experiments where only few sensors can be installed and the boundary conditions cannot be changed easily.

As future direction of this work is to reduce the computational cost and simulation time. The calculation of the complete building or the annual simulation for the evaluation of demand or comfort is a procedure that is very computationally expensive. For this, it is necessary to use reduced and simplified CFD solver, oriented specifically to buildings. This simplified programme can be implemented in thermal building simulation programmes and can be very useful for design engineers.

Author details

Alejandro Rincón Casado[1*], Magdalena Hajdukiewicz[2,3], F. Sánchez de la Flor[4] and Enrique Rodríguez Jara[4]

1 Mechanical Engineering, University of Cadiz, Cadiz, Spain

2 Department of Civil Engineering, National University of Ireland, Galway, Ireland

3 Ryan Institute, National University of Ireland, Galway, Ireland

4 Department of Machines and Thermal Motors, University of Cadiz, Cadiz, Spain

*Address all correspondence to: alejandro.rincon@uca.es

IntechOpen

© 2020 The Author(s). Licensee IntechOpen. This chapter is distributed under the terms of the Creative Commons Attribution License (http://creativecommons.org/licenses/by/3.0), which permits unrestricted use, distribution, and reproduction in any medium, provided the original work is properly cited.

References

[1] Chen Q, Srebric J. A procedure for verification, validation, and reporting of indoor environment CFD analyses. HVAC&R Research. 2002:201-216. DOI: 10.1080/10789669.2002.10391437

[2] Srebric J, Chen Q. An example of verification, validation, and reporting of indoor environment CFD analyses (RP-1133). In: ASHRAE Transactions. 2002. pp. 185-194

[3] Chen Q. Ventilation performance prediction for buildings: A method overview and recent applications. Building and Environment. 2009;**44**(4): 848-858

[4] Hajdukiewicz M, Geron M, Keane M. Formal calibration methodology for CFD models of naturally ventilated indoor environments. Building and Environment. 2013;**59**: 290-302

[5] Oberkampf W, Trucano T. Verification and validation benchmarks. Nuclear Engineering and Design. 2008; **238**(3):716-743

[6] VDI - The Association of German Engineers. VDI - 6019 Part 2- Engineering Methods for the Dimensioning of Systems for the Removal of Smoke from Buildings - Engineering Methods. Dusseldorf: Gesellschaft Technische Gebaudeausrustung; 2009

[7] Stamou A, Katsiris I. Verification of a CFD model for indoor airflow and heat transfer. Building and Environment. 2006;**41**(9):1171-1181

[8] Kim T, Kato S, Murakami S, Rho J-w. Study on indoor thermal environment of office space controlled by cooling panel system using field measurement and the numerical simulation. Building and Environment. 2005;**40**(3): 301-310

[9] Lin Z, Tian L, Yao T, Wang Q, Chow T. Experimental and numerical study of room airflow under stratum ventilation. Building and Environment. 2011;**46**(1):235-244

[10] Yongson O, Baduddin I, Zainal Z, Narayana P. Airflow analysis in an air conditioning room. Building and Environment. 2007;**42**(3):1531-1537

[11] Chafi F, Halle S. Three dimensional study for evaluating of air flow movements and thermal comfort in a model room: Experimental validation. Energy and Buildings. 2011;**43**(9): 2156-2166

[12] Rincón-Casado A, Sánchez de la Flor F, Chacón Vera E, Sánchez Ramos F. New natural convection heat transfer correlations in enclosures for building performance simulation. Engineering Applications of Computational Fluid Mechanics. 2017: 1994-2060. DOI: 10.1080/19942060. 2017.1300107

[13] Ansys I. Ansys CFX. Release V 17.0. 2017

[14] Nielsen P, Allard F, Awbi H, Davidson L, Schalin A. Computational fluid dynamics in ventilation design - REHVA guide book n. 10. International Journal of Ventilation. 2007;**6**(3): 291-293

[15] Celik I, Ghia U, Roache P, Christopher. Procedure for the estimation and reporting of uncertainty due to discretization in CFD applications. Journal of Fluids Engineering, Transactions of the ASME. 2008;**130**(7):078001

[16] Roache J. Perspective: A method for uniform reporting of grid refinement studies. Journal of Fluids Engineering. 1994;**116**(3):405-413

[17] Richardson L. The approximate arithmetical solution by finite differences of physical problems involving differential equations, with an application to the stresses in a masonry dam. Philosophical Transactions of the Royal Society of London. Series A, Containing Papers of a Mathematical or Physical Character. 1911;**210**:459-470. DOI: 10.1098/rsta.1911.0009

Chapter 2

Scaling Investigation of Low Prandtl Number Flow and Double Diffusive Heat and Mass Transfer over Inclined Walls

Mubashir O. Quadri, Matthew N. Ottah,
Olayinka Omowunmi Adewumi and Ayowole A. Oyediran

Abstract

This paper presents an essential study of scale analysis and double diffusive free convection boundary layer laminar flow of low Prandtl fluids over an inclined wall kept at uniform surface temperature. Buoyancy effect (N) was considered for an assisting flow when $N \geq 0$, which implies that the thermal and solutal forces are consolidating each other to help drive the fluid flow in the same direction. Scale analysis and similarity transformation methods are used to obtain the governing equations, and the resulting system of coupled ordinary differential equations (ODEs) is solved with the differential transform method (DTM). Results for the distributions of velocity, temperature, and concentration boundary layer of the fluid adjacent to the wall are presented. The study includes the effects of the ratio of solutal buoyancy to thermal buoyancy and important dimensionless parameters used in this work with varying angles of inclination of the wall on fluid flow and heat transfer.

Keywords: scale analysis, free convection, double diffusion, low Prandtl flow, boundary layer

1. Introduction

The impact of temperature and species concentration distribution on heat and mass transfer of fluid flow has received renewed interest to researchers and the academic community due to its multiple application areas notably in physical and chemical processes, food and manufacturing industries, geophysics, oceanography, and photosynthesis. These occurrences of thermo-solutal convection not only involve temperature variation but also concentration variation. Free convection problems driven only by temperature difference have been studied extensively by many investigators notably Akter et al. [1], Schlichting [2], Venkateswara [3], Gebhart and Pera [4], Khair and Bejan [5], and Mongruel et al. [6]. Species concentration variation sometimes plays a major role in creating the buoyancy needed in driving flow and influencing rate of heat transfer. Double diffusion has been studied by few investigators over the years. The pioneer of this area of research is the work of Gebhart and Pera in 1971 [4] where they investigated the combined buoyancy effects of thermal and mass diffusion on natural convection flow. Also, Bejan and Khair [7] carried out some analysis on heat and mass transfer by natural convection in porous media. Furthermore, the Schmidt number is the appropriate number in the concentration equation for $Pr < 1$ regime, while in the $Pr > 1$, regime Lewis number is the appropriate dimensionless number for vertical walls and this extends to inclined walls. This important criterion is sometimes omitted from heat and mass transfer studies. Allain et al. [8] also considered the problem of combined heat and mass transfer convection flows over a vertical isothermal plate. These contributors used a combination of integral and scaling laws of Bejan for their investigations. Their work was restricted to cases where two buoyancy forces aid each other; however, it was observed that heat diffusion is always more efficient than mass diffusion meaning that Lewis number is always greater than unity in most cases. It has been recommended in some previous works that more numerical or experimental results covering a wide range of Prandtl and Schmidt numbers are needed to be obtained by further investigations.

Some other research studies carried out were by Angirasa and Peterson [9], who considered free convection due to combined buoyancy forces for $N = 2$ in a thermally stratified medium, and, recently, other contributors have considered flow of power law fluids in saturated porous medium due to double diffusive free convection [10]. Other effects such as Soret and Dufour forces in a Darcy porous medium were considered by Krishna et al. [11]. The problem of mass transfer flow through an inclined plate has generated much interest from astrophysical, renewable energy system, and also hypersonic aerodynamics researchers for a number of decades [1]. It is important to note that combined heat and mass flow over an isothermal inclined wall has received little contributions from scholars [12, 13]. The key notable ones in the literature include the general model formulation of natural convection boundary layer flow over a flat plate with arbitrary inclination by Umemura and Law [14]. Their results showed that flow properties depend on both the degree of inclination and distance from the leading edge. Other investigations considered the problem of combined heat and mass transfer by MHD free convection from an inclined plate in the presence of internal heat generation of absorption [15], natural convection flow over a permeable inclined surface with variable temperature, momentum, and concentration [16], investigations on combined heat and mass transfer in hydro-magnetic dynamic boundary layer flow past an inclined plate with viscous dissipation in porous medium [17], a study on micro-polar fluid behavior in MHD-free convection with constant heat and mass flux [18] and investigations on mass transfer flow through an inclined plate with porous medium [19].

However, research conducted to critically analyze fluid behavior with the effect of species concentration and thermal diffusion on heat and mass transfer particularly for low Prandtl flows past an inclined wall is very rare. This gap has been captured in this study. The objective of this research is to investigate the effect of combined heat and species concentration involving a low Prandtl number fluid flow over an inclined wall using the method of scale analysis in formulation of the model along with the similarity transformation technique to convert partial differential equations to ordinary differential equation. The resulting dimensionless coupled and non-linear equations are solved using differential transform method. The numerics of the computation are discussed for different values of dimensionless parameters and are graphically presented.

2. Problem formulation and scale analysis

The problem of combined heat and mass transfer over a heated semi-infinite inclined solid wall is considered. The fluid is assumed to be steady, Newtonian, viscous, and incompressible. It is assumed that the wall is maintained at uniform surface temperature T_w and concentration C_w and it is immersed in fluid reservoir at rest which is kept at uniform ambient temperature T_∞ and concentration C_∞ such that $T_w > T_\infty$ and $C_w > C_\infty$. Boundary layer flow over an inclined wall driven by both thermal gradient and concentration gradient, respectively, are thereby set up due to the difference between wall values and quiescent fluid values. Hence, it is called combined heat and mass transfer phenomenon over an inclined wall (**Figure 1**).

This problem is governed by the non-linear and coupled conservation equations. Using the Boussinesq approximation and boundary layer simplifications, we have the following:

Continuity equation

$$\frac{\partial u}{\partial x} + \frac{\partial v}{\partial y} = 0, \tag{1}$$

Momentum equation

$$u\frac{\partial v}{\partial x} + v\frac{\partial v}{\partial y} = \vartheta\frac{\partial^2 v}{\partial x^2} + \rho g\beta_T(T - T_\infty)\cos\alpha + \rho g\beta_c(C - C_\infty)\cos\alpha \tag{2}$$

Energy equation

$$u\frac{\partial T}{\partial x} + v\frac{\partial T}{\partial y} = \alpha_T\frac{\partial^2 T}{\partial x^2} \tag{3}$$

Species concentration equation

$$u\frac{\partial C}{\partial x} + v\frac{\partial C}{\partial y} = D\frac{\partial^2 C}{\partial x^2} \tag{4}$$

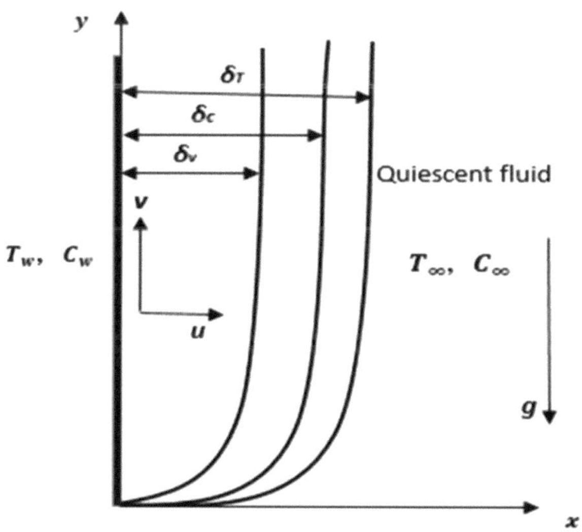

Figure 1.
Physical model of double diffusive free convection over vertical wall.

Here, u is the velocity along x-axis, v is the velocity along y-axis or along the vertical wall, T is the temperature, and C is the concentration. These equations are subject to the boundary conditions given by

$$u = 0, v = 0, T = T_w, C = C_w \; at \; x = 0 \tag{5}$$

$$u = 0, v = 0, T = T_\infty, C = C_w \; at \; x = \infty \tag{6}$$

Following the procedures as outlined by Khair and Bejan [5], it can be clearly shown that for low Prandtl number flows, the velocity, concentration, and temperature boundary layer scales (for N = 0) are:

$$\delta_v \sim HRa_*^{-\frac{1}{4}} Pr^{\frac{1}{4}} \tag{7}$$

$$\delta_T \sim H\left(Ra_*^{-\frac{1}{4}} Pr\right)^{-\frac{1}{4}} \tag{8}$$

$$\delta_C \sim H\frac{D}{\propto} Ra_*^{-1/4} Pr^{-3/4} \tag{9}$$

While vertical velocity (v) scales as.

$$v \sim \frac{\propto}{y}\left(Ra_*^{-\frac{1}{4}} Pr\right)^{1/2} \tag{10}$$

The similarity variable for velocity layer scales as

$$\eta = \frac{x}{\delta_v} = \frac{x}{y}\left(\frac{Ra_*}{Pr}\right)^{\frac{1}{4}} \tag{11}$$

And corresponding stream function ψ scales as

$$\psi \sim \alpha Ra_*^{1/4} Pr^{3/4} F(\eta) \tag{12}$$

where $F(\eta)$ is the velocity function.

Boussinesq approximation is used in Eq. (2), and the PDEs are reduced to a set of coupled ODEs using similarity variable η. It can easily be shown that for the inner layer of low Prandtl number flows, the dimensionless momentum, energy and concentration equations give:

$$\frac{-3}{4} ff'' + \frac{(f')^2}{2} = -f'''(\eta) + \theta(\eta)\cos\alpha + NC(\eta)\cos\alpha \tag{13}$$

$$\frac{3}{4} Prf\theta' = \theta'' \tag{14}$$

$$\frac{3}{4} Pr.LefC' = C'' \tag{15}$$

Eqs. (13)–(15) are solved for temperature θ, velocity f', and concentration C, respectively, subject to the boundary conditions in Eq. (16).

$$\begin{gathered} f'(0) = f(0) = 0, \theta(0) = \gamma(0) = 1 \; at \; \eta = 0 \\ f'(\infty) = \theta(\infty) = \gamma(\infty) = 0 \; at \; \eta = \infty \end{gathered} \tag{16}$$

$\overline{\xi}$ the similarity variable for concentration layer is defined as $\overline{\xi} = \frac{x}{\delta_c}, for\ \delta_c > \delta_v$ where δ_v is the thin viscous layer closest to the wall and δ_c is the concentration boundary layer. The attendant stream function is obtained as $\psi \sim DRa_*^{1/4} Pr^{-1/4} \overline{F}(\eta)$.

The ordinary differential equations governing the momentum, energy, and species concentration become:

$$\frac{-3}{4}\overline{FF}'' + \frac{\left(\overline{F}'\right)^2}{2} = -LeSc\overline{F}'''(\eta) + \overline{\theta}(\eta)\cos\alpha + N\overline{C}(\eta)\cos\alpha \tag{17}$$

$$\frac{3}{4}\overline{F}\theta' = Sc.\overline{\theta}'' \tag{18}$$

$$\frac{3}{4}\overline{F}C' = Pr.\overline{C}'' \tag{19}$$

In the outer layer, where there is inertia-buoyancy balance, ξ which is the similarity variable for thermal layer is defined as $\xi = \frac{x}{\delta_T}$, and the associated stream function obtained as $\psi \sim \alpha Ra_*^{1/4} Pr^{1/4} F(\eta)$.

The resulting dimensionless equations for low Prandtl number flow are:

$$\frac{-3}{4}FF'' + \frac{(F')^2}{2} = -PrF'''(\eta) + \tilde{\theta}(\eta)\cos\alpha + N\tilde{C}(\eta)\cos\alpha \tag{20}$$

$$\frac{3}{4}F\tilde{\theta}' = \tilde{\theta}'' \tag{21}$$

$$\frac{3}{4}LeF\tilde{C}' = \tilde{C}'' \tag{22}$$

In further works, these equations will be solved asymptotically as $Pr \to 0$ to obtain approximate analytical results.

2.1 Method of solution

The differential transform method is used to solve the non-linear similarity Eqs. (13)–(15) subject to boundary conditions in Eq. (16). The procedure to convert the PDEs to ODEs is outlined below.

$$\text{Let } Z_1 = \theta;\ Z_2 = Z_1' = \theta';\ Z_3 = F;\ Z_4 = F' = Z_3';\ Z_5 = Z_4' = F'';\ Z_6 = C;\ Z_7 = Z_6' = C'. \tag{23}$$

Such that the governing equations of motion become:

$$\left.\begin{aligned} Z_1' &= Z_2 \\ Z_2' &= -\frac{3}{4}\text{Pr}.Z_2Z_3 \\ Z_3' &= Z_4 \\ Z_4' &= Z_5 \\ Z_5' &= Z_1 cos\alpha - \frac{1}{2}Z_4^2 + \frac{3}{4}Z_3Z_5 + NZ_7 cos\alpha \\ Z_6' &= Z_7 \\ Z_7' &= -\frac{3}{4}\text{Le.Pr}.Z_3Z_7. \end{aligned}\right\} \tag{24}$$

Due to limitation of convergence of the classical DTM which is only valid near $\eta = 0$, the multi-step transformation is used. Carrying out multi-step differential transformations, we have:

$$F_i(\eta) = \sum_{i=0}^{k} \left(\frac{\eta}{Hi}\right)^i \bar{\bar{F}}_i(k)$$

$$\theta_i(\eta) = \sum_{i=0}^{k} \left(\frac{\eta}{Hi}\right)^i \bar{\bar{\theta}}_i(k)$$

$$C_i(\eta) = \sum_{i=0}^{k} \left(\frac{\eta}{Hi}\right)^i \bar{\bar{C}}_i(k) \tag{25}$$

Where i = 0, 1, 2, 3.... n indicates the ith sub-domain; k = 0, 1, 2, ... m represents the number of terms of the power series. Hi represents the sub-domain interval, and $\bar{\bar{F}}_i(k), \bar{\bar{\theta}}_i(k), \bar{\bar{C}}_i(k)$ are the transformed functions, respectively.

The transformation of the associated boundary conditions follows as:

$Z_1(0) = 1,\ Z_2(0) = a, Z_3(0) = 0, Z_4(0) = 0, Z_5(0) = b,\ Z_6(0) = 1,\ Z_7(0) = c$

where a, b, and c are obtained by solving the system of algebraic simultaneous equations, and the results obtained are shown in **Table 1**.

$$\left.\begin{aligned}
Z_1(k+1) &= \frac{Z_2\ (k)}{k+1} \\
Z_2(k+1) &= \frac{-\frac{3}{4}\mathrm{Pr}}{k+1}\sum_{i=0}^{k}(Z_3(i).Z_2(k-i)) \\
Z_3(k+1) &= \frac{Z_4\ (k)}{k+1} \\
Z_4(k+1) &= \frac{Z_5\ (k)}{k+1} \\
Z_5(k+1) &= \frac{1}{k+1}\left[Z_1(k)cos\alpha - \frac{1}{2}\sum_{i=0}^{k}Z_4(i)Z_4(k-i) + \frac{3}{4}\sum_{i=0}^{k}Z_3(i)Z_5(k-i) + N.Z_7(k)cos\alpha\right] \\
Z_6(k+1) &= \frac{Z_7\ (k)}{k+1} \\
Z_7(k+1) &= \frac{-\frac{3}{4}.\mathrm{Pr}.Le}{k+1}\sum_{i=0}^{k}[Z_3(i)Z_7(k-i)]
\end{aligned}\right\} \tag{26}$$

Parameters Domain: $Pr < 1, Le \gg 1, Le \ll 1$, and $N \geq 0$
The local Nusselt number: $Nu - [Gr_*(y)]^{\frac{1}{4}}\,(\theta')_{\eta=0}$ The Local Sherwood number: $Sh = -[Gr_*(y)]^{\frac{1}{4}}(C')_{\eta=0}$ The shearing stress on the plate: $\tau_w = \frac{\propto}{y^2} Ra_*(y)^{3/4} Pr^{1/4}(f'')_{\eta=0}$ The coefficient of skin friction: $C_f = \frac{\propto \vartheta Pr^{1/4} Ra_*(y)^{3/4}}{u^2 y^2} 2(f'')_{\eta=0}$

Table 1.
Important dimensionless parameters of interest.

3. Results and discussion

Table 1 shows the expressions for the dimensionless parameters that are of interest in this work. The solutions to Eqs. (13)–(15) subjected to Eq. (16) solved using the multi-step DTM method are presented graphically in the figures below. The results shown are for Prandtl numbers of 0.01, 0.1, 0.5, and 0.72, respectively.

Figure 2 shows the profiles of local skin friction against Prandtl number for various angles of inclination at a constant Lewis number. It could be observed from the plots that the shearing stress decreases as the Prandtl number increases for all the plate angles considered. More importantly, it is illustrated by the graphs that the local skin friction also decreases with increase in the buoyancy ratio and also with respect to the angle of inclination of the wall.

Figure 3 shows the plots of local Nusselt number against Prandtl number for various angles of inclination at a constant Lewis number. It is clearly seen from the results that the rate of heat transfer (Nusselt number) increases as the Pr increases for all the wall inclination angles. Also to note is the fact that Nusselt number increases as the buoyancy ratio is increased.

Figure 4 shows the results of how the local Sherwood number changes as the Lewis number is increased for various angles of inclination at a constant Prandtl number. It is noted from the graphs that the local Sherwood number increases as the Lewis number increases for all angles of inclination. Also, it is observed from the figures that the rate of increase of Sherwood number is dependent on N as well as the angle of inclination of the wall.

In **Figure 5**, the similarity profiles of the effect of the buoyancy ratio and the angles of inclination on the dimensionless velocities at fixed Prandtl and Lewis numbers is presented. It could be interpreted from the results for N = 0 that a maximum velocity is obtained for a vertical wall while at an angle of 60°, the minimum velocity. Also, it is clearly seen from the plots that aside from the vertical wall possessing the maximum velocity, its vertical velocity value for N = 1 is higher when compared to the case when N = 0. When the buoyancy ratio N is further increased to 1.5, the velocity for the vertical wall also increases. The trend in the figure also shows that increasing the inclination angle increases the velocity boundary layer thickness for all buoyancy ratio.

In **Figure 6**, the similarity profiles of dimensionless temperature for the wall angles of inclinations considered in this study at constant Lewis number are presented. The temperature profiles show that despite the varying buoyancy ratio, the thermal boundary layer thickness increases as the inclination angle of the wall increases.

Figure 7 presents the similarity profiles of dimensionless species concentration distributions for the angles of inclination of the wall at a constant Lewis number. It is clearly seen that as the buoyancy ratio is increased, the concentration boundary layer thickness has a decreasing trend under this same flow configuration but as the angle of inclination increases, δ_c increases.

Figure 8 shows the plots of the coupled similarity profiles of dimensionless temperature, species concentration, and velocity for wall inclination angle of 60° under the constant Lewis number of 10. It can be seen from the results that increasing N has negligible effect on the trio of velocity, concentration and temperature boundary layer thicknesses for a fixed wall angle. However, the effect of N is very noticeable in the vertical velocity values which is clearly higher when N = 1.5 compared to when N = 0.

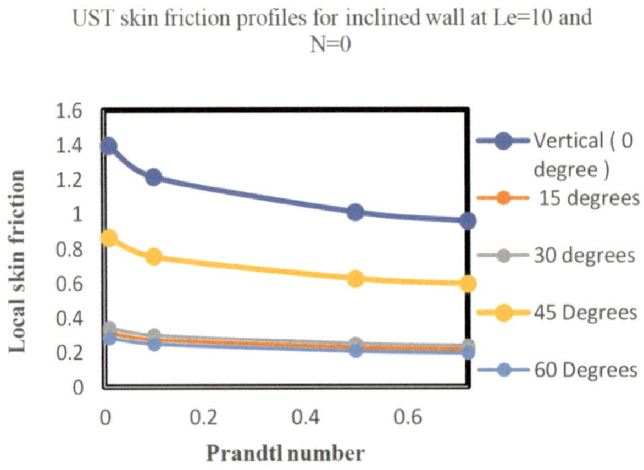

(a)

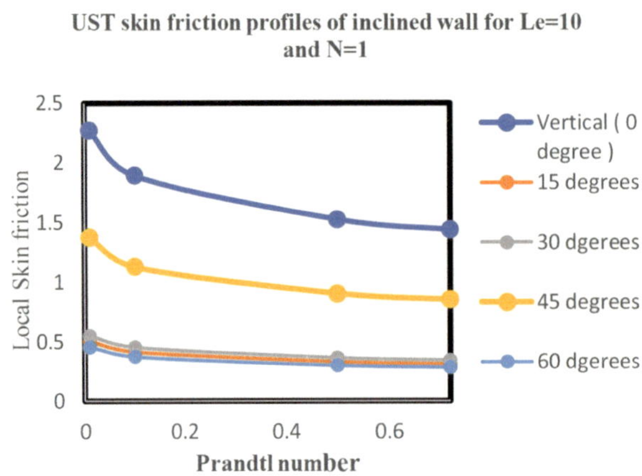

(b)

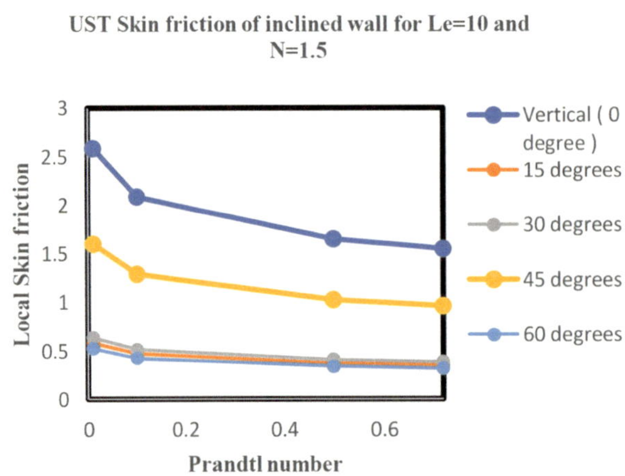

(c)

Figure 2.
Similarity profiles of effects of Prandtl number on skin friction for (a) N = 0; (b) N = 1; (c) N = 1.5 at constant Lewis number of 10.

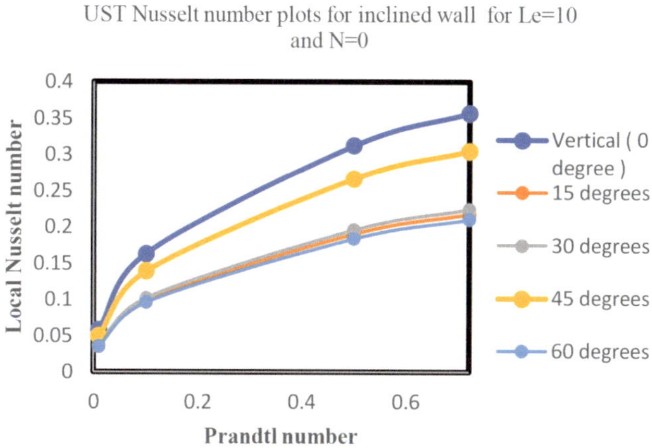

(a)

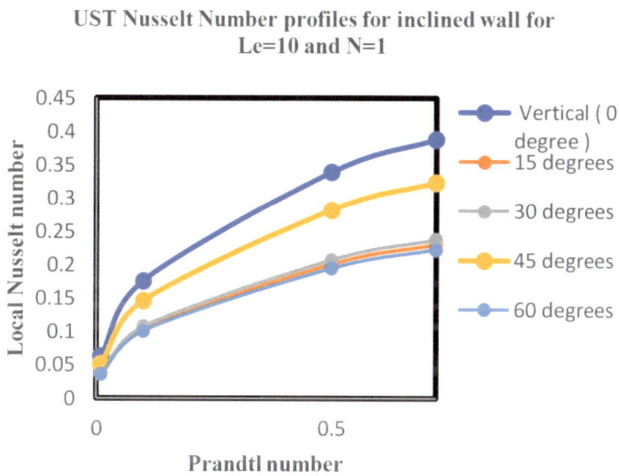

(b)

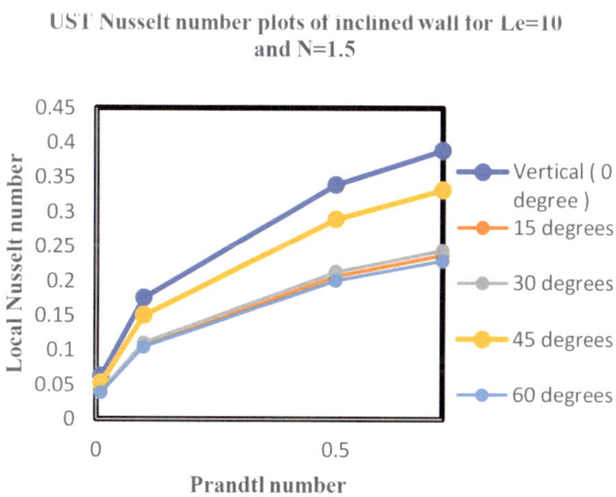

(c)

Figure 3.
Similarity profiles of effects of Prandtl number on Nusselt number for (a) N = 0; (b) N = 1; (c) N = 1.5 at constant Lewis number of 10.

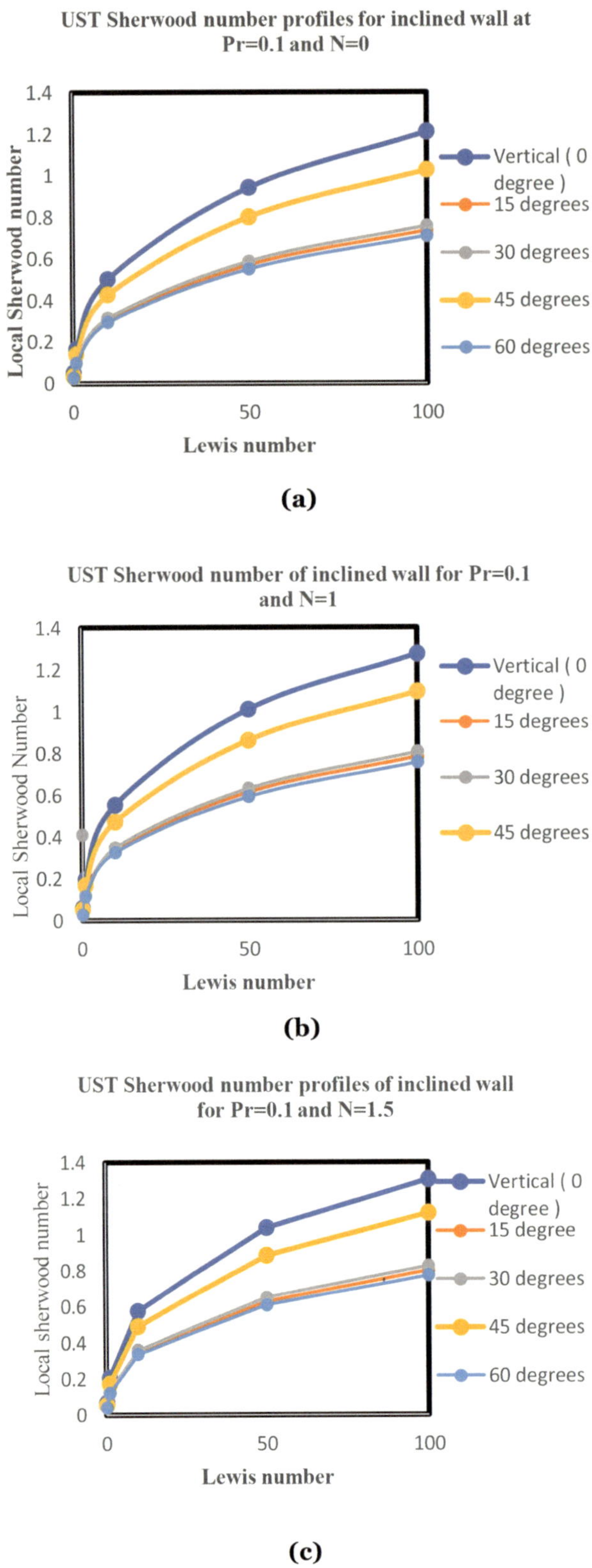

Figure 4.
Similarity profiles of effects of Lewis number on Sherwood number for (a) N = 0; (b) N = 1; (c) N = 1.5 at constant Pr of 0.1.

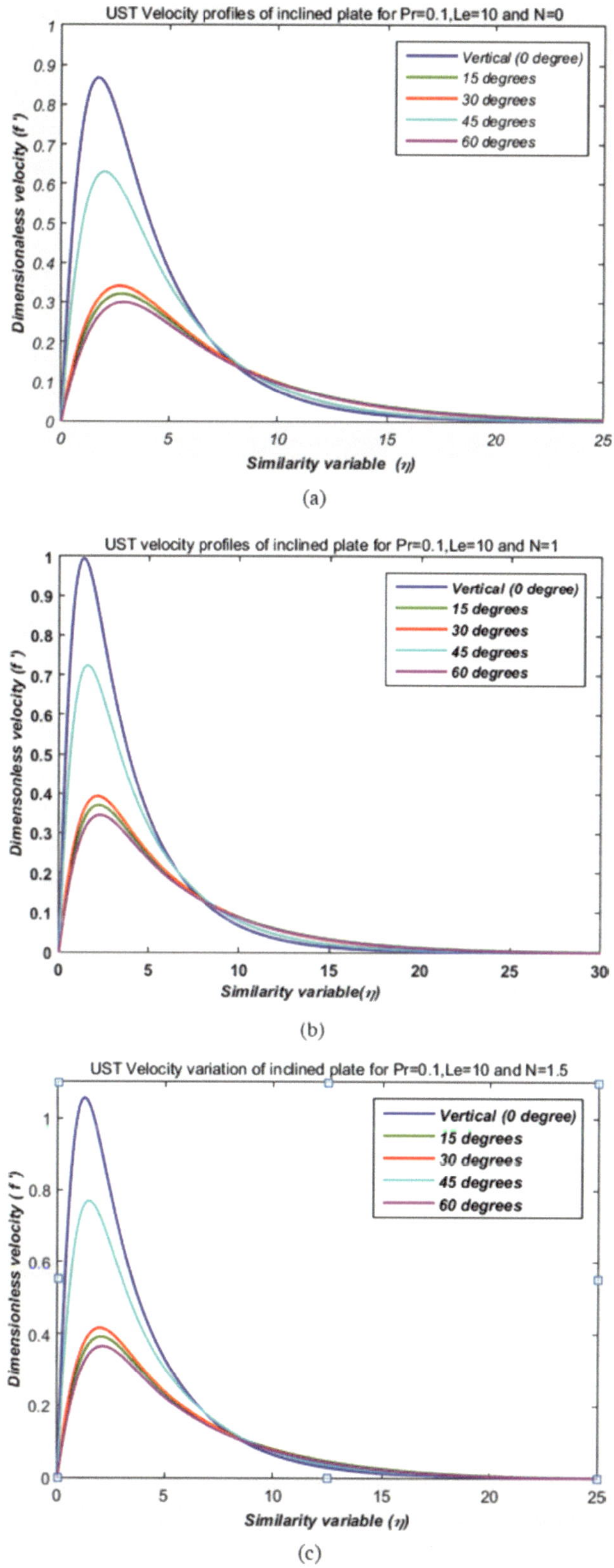

Figure 5.
Similarity profiles of dimensionless velocity for (a) $N = 0$, (b) $N = 1$, (c) $N = 1.5$ at $Pr = 0.1$ and $Le = 10$.

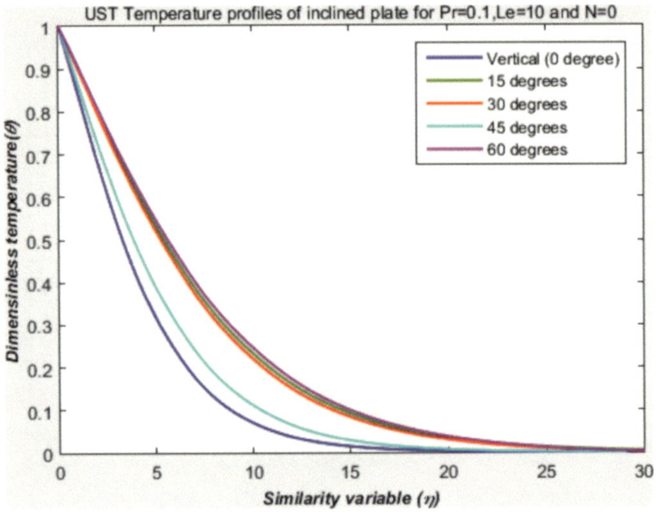

(a)

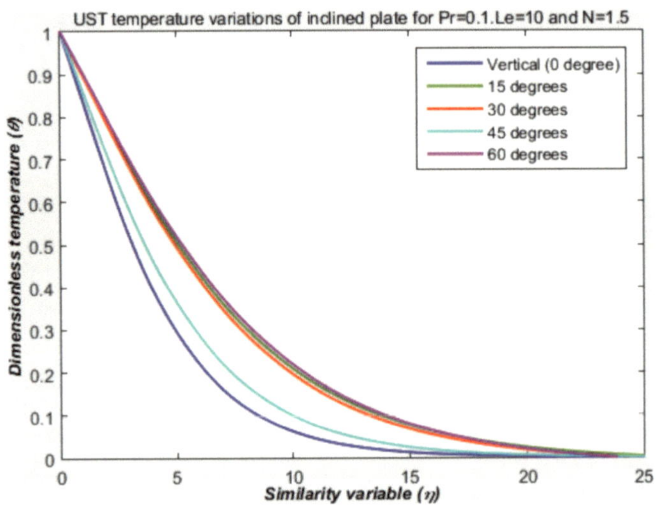

(b)

Figure 6.
Similarity profiles of dimensionless temperature for (a) $N = 0$, (b) $N = 1.5$ at $Pr = 0.1$ and $Le = 10$.

4. Conclusion

The problem of double diffusive convection and its associated boundary layer flow is of tremendous interest in academic research and various manufacturing and process industries because of its implications in energy and mass transfer efficiency in engineering and scientific applications. In this study, scale analysis and double diffusive free convection of low Prandtl fluid flow over an inclined wall is investigated in the presence of species concentration and thermal diffusion. The governing boundary layer equations obtained by scale analysis are numerically solved using differential transform method (DTM). The conclusions reached as a result of the parametric study conducted are presented below:

i. The velocity boundary layer thickness is maximum when the plate is in the vertical position, while it is minimum when the plate is at the inclined angle of 60° to the vertical for any value of Lewis number but within a certain buoyancy ratio.

ii. The thermal boundary thickness is maximum when the wall is inclined at 60° to the vertical and minimum when in a vertical position while keeping N constant.

iii. Thermal boundary thickness decreases with increase in Prandtl number for all angles of inclination while keeping both the Lewis number and buoyancy ratio constant.

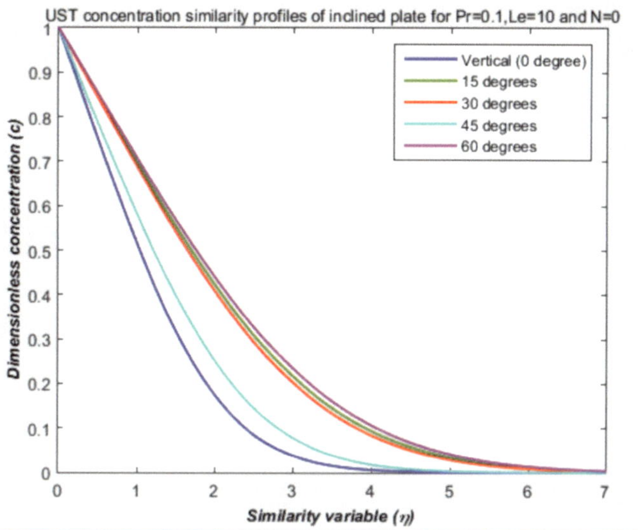

(a)

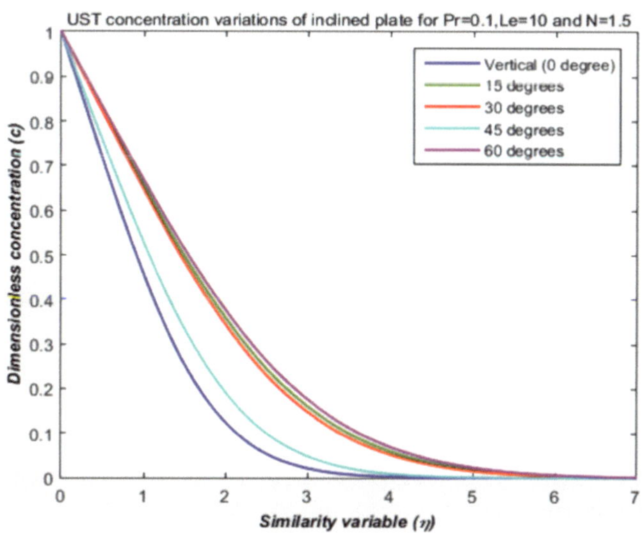

(b)

Figure 7.
Similarity profiles of dimensionless concentration for (a) $N = 0$, (b) $N = 1.5$ at $Pr = 0.1$ and $Le = 10$.

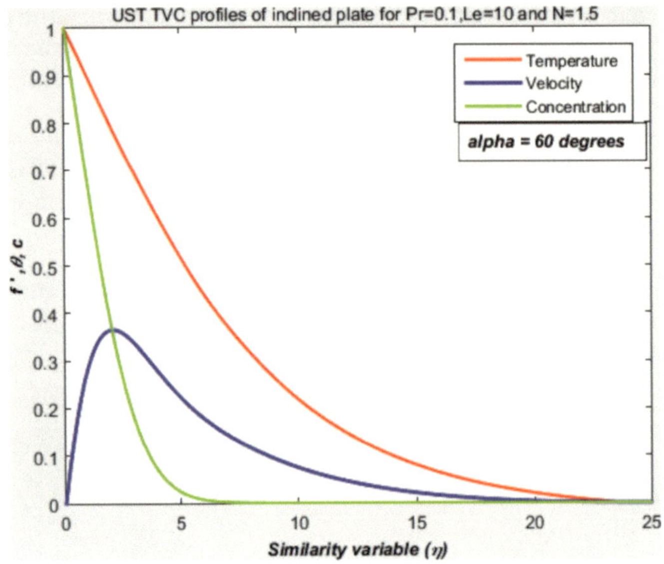

(a)

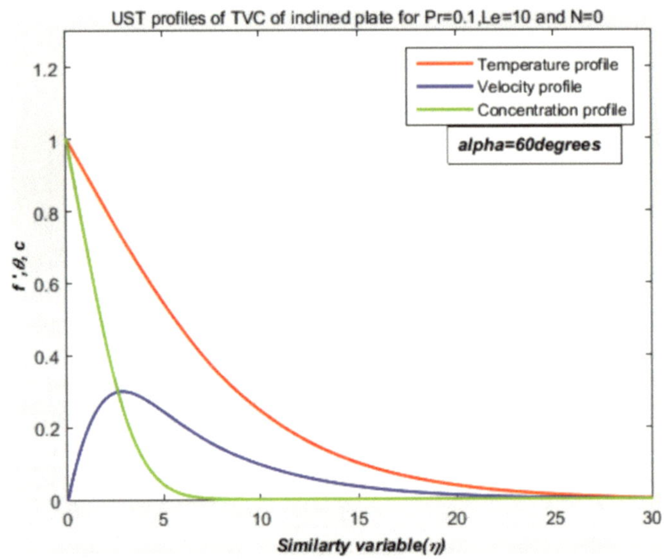

(b)

Figure 8.
Similarity profiles of dimensionless velocity, concentration, and temperature at inclination angle of 60°. (a) N = 1.5, (b) N = 0 (Pr = 0.1 and Le = 10).

iv. The concentration decreases with the increase in Lewis number for all range of values considered for N and all angles of inclination which is in agreement with Akter et al. [1]. Also, as N increases, both concentration and temperature increase while velocity decreases with increase in plate angles.

v. The velocity, concentration, and thermal boundary layer thicknesses increase with an increase in the angle of inclination of the wall.

DOI: http://dx.doi.org/10.5772/intechopen.90896

Nomenclature

C	chemical species concentration
D	chemical species diffusivity
N	buoyancy ratio
Nu	Nusselt number
g	gravity constant
k	thermal conductivity
T	temperature
Pr	Prandtl number
Ra_T	thermal Rayleigh number
Le	Lewis number
u	velocity component in x-direction
v	velocity component in y-direction
x	horizontal axis
y	vertical axis

Greek Symbols

β_T	coefficient of thermal expansion
β_C	coefficient of specie expansion
	thermal diffusivity of fluid
$c'(0)$	wall derivative of dimensionless concentration
δ_v	velocity boundary layer thickness
δ_T	thermal boundary layer thickness
δ_C	concentration boundary layer thickness
ΔC	concentration difference $(C - C_\infty)$
ΔT	temperature difference $(T - T_\infty)$
η	similarity variable
θ	dimensionless temperature
$\theta'(0)$	wall derivative of dimensionless temperature
ϑ	kinematic viscosity
ρ	density of fluid
$\theta'(0)$	constant wall dimensionless heat flux
μ	dynamic viscosity
ψ	stream function

Subscript

∞	condition at infinity
w	condition at the wall

Author details

Mubashir O. Quadri, Matthew N. Ottah, Olayinka Omowunmi Adewumi*
and Ayowole A. Oyediran
Department of Mechanical Engineering, University of Lagos, Lagos, Nigeria

*Address all correspondence to: oadewunmi@unilag.edu.ng

IntechOpen

© 2020 The Author(s). Licensee IntechOpen. This chapter is distributed under the terms of the Creative Commons Attribution License (http://creativecommons.org/licenses/by/3.0), which permits unrestricted use, distribution, and reproduction in any medium, provided the original work is properly cited.

References

[1] Akter F, Islam MM, Islam A, Khan MS, Hossain MS. Chemical reaction and thermal diffusion effects on mass transfer flow through an inclined plate. Open Journal of Fluid Dynamics. 2016;**6**:62. DOI: 10.4236/ojfd.2016.61006

[2] Schlichting H. Boundary-Layer Theory. 7th ed. New York: McGraw-Hill; 1968

[3] Rajua KV, Reddyb PBA, Suneethac S. Thermophoresis effect on a radiating inclined permeable moving plate in the presence of chemical reaction and heat absorption. International Journal of Dynamics of Fluids. 2017;**13**:89-112

[4] Gebhart B, Pera L. The nature of vertical natural convection flows resulting from the combined buoyancy effects of thermal and mass diffusion. International Journal of Heat and Mass Transfer. 1971;**14**:2025-2050. DOI: 10.1016/0017-9310(71)90026-3

[5] Khair KR, Bejan A. Mass transfer to natural convection boundary layer flow driven by heat transfer. Journal of Heat Transfer. 1985;**107**:979-981. DOI: 10.1115/1.3247535

[6] Mongruel A, Cloitre M, Allain C. Scaling of boundary-layer flows driven by double-diffusive convection. International Journal of Heat and Mass Transfer. 1996;**39**:3899-3910. DOI: 10.1016/0017-9310(96)00054-3

[7] Bejan A, Khair KR. Heat and mass transfer by natural convection in a porous medium. International Journal of Heat and Mass Transfer. 1985;**28**: 909-918. DOI: 10.1016/0017-9310(85)90272-8

[8] Allain C, Cloitre M, Mongruel A. Scaling in flows driven by heat and mass convection in a porous medium. EPL (Europhysics Letters). 1992;**20**:313. DOI: 10.1209/0295-5075/20/4/005

[9] Angirasa D, Srinivasan J. Natural convection flows due to the combined buoyancy of heat and mass diffusion in a thermally stratified medium. Journal of Heat Transfer. 1989;**111**:657-663. DOI: 10.1115/1.3250733

[10] Angirasa D, Peterson GP, Pop I. Combined heat and mass transfer by natural convection with opposing buoyancy effects in a fluid saturated porous medium. International Journal of Heat and Mass Transfer. 1997;**40**: 2755-2773. DOI: 10.1016/S0017-9310(96)00354-7

[11] Murthy SK, Kumar BR, Chandra P, Sangwan V, Nigam M. A study of double diffusive free convection from a corrugated vertical surface in a Darcy porous medium under Soret and Dufour effects. Journal of Heat Transfer. 2011; **133**:092601. DOI: 10.1115/1.4003813

[12] Ferdows M, Khan MS, Bég OA, Azad MAK, Alam MM. Numerical study of transient magnetohydrodynamic radiative free convection nanofluid flow from a stretching permeable surface. Proceedings of the Institution of Mechanical Engineers, Part E: Journal of Process Mechanical Engineering. 2014; **228**:181-196. DOI: 10.1177/0954408913493406

[13] Khan MS, Karim I, Ali LE, Islam A. Unsteady MHD free convection boundary-layer flow of a nanofluid along a stretching sheet with thermal radiation and viscous dissipation effects. International Nano Letters. 2012;**2**:24. DOI: 10.1186/2228-5326-2-24

[14] Umemura A, Law CK. Natural-convection boundary-layer flow over a heated plate with arbitrary inclination. Journal of Fluid Mechanics. 1990;**219**: 571-584. DOI: 10.1017/S0022112090003081

[15] Chamka A, Khaled ARA. Simultaneously heat and mass transfer

in free convection. Industrial Engineering Chemical. 2001;**49**: 961-968. DOI: 10.1021/ie50570a025

[16] Reddy MG, Reddy NB. Mass transfer and heat generation effects on MHD free convection flow past an inclined vertical surface in a porous medium. Journal of Applied Fluid Mechanics. 2011;**4**:7-11

[17] Singh PK. Heat and mass transfer in MHD boundary layer flow past an inclined plate with viscous dissipation in porous medium. International Journal of Scientific and Engineering Research. 2012;**3**:1-11

[18] Ali LE, Islam A, Islam N. Investigate micropolar fluid behavior on MHD free convection and mass transfer flow with constant heat and mass fluxes by finite difference method. American Journal of Applied Mathematics. 2015;**3**:157-168. DOI: 10.11648/j.ajam.20150303.23

[19] Islam M, Akter F, Islam A. Mass transfer flow through an inclined plate with porous medium. American Journal of Applied Mathematics. 2015;**3**:215-220. DOI: 10.11648/j.ajam.20150305.12

Chapter 3

Bingham Fluid Simulation in Porous Media with Lattice Boltzmann Method

José Luis Velázquez Ortega

Abstract

Generate a lattice Boltzmann model (LBM), which allows to simulate the behavior of a Bingham fluid through a rectangular channel with the D2Q9 model. For this purpose, a relaxation parameter is proposed based on the rheological parameters of the Bingham model. The validation will be carried out with the solution of the movement equation, and velocity profiles will be obtained for three different Bingham numbers (Bn). Other simulations will be made in a rectangular channel in the presence of arbitrarily and randomly generated porous media. The main objective is to propose a method to predict the behavior of non-Newtonian fluids (Bingham fluid) through porous media, which have many applications in the chemical industry avoiding at the same time the expensive experimentation of these systems, with predicting models.

Keywords: lattice Boltzmann model, non-Newtonian fluids, Bingham fluid, porous media, velocity profiles

1. Introduction

A continuous medium is characterized by the fact that its atoms or molecules are so close together, in such way that they could be considered macroscopically as a homogeneous mass, whose behavior can be foreseen without considering the movement of each of its elementary particles that compose it. In this sense it is assumed that there are no gaps or separations between the particles.

The movement of fluids can have a wide variety of behaviors for both simple and complex flows (biological and food systems). In addition to this, using Reynolds number, we can know if the flow has turbulent or laminar regime.

Mass conservation is the basic principle of fluid movement, which requires that when the fluid is in motion, it moves so that the mass is preserved. The movement of fluids is governed in general by the continuity equation.

$$\frac{\partial \rho}{\partial t} + \nabla \cdot \left(\rho \vec{v}\right) = 0 \quad (1)$$

On the other hand, the Navier-Stokes equation, which in general terms corresponds to the application of Newton's second law of classical mechanics to fluid movement, is described as follows:

$$\rho\left[\frac{\partial\vec{v}}{\partial t}+\vec{v}\nabla\cdot\vec{v}\right]=-\nabla P+\mu\nabla^2\vec{v}+\rho\vec{g} \tag{2}$$

for a fluid with velocity $\vec{v}$, density ρ, pressure p, and kinematic shear viscosity μ. The Navier–Stokes equation is a second-order partial differential equation, which can have an analytical solution only for a small number of cases.

There are basically three types of fluids from the point of view of fluid dynamics, which are called Newtonians, non-Newtonians, and viscoelastic.

1.1 Newtonian fluids

In this type of fluid, the shear stress or shear force per unit area is proportional to the viscosity gradient:

$$\tau_{yx}=\mu\left(\frac{dv_x}{dy}\right) \tag{3}$$

In Eq. (3), τ_{yx} is the shear stress, (dv_x/dy) is the shear rate, and μ is the viscosity, and its graphic representation can be seen in **Figure 1**.

All gases, liquid water, and liquids of single molecules (ammonia, alcohol, benzene, petroleum, chloroform, butane, etc.) are Newtonian. Many food materials such as milk, apple juice, orange juice, wine, and beer have a Newtonian behavior [1].

1.2 Non-Newtonian fluids

When the relationship between shear stress and shear rate is not linear, the fluid is called non-Newtonian. There are many these types of fluids, and their behaviors are shown in **Figure 1**.

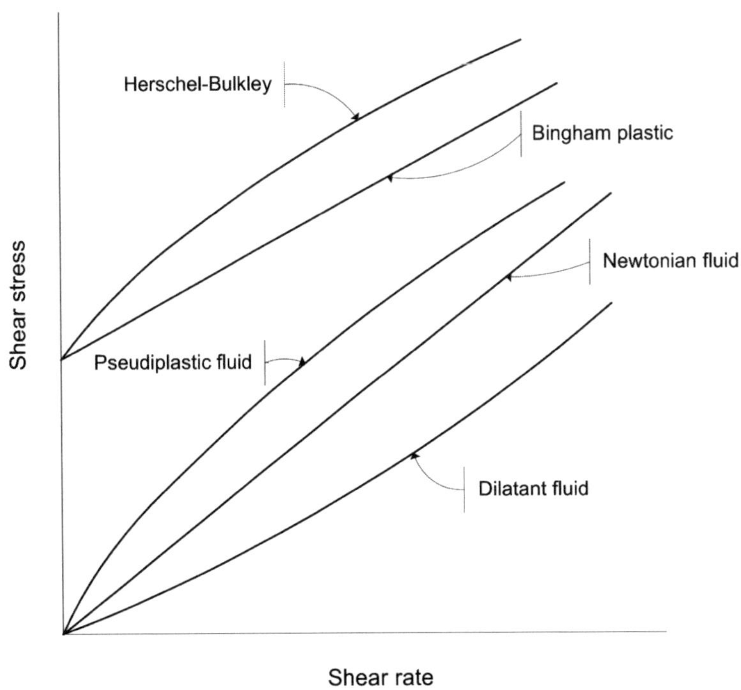

Figure 1.
Types of times-independent flow behavior [2].

In the case of these fluids, the viscosity is no longer constant; therefore, the relationship between the shear stress and the shear rate of the fluid is no longer linear. For this reason, a new term is now introduced and is known as apparent viscosity or also known as shear rate-dependent viscosity [3].

Some fluids have a yield stress, from which the fluid begins to move. Below this tension the shear rate would be zero. This relationship is not linear or, if it is, it does not pass through the origin [4]. Complex mixtures are considered non-Newtonian fluids: grouts, pastes, gels, polymer solutions, etc. Most non-Newtonian fluids are mixtures with constituents of very different sizes. For example, toothpaste is composed of solid particles suspended in an aqueous solution of several polymers. Solid particles are much larger than water molecules, and polymer molecules are much larger than water molecules.

Much of the research that is carried out in the field of non-Newtonian fluids has been focused in the measure of its shear stress-shear strain curves and to look for mathematical descriptions of these curves. The study of the behavior of the flow of materials is called rheology (a term that originates from Greek words that give the meaning of "the study of flow"); thus, diagrams such as the one shown in **Figure 1** are often called rheograms.

In the case of Bingham fluids, sometimes called Bingham plastics, they resist a small shear force indefinitely, but they flow easily under large shear stresses. In other words, at low stresses the plastic viscosity of Bingham is infinite, and at greater stresses the viscosity decreases with the increase in the velocity gradient. Examples are bread dough, toothpaste, apple sauce, some paints, plastics, mayonnaise, ketchup, aleas, and some grouts [5, 6].

2. Bingham model

Eugene C. Bingham described the paintings with this model in 1919, published in his book *Fluidity and Plasticity* in 1922. The model was analyzed by Oldroyd (1947), Reiner (1958), and Prager (1961).

The main feature of the Bingham model is the yield stress, necessary for the fluid to deform or flow. Above this minimum yield stress, the fluid begins to move. If this yield stress is not exceeded, the fluid behaves like a rigid or quasi-rigid body, with zero shear rate.

The relationship between the shear stress and the velocity gradient is linear, but it does not go through the origin for a Bingham plastic (**Figure 1**); its mathematical model is given by

$$\tau_{yx} = \tau_0 + \mu_B \overbrace{\frac{dv_x}{dy}}^{\dot{\gamma}} \quad \text{if} \ \left|\tau_{yx}\right| > \tau_0 \tag{4}$$

$$\frac{dv_x}{dy} = 0; \ \text{if} \ \left|\tau_{yx}\right| < \tau_0$$

$$\mu(\dot{\gamma}) = \mu_B + \frac{\tau_0}{\dot{\gamma}} \quad \text{for} \ \tau_{yx} > \tau_0 \tag{5}$$

$$\dot{\gamma} = 0; \ \text{for} \ \tau_{yx} \leq \tau_0$$

where τ_0 is the yield stress, μ_B is the plastic viscosity of Bingham, and $\mu(\dot{\gamma})$ is the apparent viscosity, which decreases with the increase in the magnitude of the shear rate $\dot{\gamma}$.

Other examples of Bingham-type fluids in foods are tomato sauce, whipped cream, whipped egg white, margarine, and mustard-type condiments [7, 8].

2.1 Analytical solution of the equation of motion in the case of a Bingham-type fluid

In this section, we will obtain a mathematical expression that shows how to quantify the velocity profile for a non-Newtonian fluid for the Bingham model.

In the case of a Poiseuille flow, the flow of a Bingham-type fluid in the x-direction is considered, between two plates separated by a distance 2H, taking into account the steady-state conditions, constant cross section, absence of gravitational effects and isothermal flow, and being incompressible, such as the one shown in **Figure 2**.

From the equation of motion, we can obtain the stress profile, as well as the velocity profile:

$$\rho\left[\frac{\partial \vec{v}}{\partial t} + \vec{v}\cdot\nabla\vec{v}\right] = \nabla\cdot\bar{\bar{\tau}} - \nabla P + \rho\vec{g} \tag{6}$$

The pressure gradient effect is considered as a favorable driving force for fluid movement. Usually, the pressure decreases at a constant rate from the initial end to the end in the x direction.

$$\nabla P \cong \frac{\Delta P}{\Delta z} = \frac{(P_L - P_0)}{L} = \frac{\Delta P}{L} \tag{7}$$

The component of the flow density tensor of the amount of movement of Eq. (6) is τ_{yx}; therefore, considering the above conditions, we have the following equation to solve:

$$\frac{d\tau_{yx}}{dy} = \left(\frac{\Delta P}{L}\right) \tag{8}$$

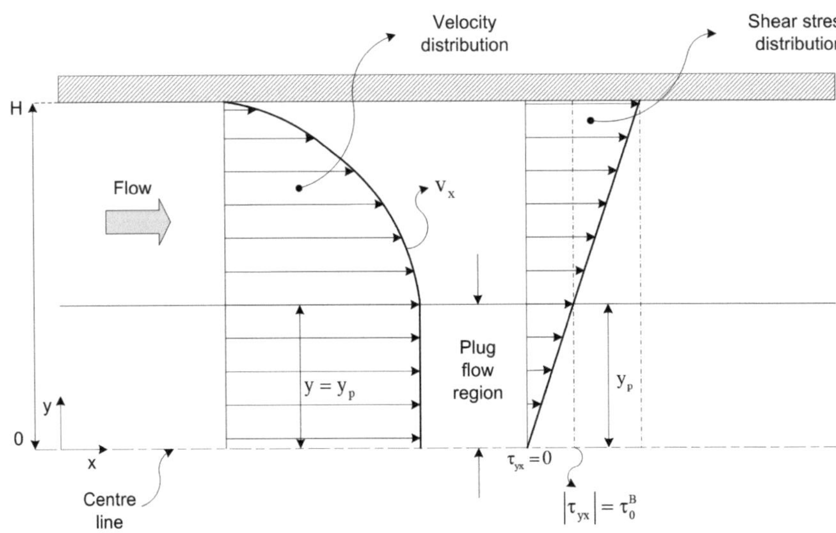

Figure 2.
Flow of a Bingham fluid between two plates one half view [2].

Making a separation of variables, in addition to performing the corresponding integrals in Eq. (8), we obtain

$$\tau_{yx} = \left(\frac{\Delta P}{L}\right) y + c_1 \tag{9}$$

The integration constant is zero, when $\tau_{yx} = 0$ at $y = 0$, i.e., at the center of the plates, the shear stress is minimal. Therefore, the shear stress profile is

$$\tau_{yx} = \left(\frac{\Delta P}{L}\right) y \tag{10}$$

Equating Eq. (10) with the Bingham model, we obtain

$$\overbrace{\mu_B \left(\frac{dv_x}{dy}\right) + \tau_0}^{\text{Bingham model}} = \left(\frac{\Delta P}{L}\right) y \tag{11}$$

Making a second separation of variables, in addition to performing the corresponding integrals, we have

$$v_x = \left(\frac{\Delta P}{L}\right) \frac{1}{\mu_B} \frac{y^2}{2} - \frac{\tau_0}{\mu_B} y + c_2 \tag{12}$$

In **Figure 2**, it is shown that velocity is zero on the plates; i.e., $v_x = 0$ at $y = \pm H$. Using this condition, the value of c_2 is obtained. Substituting in Eq. (12) we obtain

$$v_x = \frac{\tau_0}{\mu_B} H\left(1 - \frac{y}{H}\right) + \left(\frac{\Delta P}{L}\right) \frac{H^2}{2\mu_B} \left[\left(\frac{y}{H}\right)^2 - 1\right] \tag{13}$$

To know the velocity profile in the region of the plug flow, the condition for the yield stress is proposed according to Eq. (10), when $y = y_0$:

$$y = y_0; \qquad \tau_0 = \left(\frac{\Delta P}{L}\right) y_0 \tag{14}$$

Substituting Eq. (14) into Eq. (13), the velocity in the plug flow region is obtained:

$$v_0 = \left(\frac{\Delta P}{L}\right) \frac{H^2}{2\mu_B} \left(1 - \frac{y_0}{H}\right)^2 \tag{15}$$

Commonly, velocity profiles are usually represented with on the Bingham number, which is defined as

$$Bn = \frac{\tau_0}{\mu_B} \frac{H}{v} \tag{16}$$

In Eq. (16), v is a characteristic velocity. Dividing Eq. (13) by this velocity,

$$\frac{v_x}{v} = Bn\left(1 - \frac{y}{H}\right) + \left(\frac{\Delta P}{L}\right) \frac{H^2}{2v\mu_B} \left[\left(\frac{y}{H}\right)^2 - 1\right] \tag{17}$$

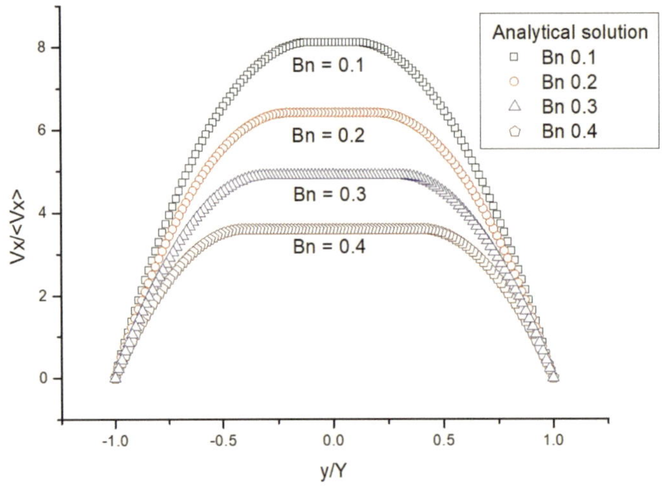

Figure 3.
Velocity profile for a Bingham fluid; Bn = 0.1, 0.2, 0.3, and 0.4 [4].

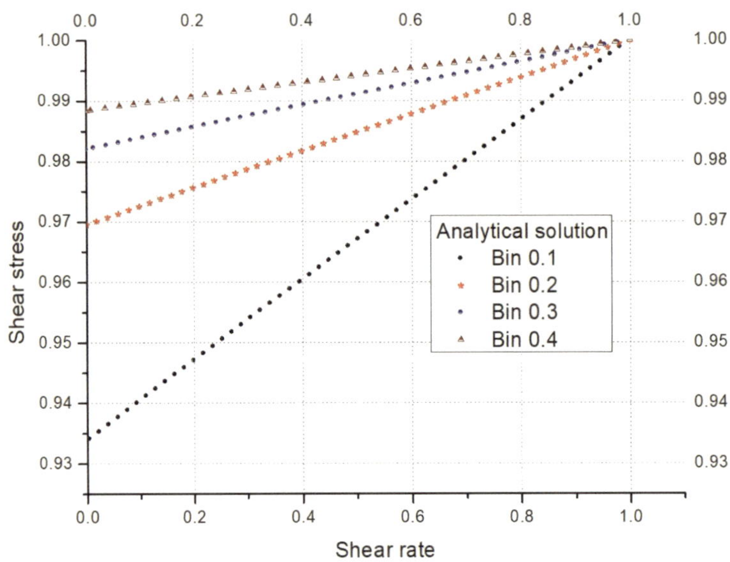

Figure 4.
Shear stress vs. normalized shear rate with analytical solution.

is obtained. Eq. (17) represents the velocity profile for the Poiseuille flow between two parallel plates, in the case of a Bingham-type fluid, and its graphic representation is that shown in **Figure 3**, for values of Bn = 0.1, 0.2, 0.3, and 0.4, with dimensionless values [4].

The graph of the shear stress vs. normalized shear rate (rheogram) is shown in **Figure 4**.

3. Lattice Boltzmann Bhatnagar-Gross-Krook (BGK)

The Lattice Boltzmann Method (LBM) generally consists of a discrete lattice; each site (node) is represented by the distribution function, which is defined by the

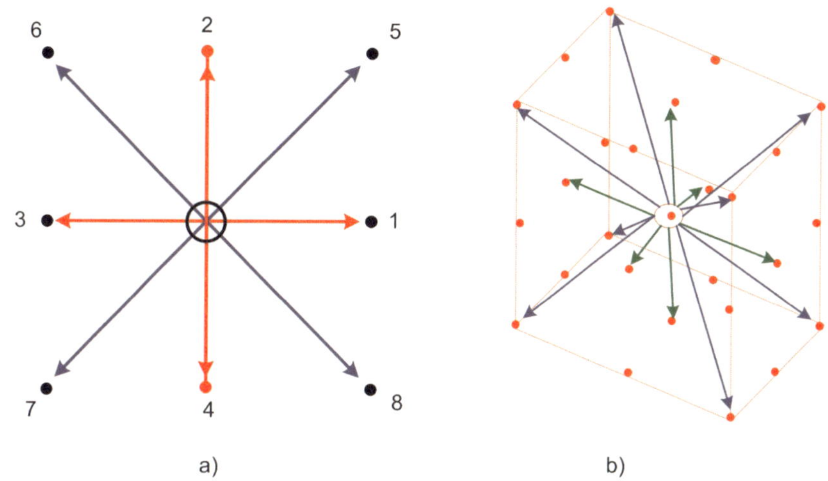

Figure 5.
Allowed directions for particle movement. (a) Model D2Q9, (b) model D2Q15.

velocity of a particle and limited to a discrete group of allowed velocities. During each time step, the movement or jump of particle to nearby lattice sites, along its direction of movement, a collision with another particle can occur when they reach the same site. The result of the collisions is determined by means of the solution of the kinetic equation of Boltzmann for the new function of distribution of the particle to that site, and in this way the function of distribution of the particle is updated [9, 10].

There are different lattice models, which are given by DmQn; m indicates the dimension and n the permitted velocity, thus, the D2Q9 model (two-dimensional with nine speed directions), of which four sites correspond to nearby neighbors (right, left, up, and down), four other points of the lattice vectors along the diagonal faces of the following sites, along these diagonals. In this way the particles can travel in eight directions for each lattice site. The circle in the center of the square represents the vector, which has a value of zero and represents particles that have no movement, that is, particles at rest. Then, there are a total of nine real numbers that describe the distribution function of the particle for a lattice site (see **Figure 5**).

The process of the propagation and collision of particles generally occurs in two stages: the first is to denote the advance of the particles to the next lattice site along the directions of movement; this is for each time step Δt. The second stage is to simulate the collisions of the particles, which causes them to propagate in different directions at each lattice site [11, 12]. These stages can be described through the discretized Boltzmann equation on a lattice.

$$f_i\left(\vec{x} + \vec{e}_i, t + 1\right) - f_i\left(\vec{x}, t\right) = -\Omega \tag{18}$$

To simplify Eq. (18), the BGK approximation is usually used; this approximation replaces the term Ω:

$$\Omega = \frac{f^{eq} - f}{\tau} \tag{19}$$

This operator models the effect of the collision as a relaxation of the distribution function towards the Maxwell equilibrium distribution. The parameter τ (relaxation time) has dimensions of time and controls the frequency with which the distribution function relaxes to reach equilibrium, that is, this time determines the rate at which the fluctuations in the system lead to this state of equilibrium [13].

$$f_i\left(\vec{x}+\vec{e}_i, t+1\right) - f_i\left(\vec{x}, t\right) = -\frac{1}{\tau}\left[f_i\left(\vec{x}, t\right) - f_i^{eq}\left(\vec{x}, t\right)\right] \quad (20)$$

The macroscopic variables (mass (ρ) and velocity $\left(\vec{u}\right)$) can be calculated directly from the values of the distribution function as

$$\begin{aligned} \rho\left(\vec{x}, t\right) &= \sum_{i=1}^{n} f_i\left(\vec{x}, t\right) \\ \vec{u}\left(\vec{x}, t\right) &= \frac{1}{\rho\left(\vec{x}, t\right)} \sum_{i=1}^{n} f_i\left(\vec{x}, t\right)\vec{e}_i \end{aligned} \quad (21)$$

In the case of a Newtonian fluid, the ratio of kinematic viscosity and relaxation time is given by [13] $\upsilon = 1/3\ (\tau\text{-}1/2)$.

4. Simulation of Bingham fluids with the lattice Boltzmann method

For the simulations of the Bingham fluid with the Lattice Boltzmann Method, a modification was made to the LBGK approach presented by Wang and Ho [14] and Tang et al. [15], for a D2Q9 model, which consists in proposing the relaxation parameter τ based on the apparent viscosity:

$$\tau = 3\left[\mu_B + \tau_0(\dot{\gamma})\right] + \frac{1}{2} \quad (22)$$

In Eq. (22), μ_B, τ_0, and $\dot{\gamma}$ are the Bingham viscosity, yield stress, and shear rate, respectively. Parameter τ was used in the Lattice Boltzmann equation (Eq. (20)). The simulations were carried out on a 64×64 lattice, using "bounce back" conditions on the solid walls to ensure that the velocities are zero and periodic boundary conditions at the fluid inlet and outlet, so that the nodes located in the border will have their neighboring nodes on the opposite border. The steady state was reached at 360,000 time steps.

The validation of the proposal in the LBM was performed by comparing the results of the analytical solutions for a Poiseuille flow between two separate plates a distance 2H, shown in **Figure 2** using Eq. (17). The used conditions were pressure force = 2.66E-2, yield stress = 2.00E-5, and Bingham viscosity of 0.4 for a Bin = 0.1; pressure force = 5.83E-3, yield stress = 1.10E-5, and Bingham viscosity of 0.08 for a Bin = 0.2; pressure force = 5.19E-3, yield stress = 1.40E-5, and Bingham viscosity of 0.07 for a Bin = 0.3; and pressure force = 1.88E-3, yield stress = 6.50E-6, and Bingham viscosity of 0.025 for a Bin = 0.4.

Simulations were performed in porous media, applying the LBM in the case of deterministic porous medium and random porous media. A modification of a "Box-Muller method," which is a random number generator, was inserted in random porous media, and blocks were inserted arbitrarily in the lattice for deterministic porous media [16, 17].

An alternative way or method is proposed for obtaining local permeabilities for deterministic and random porous media. This one consists in a modification of

Darcy's law; for this the permeability is calculated based on the apparent viscosity according to the following equation:

$$K = \left(\mu_B + \frac{\tau_0}{\dot{\gamma}} \right) v / \left({}^{dP}/_{dx} \right) \tag{23}$$

In Eq. (23), K is the permeability, v is the velocity, and dP/dx is a pressure force.

The discrete macroscopic pressure P is given by the state equation that relates the discrete density to the pressure $P = c_s^2 \rho$, where c_s is the speed of sound and ρ is the density that is calculated through Eq. (21).

The relationship is valid for incompressible fluid simulations and is only allowed to fluctuate locally around a fixed value [18]. Hidemitsu Hayashi proposed two LBMs for the flow generated by the pressure gradient (FGPG) and the flow driven by an external force (FDEF), which are consistent with each other [19].

The criterion used by Wang and Ho [14] was taken, for which yielding occurs when the magnitude of the extra shear stress tensor exceeded the yield stress, τ_0, i.e., be yielded when $|\tau_{yx}| \rangle \tau_o$ and unyielded if $|\tau_{yx}| \le \tau_o$.

Figure 6 shows the validation of the velocity profiles with the analytical solution and the simulations with LBM. The error between both solutions was less than 2.0%.

For the development of the work, three porous media with a deterministic structure and nine random were proposed; in each of them all the simulations were performed for three Bingham numbers (0.2, 0.3, and 0.4).

In **Figures 7–15**, the speed patterns for all simulations are shown; for this the values of the yield stress, the pressure forces, and the viscosities were varied.

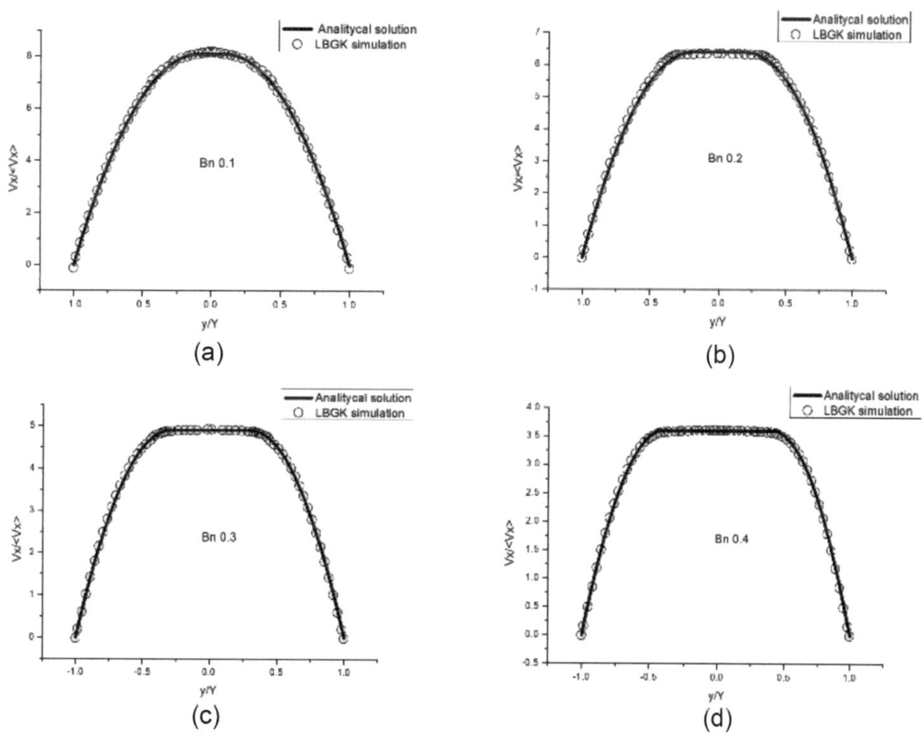

Figure 6.
Comparison of velocity profiles with different Bingham numbers for the analytical solutions and the proposed LBM. Normalized velocity profiles for (a) Bin = 0.1, (b) Bin = 0.2, (c) Bin = 0.3, (d) Bin = 0.4.

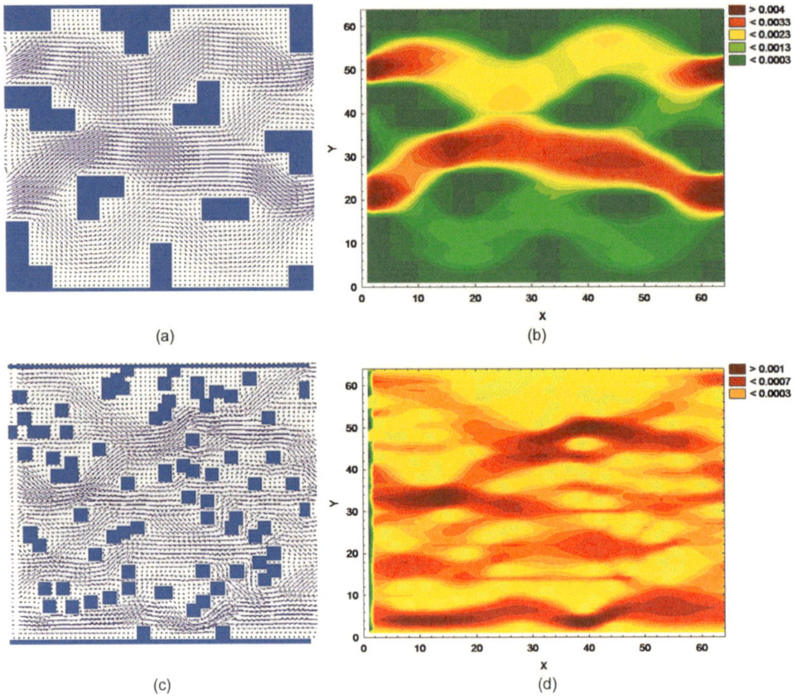

Figure 7.
Bingham = 0.2, yield stress = 1.1E-5, pressure force = 5.83E-3, and viscosity = 0.08. Porosity 81.68%, deterministic porous medium: (a) vectorized flow, (b) velocity patterns. Porosity 81.62%, random porous medium: (c) vectorized flow, (d) velocity patterns.

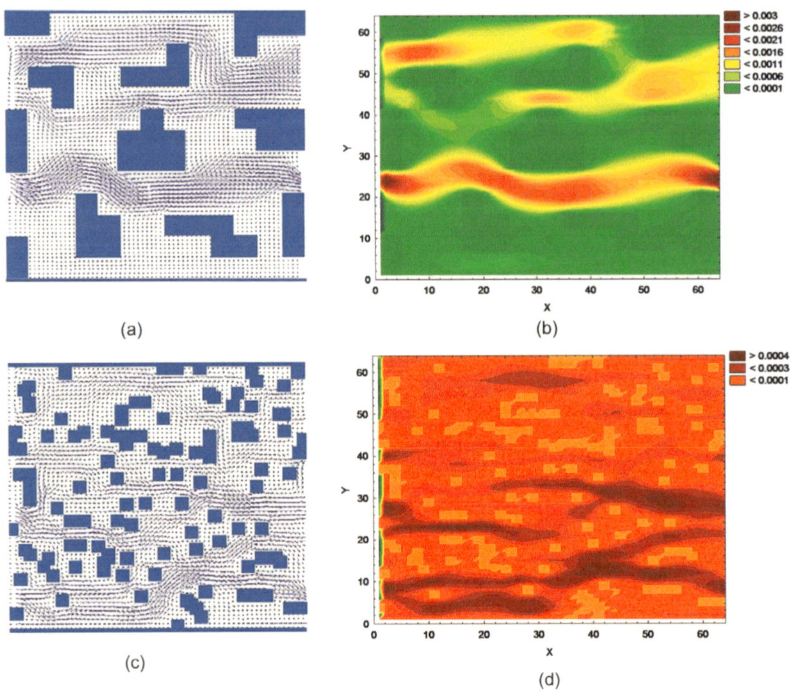

Figure 8.
Bingham = 0.2, yield stress = 1.1E-5, pressure force = 5.83E-3, and viscosity = 0.08. Porosity 73.75%, deterministic porous medium: (a) vectorized flow, (b) velocity patterns. Porosity 73.68%, random porous medium: (c) vectorized flow, (d) velocity patterns.

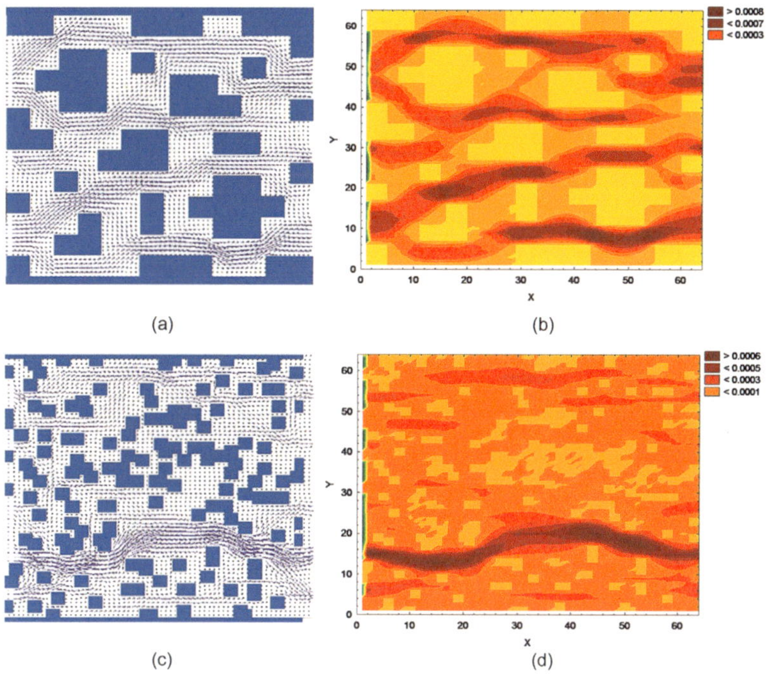

Figure 9.
Bingham = 0.2, yield stress = 1.1E-5, pressure force = 5.83E-3, and viscosity = 0.08. Porosity 65.82%, deterministic porous medium: (a) vectorized flow, (b) velocity patterns. Porosity 65.75%, random porous medium: (c) vectorized flow, (d) velocity patterns.

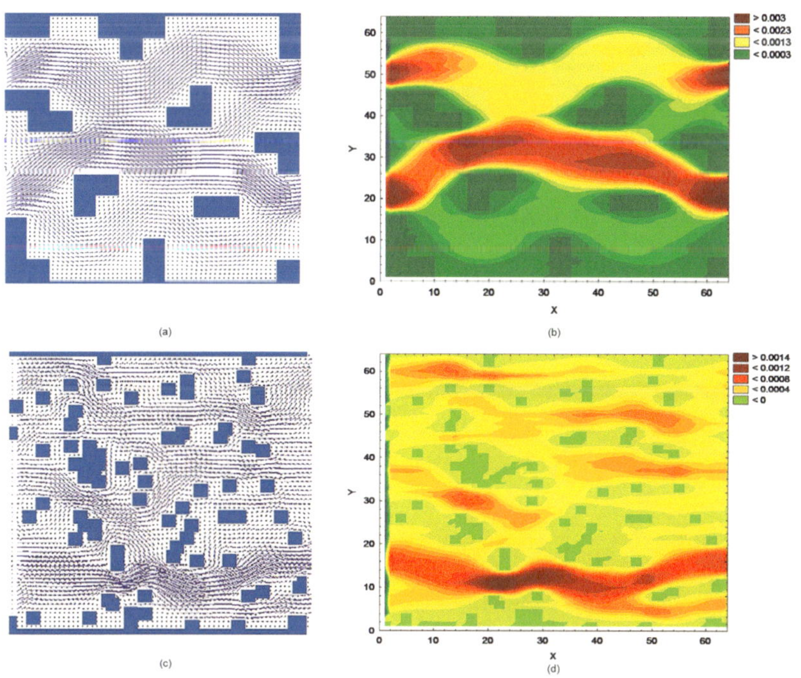

Figure 10.
Bingham = 0.3, yield stress = 1.4E-5, pressure force = 5.19E-3, and viscosity = 0.07. Porosity 81.68%, deterministic porous medium: (a) vectorized flow, (b) velocity patterns. Porosity 81.62%, random porous medium: (c) vectorized flow, (d) velocity patterns.

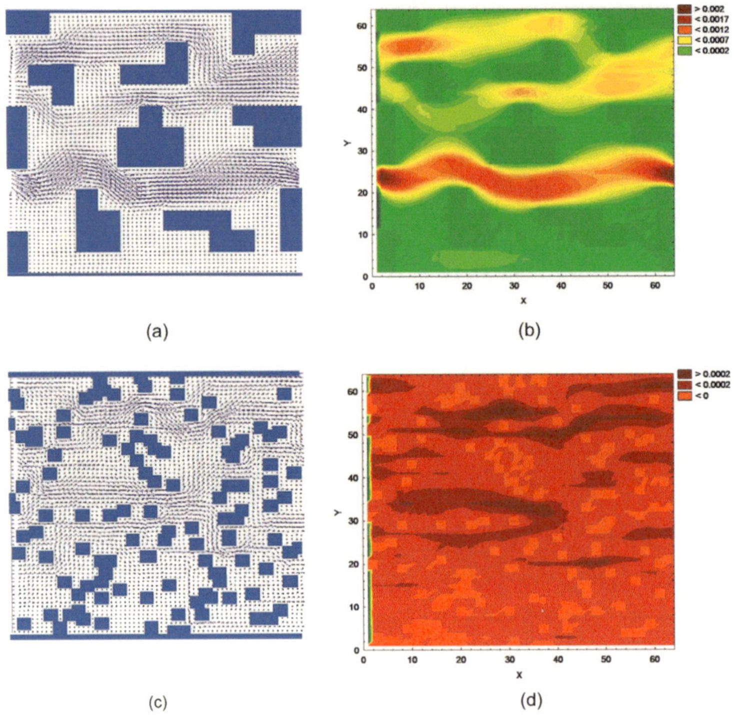

Figure 11.
Bingham = 0.3, yield stress = 1.4E-5, pressure force = 5.19E-3, and viscosity = 0.07. Porosity 73.75%, deterministic porous medium: (a) vectorized flow, (b) velocity patterns. Porosity 73.68%, random porous medium: (c) vectorized flow, (d) velocity patterns.

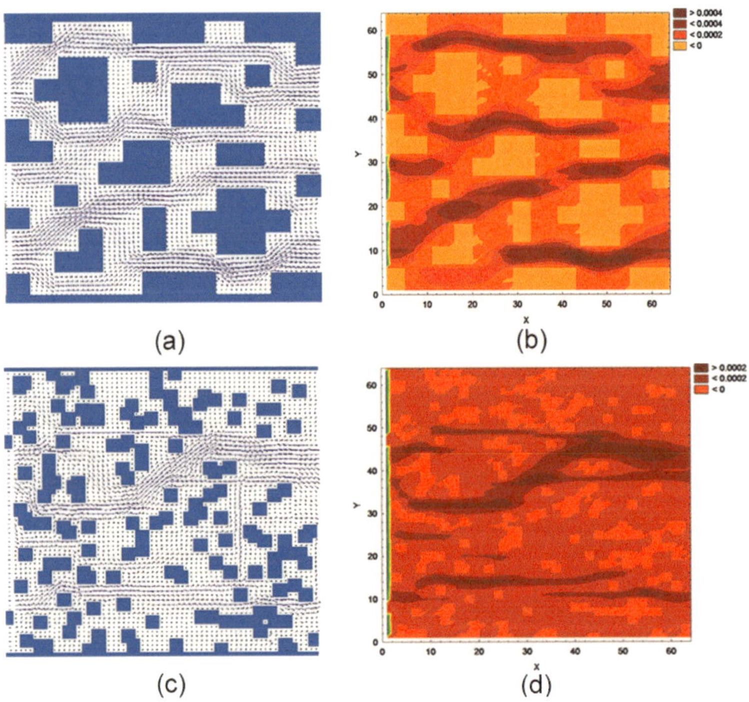

Figure 12.
Bingham = 0.3, yield stress = 1.4E-5, pressure force = 5.19E-3, and viscosity = 0.07. Porosity 65.82%, deterministic porous medium: (a) vectorized flow, (b) velocity patterns. Porosity 65.75%, random porous medium: (c) vectorized flow, (d) velocity patterns.

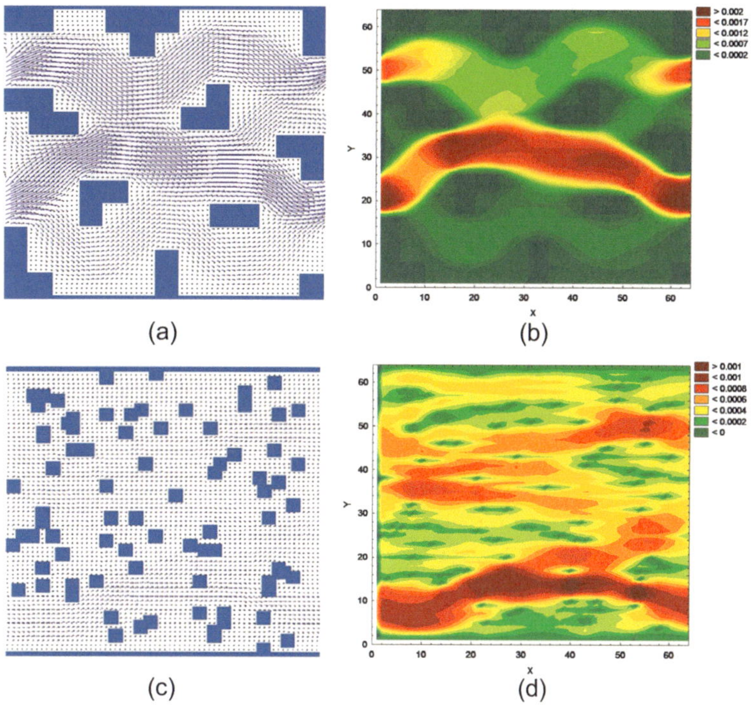

Figure 13.
Bingham = 0.4, yield stress = 6.5E-6, pressure force = 1.88E-3, and viscosity = 0.025. Porosity 81.68%, deterministic porous medium: (a) vectorized flow, (b) velocity patterns. Porosity 81.62%, random porous medium: (c) vectorized flow, (d) velocity patterns.

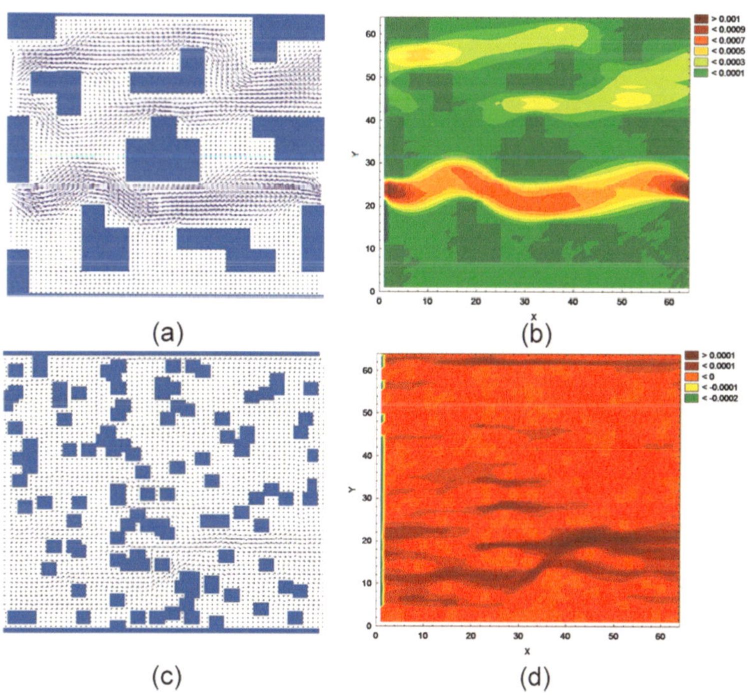

Figure 14.
Bingham = 0.4, yield stress = 6.5E-6, pressure force = 1.88E-3, and viscosity = 0.025. Porosity 73.75%, deterministic porous medium: (a) vectorized flow, (b) velocity patterns. Porosity 73.68%, random porous medium: (c) vectorized flow, (d) velocity patterns.

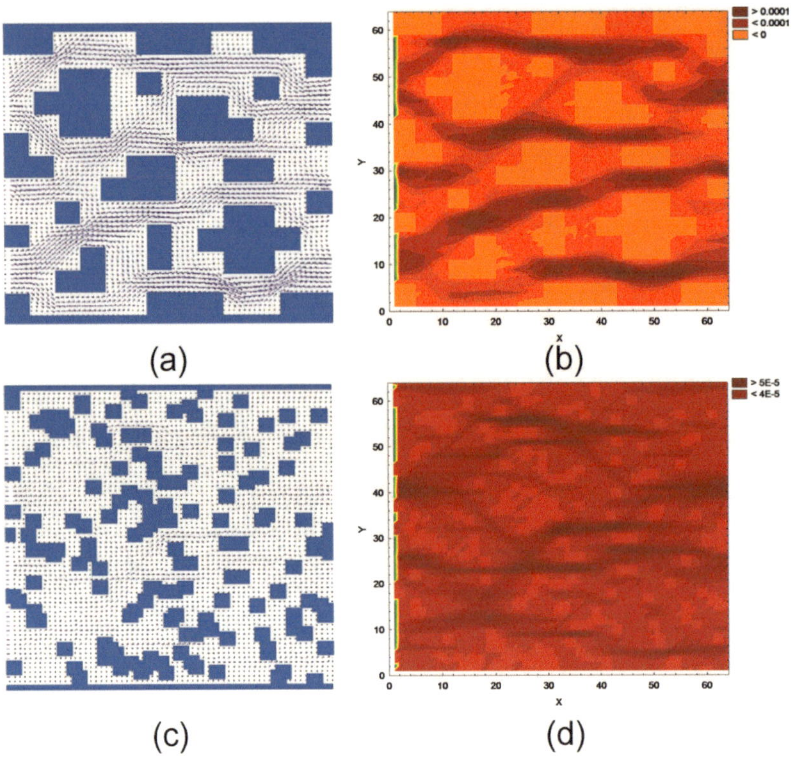

Figure 15.
Bingham = 0.4, yield stress = 6.5E-6, pressure force = 1.88E-3, and viscosity = 0.025. Porosity 65.82%, deterministic porous medium: (a) vectorized flow, (b) velocity patterns. Porosity 65.75%, random porous medium: (c) vectorized flow, (d) velocity patterns.

In **Figures 7–15** the decrease in porosity corresponds to a decrease in velocity. It can be noted that for a Bingham number of 0.2 (**Figures 7–9**), higher velocities are shown in deterministic porous media than random media. But in the latter, there are more places to pass the fluid. These behaviors are presented in the simulations for the three Bingham numbers used (0.2, 0.2, and 0.3). Comparing the Bingham number of 0.2 with that of 0.4, a decrease in the velocity in the deterministic porous media is observed as well as the random ones according to the conditions handled; this is due to the decrease of the initial effort, the pressure force, and the viscosity.

Local permeabilities were simulated based on apparent viscosities for all deterministic and random porous media. In **Figures 16–19**, only some of the results of local permeabilities for deterministic porous media 81.68 and 65.82% with Bingham numbers of 0.2 and 0.4, respectively, are shown. Likewise, only the result of the simulations for two random porous media 81.62 and 65.75% for Bingham numbers of 0.2 and 0.4, respectively, are shown.

Figures 16–19 show the zones in blue of the local permeabilities. It is remarkable that the blue areas predominate in random porous media. By comparing the Bingham number of 0.2 for the two deterministic and random porous media according to **Figures 16** and **17** with that of 0.4 of **Figures 18** and **19**, an increase in permeabilities can be seen.

Finally, pressures for all porous media were simulated. **Figures 20–25** show some of the results obtained in the case of deterministic porous media with porosities of 81.68, 73.75, and 65.82% for Bingham numbers 0.2, 0.3, and 0.4, respectively. Similarly, the results are presented for random porous media with porosities of 81.62, 73.68, and 65.75% for Bingham numbers 0.2, 0.3, and 0.4, respectively.

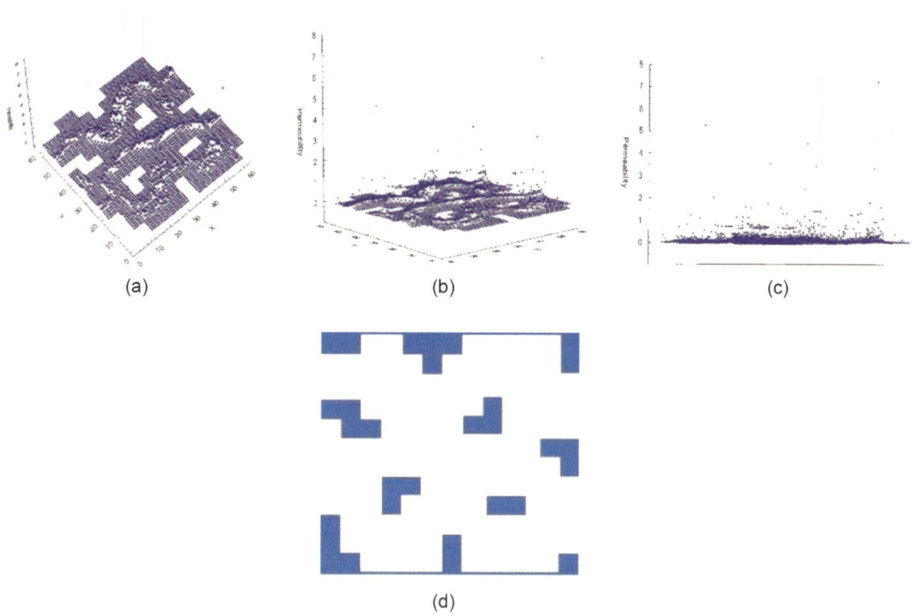

Figure 16.
Local permeabilities for deterministic porous media with porosity of 81.68%, Bingham = 0.2, yield stress = 1.1E-5, pressure force = 5.83E-3 and viscosity = 0.08. (a–c) The rough numerical date of permeability, (d) pattern porous media.

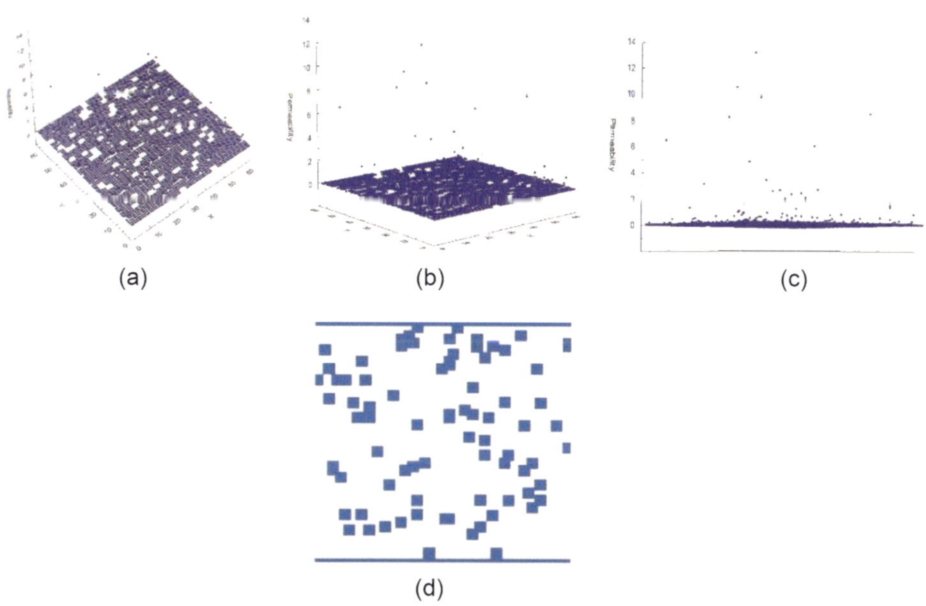

Figure 17.
Local permeabilities for random porous media with porosity of 81.62%, Bingham = 0.2, yield stress = 1.1E-5, pressure force = 5.83E-3 and viscosity = 0.08; (a–c) The rough numerical date of permeability, (d) pattern porous media.

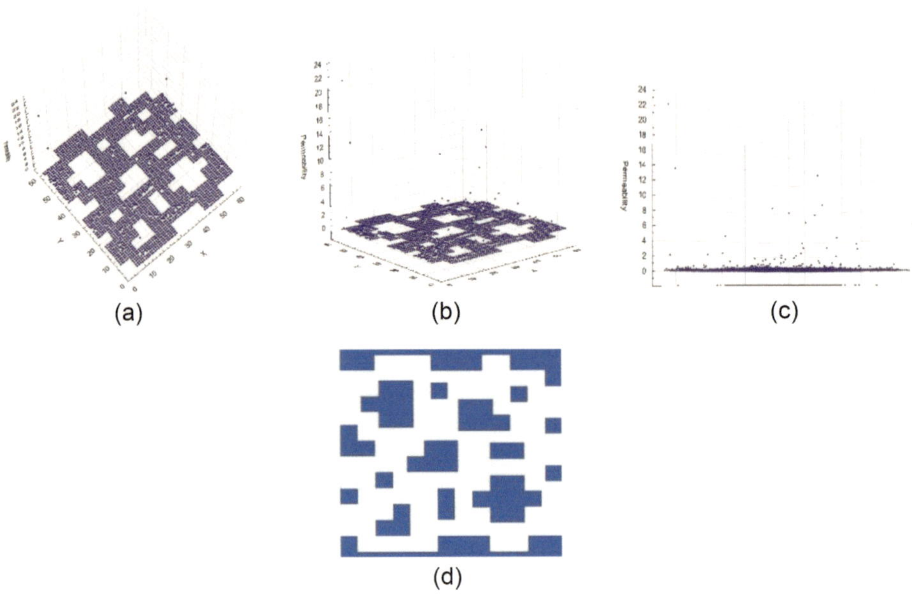

Figure 18.
Local permeabilities for deterministic porous media with porosity of 65.82%, Bingham = 0.4, yield stress = 6.5E-6, pressure force = 1.88-3 and viscosity = 0.025; (a–c) the rough numerical date of permeability, (d) pattern porous media.

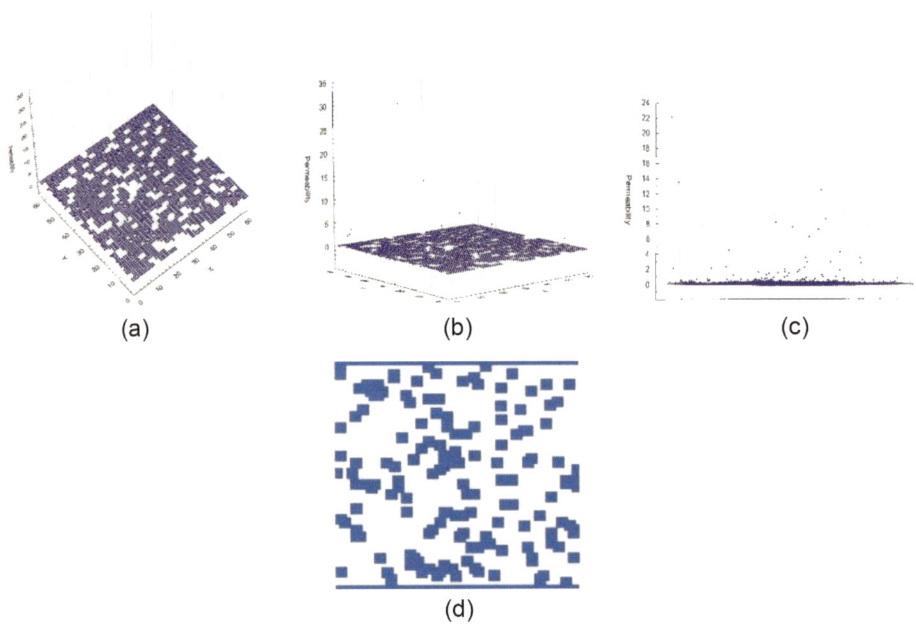

Figure 19.
Local permeabilities for random porous media with porosity of 65.75%, Bingham = 0.4, yield stress = 6.5E-6, pressure force = 1.88-3 and viscosity = 0.025; (a–c) The rough numerical date of permeability, (d) pattern porous media.

Figures 20–25, you can see in all cases the difference in pressures, higher pressure in the red colors, and less pressure in the green colors, in addition to observing the pressures in the different zones.

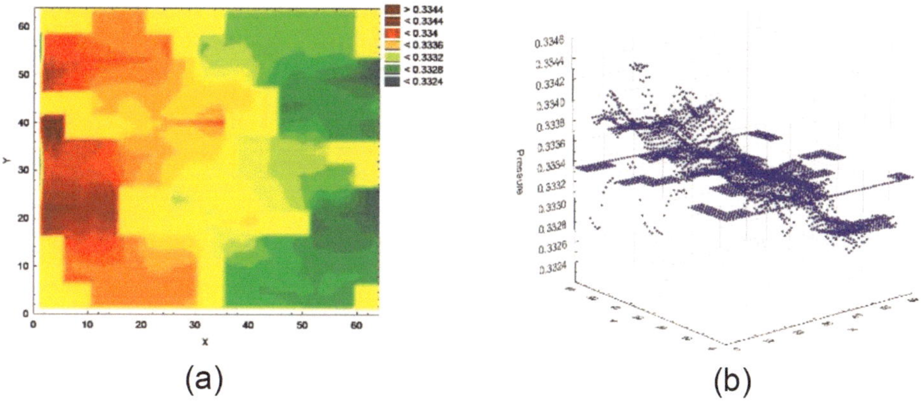

Figure 20.
Pressures for deterministic porous medium with porosity of 81.68%, Bingham = 0.2, yield stress = 1.1E-5, pressure force = 5.83E-3 and Viscosity = 0.08. (a) Pressure distribution, (b) the rough numerical date of pressure.

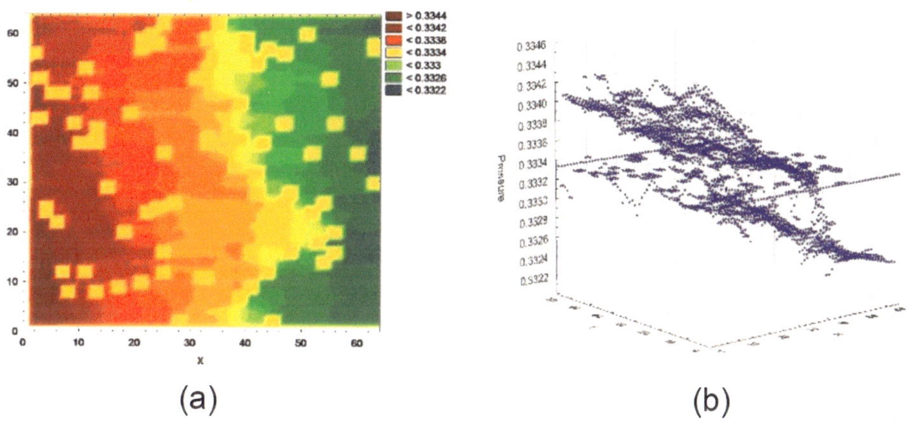

Figure 21.
Pressures for random porous medium with porosity of 81.62%, Bingham = 0.2, yield stress = 1.1E-5, pressure force = 5.83E-3 and Viscosity = 0.08. (a) Pressure distribution, (b) the rough numerical date of pressure.

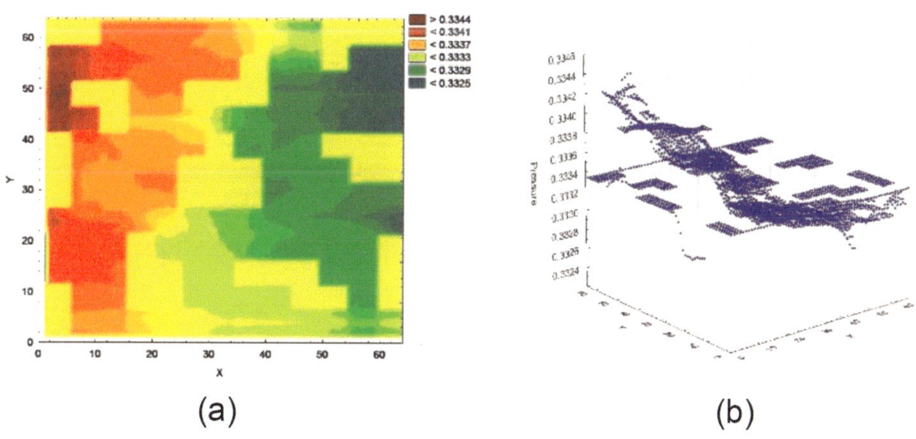

Figure 22.
Pressures for deterministic porous medium with porosity of 73.75%, Bingham = 0.3, yield stress = 1.4E-5, pressure force = 5.19E-3 and Viscosity = 0.07. (a) Pressure distribution, (b) the rough numerical date of pressure.

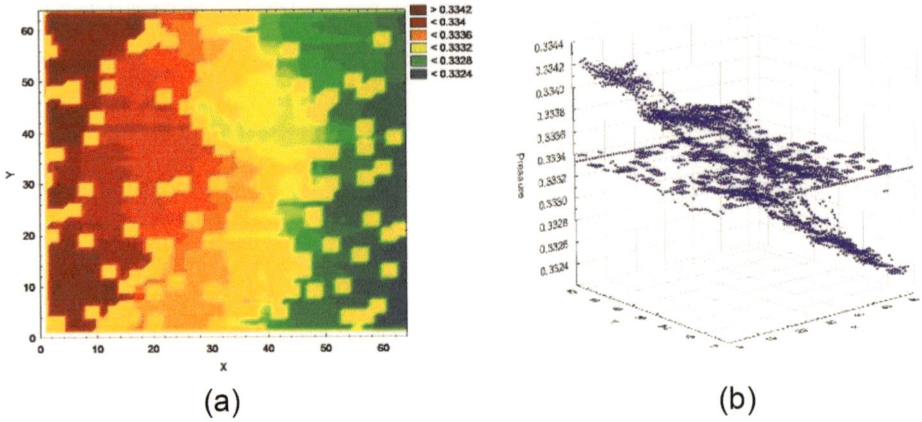

Figure 23.
Pressures for random porous medium with porosity of 73.68%, Bingham = 0.3, yield stress = 1.4E-5, pressure force = 5.19E-3 and Viscosity = 0.07. (a) Pressure distribution, (b) the rough numerical date of pressure.

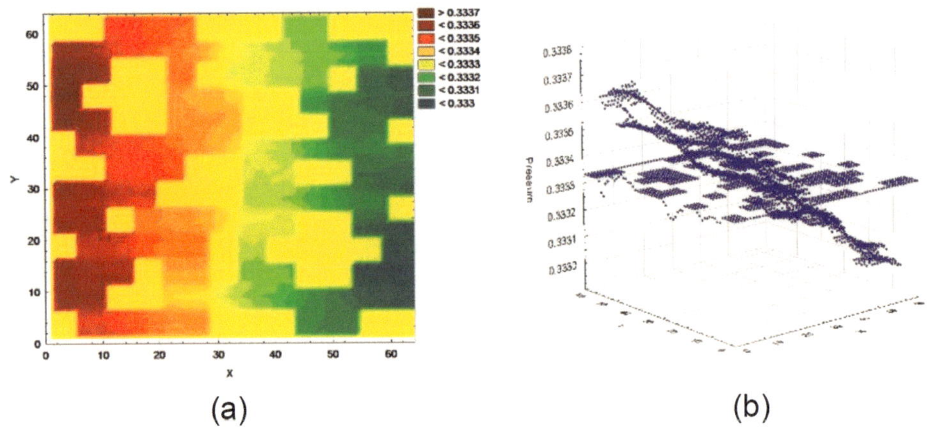

Figure 24.
Pressures for deterministic porous medium with porosity of 65.82%, Bingham = 0.4, yield stress = 6.5E-6, pressure force = 1.88E-3 and Viscosity = 0.025. (a) Pressure distribution, (b) the rough numerical date of pressure.

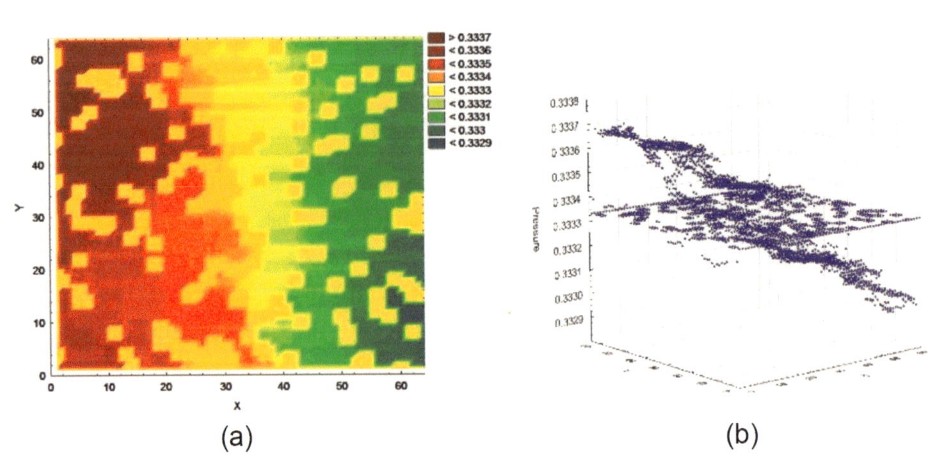

Figure 25.
Pressures for random porous medium with porosity of 65.75%, Bingham = 0.4, yield stress = 6.5E-6, pressure force = 1.88E-3 and Viscosity = 0.025. (a) Pressure distribution, (b) the rough numerical date of pressure.

5. Conclusions

In the present work, the Lattice Boltzmann Method was applied to a problem of the flow of a non-Newtonian Bingham-type fluid between two plates, in the case of a Poiseuille flow.

This method is an alternative to the conventional ones used in computational fluid mechanics, its programming is not complicated, and today it is applied to many engineering problems.

Validations were carried out with the analytical solution of the velocity profiles for the case of a Poiseuille flow and the simulations with Lattice Boltzmann, for the case of Bingham-type fluids, for values of the Bingham number (Bin) of 0.1, 0.2, 0.3, and 0.4. The results of all the simulations were quite acceptable, since the percentage of error between both results did not exceed 2.0%.

The LBM proves to be kind for simulations with small lattices, such as the one used in the present work 64×64. All simulations were performed in a laminar regime.

Three deterministic porous media with porosities of 81.68, 75.75, and 65.82% and nine randomized ones with porosities of 81.62, 73.68, and 65.75% were proposed for three Bingham numbers (0.2, 0.3, and 0.4), to perform all simulations. In them the pressure forces, yield stress, and viscosities were varied.

Profiles of velocities, permeabilities, and local pressures were obtained, in all cases the results and behaviors were acceptable for all porous media, and the three Bingham numbers, although only some of the results obtained, were presented at work.

The LBM with the necessary restrictions allows to perfectly simulate the behavior of fluids, as is the case of the Bingham type; the importance of this is the application of multiple industrial processes, in the displacement of fluids reducing costs and time.

Finally, it would be convenient to perform simulations with turbulent flows to verify the goodness of the method with this type of fluid, in which its description is more complex.

Acknowledgements

The present work was developed under the sponsorship of the Facultad de Estudios Superiores Cuautitlán-Universidad Nacional Autónoma de México.

Conflict of interest

The author declares no conflicts of interest regarding the publication of this paper.

Author details

José Luis Velázquez Ortega
Facultad de Estudios Superiores Cuautitlán – UNAM, Cuautitlán Izcalli,
Estado de México, México

*Address all correspondence to: siulj@unam.mx

IntechOpen

© 2019 The Author(s). Licensee IntechOpen. This chapter is distributed under the terms of the Creative Commons Attribution License (http://creativecommons.org/licenses/by/3.0), which permits unrestricted use, distribution, and reproduction in any medium, provided the original work is properly cited. CC BY

References

[1] Pérez V. La Ecuación de redes de Boltzmann: Una alternativa para la simulación de fluidos Newtonianos y de Ostwald de Waele [Thesis]. Cuautitlán: Universidad Nacional Autónoma de México; 2016

[2] Chhabra RP, Richardson JF. Non-Newtonian Flow in the Process Industries, Fundamentals and Engineering Applications. 2nd ed. Burlington: Butterworth-Heinemann; 2008. 518p

[3] León GE. Simulación de fluidos pseudoplásticos con métodos actuales de la dinámica de fluidos computacionales y su comparación con datos experimentales [Thesis]. Cuautitlán: Universidad Nacional Autónoma de México; 2017

[4] Guerrero V. Simulación del perfil de velocidades para un fluido tipo Bingham en canales rectangulares empleando redes de Boltzmann [Thesis]. Cuautitlán: Universidad Nacional Autónoma de México; 2018

[5] De Nevers N. Mecánica de Fluidos para Ingenieros Químicos. 1st ed. México: CECSA; 2006

[6] Moreno E, Cervera M. Elementos finitos mixtos para flujos confinados de Bingham y de Herschel-Bulkley. Parte I: Formulación. Revista Internacional de Metodos Numericos para Calculo y Diseno en Ingenieria. 2016;**32**:100-109. DOI: 10.1016/j.rimni.2015.02.004

[7] Moreno E, Cervera M. Elementos Finitos Mixtos Estabilizados para Flujos Viscoplasticos, Formulación y Aplicaciones. 1ª ed. Gran Capitán: Centro Internacional de Métodos Numéricos en Ingeniería; 2014. 319p

[8] Bird R, Stewart W, Lightfoot E. Fenómenos de Transporte. 2nd ed. México: Limusa Wiley; 2006. 1062p

[9] Chen S, Doolen G. Lattice Boltzmann method for fluid flows. Annual Review of Fluid Mechanics. 1998;**30**:329-364. DOI: 10.1146/annurev.fluid.30.1.329

[10] Wilke J. Cache optimizations for the lattice Boltzmann method in 2D [Thesis]. Erlangen: Friedrich-Alexander-Universität Erlangen-Nürnberg; 2003

[11] Chen S, Doolen G, Eggert K. Lattice Boltzmann versatile tool for multiphase, fluid dynamics and other complicated flows. Los Alamos Science. 1994;**20**: 100-111

[12] ThÄurey N. A single-phase free-surface lattice Boltzmann [Thesis]. Erlangen: Friedrich-Alexander-Universität Erlangen-Nürnberg; 2003

[13] Usman R, Alireza E, Torsten M. The lattice-Boltzmann method on optimal sampling lattices. IEEE Transactions on Visualization & Computer Graphics. 2009;**15**:630-641

[14] Wang C, Ho J. Lattice Boltzmann modeling of Bingham plastics. Physica A: Statistical Mechanics and its Applications. 2008;**387**:4740-4748. DOI: 10.1016/j.physa.2008.04.008

[15] Tang G, Wang S, Ye P, Tao W. Bingham fluid simulation with the incompressible lattice Boltzmann model. Journal of Non-Newtonian Fluid Mechanics. 2011;**166**:145-151. DOI: 10.1016/j.jnnfm.2010.11.005

[16] Kloeden P, Platen E. Numerical Solution of Stochastic Differential Equations. New York: Springer-Verlag; 1999. DOI: 10.1007/978-3-662-12616-5. 636p

[17] Golder E, Settle J. The box-Muller method for generating pseudo-random

normal deviates. Applied Statistics. 1976;**25**:12-20. DOI: 10.2307/2346513

[18] Carrasco B. Aplication of the Lattice Boltzmann Method to issues of coolant flows in nuclear reactors [Thesis]. Barcelona-Mùnich: Universitat Politècnica de Catalunya, Escuela Técnica Superior de Ingeniería Industrial de Barcelona-Techische Universität München; 2013

[19] Hayashi H. Lattice Boltzmann method and its application to flow analysis in porous media. R&D Review of Toyota CRDL. 2003;**38**:17-25

Chapter 4

Interface Treatment for Conjugate Conditions in the Lattice Boltzmann Method for the Convection Diffusion Equation

David Korba and Like Li

Abstract

The lattice Boltzmann method (LBM) has emerged as an attractive numerical method for fluid flows and thermal and mass transport. For LBM modeling of transport between different phases or materials of distinct properties, effective treatment for the conjugate conditions at the interface is required. Recognizing the benefit of satisfying the conjugate conditions in each time step without iterative computations using LBM, various interface schemes have been proposed in the last decade. This chapter provides a review of those interface schemes, with a focus on the comparison of numerical accuracy and convergence orders. It is shown that in order to preserve the second-order accuracy in LBM, the local interface geometry must be considered; and the modified geometry-ignored interface schemes result in degraded convergence orders and/or much higher error magnitude. It is also verified that with appropriate interface schemes, interfacial transport with scalar and flux jumps can be effectively modeled.

Keywords: conjugate conditions, boundary conditions, heat and mass transfer, lattice Boltzmann, numerical accuracy

1. Introduction

Heat and mass transfer between multi-phases or different materials with interfacial conjugate conditions is frequently encountered in fundamental sciences and numerous engineering applications involving fluid dynamics, thermal transport, materials sciences, and chemical reactions. Examples are cooling of turbine blades, heat exchangers and electronic devices, thermal insulation on heat pipes and chemical reactors, heat conduction in composite materials, and heat and mass transfer between solid particles and their surrounding fluids [1–8], to name a few. The most well-known conjugate conditions include the continuity of both the temperature (concentration) and the heat (mass) flux at the interface. Other conjugate conditions, such as with temperature (concentration) jumps and/or flux discontinuities [9], and Henry's law relationship [10], are also noticed at fluid-solid interfaces or interfaces of two solids or fluids of different thermal (mass diffusion) properties.

The non-smoothness or discontinuities/jumps in the physical or transport properties, and consequently in the distribution of the temperature (concentration) field

IntechOpen

across the interface, pose a great challenge to any numerical method applied to solve the interface problems. Development of accurate and efficient numerical schemes to treat the interface conditions has attracted much attention in the literature, such as the immersed boundary method (IBM) [11, 12], the immersed interface method (IIM) [13, 14], the ghost fluid method (GFM) [15, 16], the sharp interface Cartesian grid method [17, 18], and the matched interface and boundary (MIB) method [19]. Most of these methods are formulated in the finite-difference, finite-volume or finite-element frameworks.

When applying those traditional numerical methods, a popular approach to implement the conjugate conditions is to employ iterative schemes, in which a Dirichlet interface condition is imposed for one phase or material and a Neumann interface condition for the other. The heat and mass transfer in each phase is separately solved, and the continuity or prescribed jump condition at the interface could be satisfied after multiple iterations. For conjugate transport with complex interface geometry, the iterative schemes would become difficult to implement and they normally necessitate a considerable amount of computational effort.

The lattice Boltzmann method (LBM), which has emerged as an attractive alternative numerical method for modeling fluid flows and heat mass transfer (see [20–22] and Refs. therein), has been demonstrated to be an effective and efficient numerical approach for conjugate interface conditions in tandem with the convection diffusion equation (CDE) [9, 23]. In this chapter, we present a critical review of the various interface schemes proposed in the literature, with a focus on the comparison of numerical accuracy.

The well-known features of the LBM method include its explicit algorithm, ease in implementation, capability to treat complex geometry, and compatibility with parallel computing [20, 21]. Boundary condition treatment is essential to the integrity of LBM since the kinetic theory-based method deals directly with the microscopic distribution functions (DFs) rather than the macroscopic conservation equations. Earlier LB models treat the collision effects with a single-relaxation-time (SRT) approximation, commonly referred to as the Bhatnagar-Gross-Krook (BGK) model [24–26]. However, the SRT model is limited such that it can only describe isotropic diffusion [20]. In recent years, models such as the two-relaxation-time (TRT) [27, 28] and multiple-relaxation-time (MRT) [20, 29, 30] LB models have been proposed that can handle anisotropic diffusion. Representative LB models proposed in the literature include the general BGK model by Shi and Guo [31] for the nonlinear CDE, the D3Q7/D2Q5 MRT models by Yoshida and Nagaoka [20] for the general convection anisotropic diffusion equation, and the D1Q3/D2Q9/D3Q19 MRT models by Chai and Zhao [30] for the general nonlinear convection anisotropic diffusion equation, to name a few. The MRT models have improved numerical accuracy and stability compared to the SRT models [20, 28, 30]. The D3Q7/D2Q5 model proposed in [20] is used for this review, as it preserves second order spatial accuracy when recovering the general CDE following an asymptotic analysis. Based on the D3Q7/D2Q5 LB models, Li et al. [21] proposed second-order accurate boundary treatments for both the Dirichlet and Neumann conditions; they have also established a general framework for heat and mass transfer simulations with direct extension to curved boundary situations. In their framework, explicit analytical expressions were developed to relate the macroscopic quantities, such as boundary temperature (concentration) and their fluxes, and interior temperature (concentration) gradients, to the microscopic DFs in the LB model.

The first work that explicitly addressed the fluid-solid interface condition in LBM was conducted by Wang et al. [6]. They proposed a simple "half lattice division" (HLD) treatment in which no special treatment is required and the temperature and flux continuity condition at the interface was automatically satisfied for steady cases. Improvement of this HLD based scheme was conducted by several groups (see a review in Ref. [23]) for unsteady cases. Importantly, with the interface treated as a shared boundary between the adjacent domains, the boundary conditions by Li et al. [21] were applied to interface conditions and particular interface schemes were proposed and verified in [23] for standard conjugate conditions, and in [9] for conjugate heat and mass transfer with interfacial jump conditions. This idea of developing analytical relationships for the DFs to satisfy the conjugate conditions was also extended to handle general interface conditions in [32, 33]. In all the previous schemes in [9, 23, 32, 33], the local geometry was taken into account and the second-order accuracy of the LBM solution can be preserved.

There is another category of interface schemes that has attracted interest in the LBM community. In those schemes, additional source terms [34], alternative LBE formulations [35, 36], or modified equilibrium DFs [37, 38], were proposed to handle the conjugate conditions. The main motivation for those schemes is to avoid the consideration of the interface geometry or topology, which can be a challenge in complex systems such as porous media. As pointed out in [33], however, these schemes usually suffer from degraded numerical accuracy and/or convergence orders. This perspective will be illustrated in detail in this chapter.

The rest of this chapter is organized as follows. In Section 2, the different types of conjugate conditions are presented. The LB model for the general CDE is briefly described in Section 3. In Section 4, the representative interface schemes are summarized. Discussion on the numerical accuracy of the selected interface schemes is provided in Section 5 with representative numerical examples. Conclusions and outlook are given in Section 6.

2. Conjugate conditions in heat and mass transfer

In order to define the conjugate conditions, we begin by defining two domains 1 and 2, as shown in **Figure 1**. The conjugate interface conditions, from a heat and mass transfer perspective, can be defined as:

$$\Phi_{df} - \Phi_{ds} + \phi_{\text{jump}} \tag{1a}$$

$$\mathbf{n}\cdot\left(k\nabla\phi+\rho c_p\mathbf{u}\phi\right)_f=\mathbf{n}\cdot\left(k\nabla\phi+\rho c_p\mathbf{u}\phi\right)_s+\mathbf{n}\cdot\mathbf{q}_{\text{jump}}^T \text{ in heat transfer,} \tag{1b}$$

$$\mathbf{n}\cdot\left(D_m\nabla\phi+\mathbf{u}\phi\right)_f=\mathbf{n}\cdot\left(D_m\nabla\phi+\mathbf{u}\phi\right)_s+\mathbf{n}\cdot\mathbf{q}_{\text{jump}}^C \text{ in mass transfer,} \tag{1c}$$

where $\mathbf{n}$ represents the normal direction, ϕ the macroscopic scalar variable of interest (temperature or concentration), k the thermal conductivity, ρ the density, c_p the heat capacity, $\mathbf{u}$ the velocity vector, D_m the mass diffusivity, and ϕ_{jump} and $\mathbf{q}_{\text{jump}}$ the jump conditions at the interface.

Some examples of jump conditions can be found in cases such as concentration jumps (Henry's law [10]) or temperature jumps at the interface [9]. Eqs. (1a)–(1c) reduce to the standard conjugate conditions in [23] with no jumps and zero normal velocity; they can also be extended to yield two general relationships between interfacial scalar values and their fluxes as in [32, 33].

3. Lattice Boltzmann model for the general CDE

The governing heat and mass transfer equation within each domain can be written as a general convection diffusion equation as:

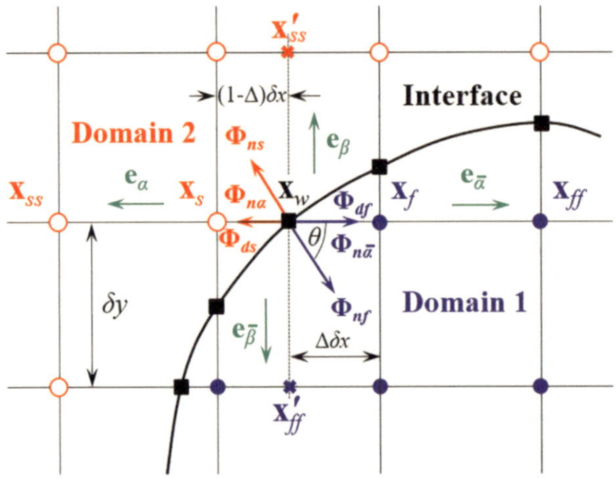

Figure 1.
Illustration of the local geometry of an interface in the lattice (filled circles: lattice nodes in domain 1, filled squares: interface nodes, and open circles: lattice nodes in domain 2). With permission from [9].

$$\frac{\partial \phi}{\partial t} + \frac{\partial}{\partial x_j}(v_j \phi) = \frac{\partial}{\partial x_j}\left(D_{ij}\frac{\partial \phi}{\partial x_j}\right) + G \tag{2}$$

where D_{ij} represents the diffusion coefficient, and G is the general source term.

For fluid flow simulations, the D2Q9/D3Q19 LB models are the most popular selections due to their accuracy and robustness [39]. While for the scalar CDE (2), the D2Q5/D3Q7 LB models are most widely used [20, 40]. To recover the CDE to second-order accuracy, the evolution equation follows

$$g_\alpha(\mathbf{x} + \mathbf{e}_\alpha \delta t, t + \delta t) - g_\alpha(\mathbf{x}, t) = \left[\mathbf{L} \cdot (\mathbf{g} - \mathbf{g}^{eq})(\mathbf{x}, t)\right]_\alpha + \omega_\alpha G(\mathbf{x}, t)\delta t \tag{3}$$

where the microscopic distribution function, $g_\alpha(\mathbf{x}, t) \equiv g(\mathbf{x}, \xi_\alpha, t)$, is defined in the discrete velocity space, ξ is the particle velocity vector that is discretized to a small set of discrete velocities $\{\xi_\alpha | \alpha = 0, 1, \ldots, m-1\}$, $\mathbf{e}_\alpha$ is the αth discrete velocity vector, δt is the time step, $\mathbf{L}$ is the collision operator, $g_\alpha^{eq}(\mathbf{x}, t)$ is the equilibrium distribution function, and ω_α is the weight coefficient. The macroscopic scalar variable is obtained from

$$\phi(\mathbf{x}, t) = \sum_{\alpha=0}^{m-1} g_\alpha(\mathbf{x}, t) \tag{4}$$

The equilibrium distribution function can be defined as [20, 40]

$$g_\alpha^{eq} = \omega_\alpha \phi \left(1 + \frac{\mathbf{e}_\alpha \cdot \mathbf{u}}{c_s^2}\right) \tag{5}$$

where c_s is the speed of sound with $c_s = c/\sqrt{3} = (\delta x/\delta t)/\sqrt{3} = 1/\sqrt{3}$.

When using the multiple-relaxation-time (MRT) collision operator, and the collision-streaming process for efficient computations, Eq. (3) is split into two steps:

Collision step

$$\hat{g}_\alpha(\mathbf{x}, t) = g_\alpha(\mathbf{x}, t) - \left[\mathbf{M}^{-1}\mathbf{S}\left(\mathbf{m}(\mathbf{x}, t) - \mathbf{m}^{(eq)}(\mathbf{x}, t)\right)\right]_\alpha + \omega_\alpha G(\mathbf{x}, t)\delta t, \text{ and} \tag{6}$$

Streaming step

$$g_\alpha(\mathbf{x} + \mathbf{e}_\alpha \delta t, t + \delta t) = \hat{g}_\alpha(\mathbf{x}, t). \tag{7}$$

In the above, $\mathbf{M}$ is a matrix to transform the distribution functions $\mathbf{g}^{(\mathrm{eq})}$ to their moments $\mathbf{m}^{(\mathrm{eq})}$ by the relation $\mathbf{m}^{(\mathrm{eq})} = \mathbf{M}\mathbf{g}^{(\mathrm{eq})}$. $\mathbf{S}$ is a matrix of relaxation coefficients τ_{ij}. In the D3Q7 model, the equilibrium moments of the distribution functions are $\mathbf{m}^{(\mathrm{eq})} = (\phi, u\phi, v\phi, w\phi, a\phi, 0, 0)^T$, where u, v, and w are the macroscopic velocity components, and a is a constant related to the weight coefficients. For details about the matrices and the constants in the LB models the reader can refer to [20, 32].

4. Interface schemes for conjugate conditions

One unique feature of the LBM method is that both the Dirichlet-type boundary value and the Neumann-type boundary flux, i.e., temperature/concentration gradient, can be obtained from a simple moment of the distribution functions with appropriate boundary schemes [20, 21]. It eliminates finite-difference type approximation schemes for the flux. This idea can also be applied to construct interface schemes by treating the interface as a shared boundary between the two adjacent domains [9, 23, 32, 33]. We consider the basic situation with zero convective flux ($u_n = 0$) and the normal of the interface parallel to the discrete velocity vector, i.e., parallel straight interface; the interface scheme for more general situations such as curved geometry can be similarly constructed as shown in [9, 23, 32, 33]. The conjugate conditions in Eqs. (1b) and (1c) thus reduce to (see **Figure 1**)

$$\Phi_{nf} = -D_f \frac{\partial \phi_f}{\partial n_f} = \sigma D_s \frac{\partial \phi_s}{\partial n_s} + q_{\mathrm{jump}} = -\sigma \Phi_{ns} + q_{\mathrm{jump}}, \tag{8}$$

with $\sigma = 1$, $q_{\mathrm{jump}} = q^{\mathrm{C}}_{\mathrm{jump}}$ in mass transfer, and $\sigma = \frac{(\rho c_p)_s}{(\rho c_p)_f}$, $q_{\mathrm{jump}} = q^{\mathrm{T}}_{\mathrm{jump}} / (\rho c_p)_f$ in heat transfer.

Depending on whether the local interface geometry is considered, the interface schemes fall into two different categories, as discussed in Sections 4.1 and 4.2.

4.1 Interpolation-based interface schemes

According to the second-order interpolation-based boundary schemes developed in [21], the relationships between the distribution functions and the interfacial ϕ values and their fluxes for each domains can be obtained: for the Dirichlet condition treatment,

$$g_{\bar{\alpha}}(\mathbf{x}_f, t + \delta t) = c_{d1}\hat{g}_\alpha(\mathbf{x}_f, t) + c_{d2}\hat{g}_\alpha(\mathbf{x}_{ff}, t) + c_{d3}\hat{g}_{\bar{\alpha}}(\mathbf{x}_f, t) + c_{d4}\varepsilon_D \Phi_{df}, \tag{9a}$$

$$g_\alpha(\mathbf{x}_s, t + \delta t) = c^*_{d1}\hat{g}_{\bar{\alpha}}(\mathbf{x}_s, t) + c^*_{d2}\hat{g}_{\bar{\alpha}}(\mathbf{x}_{ss}, t) + c^*_{d3}\hat{g}_\alpha(\mathbf{x}_s, t) + c^*_{d4}\varepsilon_D \Phi_{ds}. \tag{9b}$$

Similarly, for the Neumann condition treatment,

$$g_{\bar{\alpha}}(\mathbf{x}_f, t + \delta t) = c_{n1}\hat{g}_\alpha(\mathbf{x}_f, t) + c_{n2}\hat{g}_\alpha(\mathbf{x}_{ff}, t) + c_{n3}\hat{g}_{\bar{\alpha}}(\mathbf{x}_f, t) + c_{n4}(\delta t/\delta x)\Phi_{n\bar{\alpha}}, \tag{10a}$$

$$g_\alpha(\mathbf{x}_s, t + \delta t) = c^*_{n1}\hat{g}_{\bar{\alpha}}(\mathbf{x}_s, t) + c^*_{n2}\hat{g}_{\bar{\alpha}}(\mathbf{x}_{ss}, t) + c^*_{n3}\hat{g}_\alpha(\mathbf{x}_s, t) + c^*_{n4}(\delta t/\delta x)\Phi_{n\alpha}, \tag{10b}$$

where $\Phi_{n\bar{\alpha}}$ and $\Phi_{n\alpha}$ are the respective interfacial fluxes along the discrete lattice velocity directions $\mathbf{e}_{\bar{\alpha}}$ and $\mathbf{e}_{\alpha}$, $\mathbf{x}_f$ and $\mathbf{x}_{ff}$ are the first and second interior lattice nodes along $\mathbf{e}_{\bar{\alpha}}$ direction in Domain 1 ($\mathbf{x}_{ff} = \mathbf{x}_f + \mathbf{e}_{\bar{\alpha}}\,\delta t$), and $\mathbf{x}_s$ and $\mathbf{x}_{ss}$ are the lattice nodes along $\mathbf{e}_{\alpha}$ direction in Domain 2 ($\mathbf{x}_{ss} = \mathbf{x}_s + \mathbf{e}_{\alpha}\,\delta t = \mathbf{x}_s - \mathbf{e}_{\bar{\alpha}}\,\delta t$), respectively (see **Figure 1**). All the coefficients in Eqs. (9) and (10) are only related to the local geometry as denoted by the link intersection fraction, Δ, at the interface [21, 23].

When $\mathbf{e}_{\bar{\alpha}}$ is aligned with the interface normal directions, $\Phi_{n\bar{\alpha}} = \Phi_{nf}$ and $\Phi_{n\alpha} = \Phi_{ns}$ are readily noticed. Hence, Eqs. (1), (8), (9a), (9b), (10a) and (10b) constitute a linear system of six equations, and the six unknowns $g_{\bar{\alpha}}(\mathbf{x}_f, t+\delta t)$, $g_{\alpha}(\mathbf{x}_s, t+\delta t)$, Φ_{df}, Φ_{ds}, Φ_{nf}, and Φ_{ns} can be analytically solved. The interface scheme thus becomes [9]:

$$\begin{aligned} g_{\bar{\alpha}}(\mathbf{x}_f, t+\delta t) = {} & A_1^f \hat{g}_{\alpha}(\mathbf{x}_f, t) + A_2^f \hat{g}_{\alpha}(\mathbf{x}_{ff}, t) + A_3^f \hat{g}_{\bar{\alpha}}(\mathbf{x}_f, t) \\ & + B_1^f \hat{g}_{\bar{\alpha}}(\mathbf{x}_s, t) + B_2^f \hat{g}_{\bar{\alpha}}(\mathbf{x}_{ss}, t) + B_3^f \hat{g}_{\alpha}(\mathbf{x}_s, t), \\ & + \chi_q q_{\text{jump}} + \chi_t \phi_{\text{jump}}, \end{aligned} \tag{11a}$$

$$\begin{aligned} g_{\alpha}(\mathbf{x}_s, t+\delta t) = {} & A_1^s \hat{g}_{\bar{\alpha}}(\mathbf{x}_s, t) + A_2^s \hat{g}_{\bar{\alpha}}(\mathbf{x}_{ss}, t) + A_3^s \hat{g}_{\alpha}(\mathbf{x}_s, t) \\ & + B_1^s \hat{g}_{\alpha}(\mathbf{x}_f, t) + B_2^s \hat{g}_{\alpha}(\mathbf{x}_{ff}, t) + B_3^s \hat{g}_{\bar{\alpha}}(\mathbf{x}_f, t). \\ & + \gamma_q q_{\text{jump}} + \gamma_t \phi_{\text{jump}}. \end{aligned} \tag{11b}$$

The coefficients in Eqs. (11a) and (11b) are now determined by the geometry fraction Δ and the property ratio σ. It is worth noting that there is an adjustable parameter in those coefficients since the second-order Dirichlet boundary scheme allows one adjustable parameter, as shown in [21], where three particular Dirichlet schemes were also presented:

$$\text{Scheme 1}: c_{d1} = \begin{cases} -2\Delta, & (0 \le \Delta \le 0.5), \\ -\dfrac{1}{2\Delta}, & (\Delta > 0.5), \end{cases} \tag{12a}$$

$$\text{Scheme 2}: c_{d1} = -2(1-\Delta), \text{ and} \tag{12b}$$

$$\text{Scheme 3}: c_{d1} = -1. \tag{12c}$$

The corresponding interface schemes, can thus also be obtained. Those schemes will be numerically verified in Section 5.2 for a test case including interfacial jump conditions.

When the straight interface is located "halfway" between the lattice nodes ($\Delta = 0.5$), the unknown DFs can be calculated by only knowing two single-node post-collision DFs, i.e., without interpolation:

$$g_{\bar{\alpha}}(\mathbf{x}_f, t+\delta t) = \left(\frac{1-\sigma}{1+\sigma}\right)\hat{g}_{\alpha}(\mathbf{x}_f, t) + \left(\frac{2\sigma}{1+\sigma}\right)\hat{g}_{\bar{\alpha}}(\mathbf{x}_s, t) + \frac{q_{\text{jump}}}{1+\sigma} + \frac{\varepsilon_D \sigma \phi_{\text{jump}}}{1+\sigma}, \tag{13a}$$

$$g_{\alpha}(\mathbf{x}_s, t+\delta t) = -\left(\frac{1-\sigma}{1+\sigma}\right)\hat{g}_{\bar{\alpha}}(\mathbf{x}_s, t) + \left(\frac{2}{1+\sigma}\right)\hat{g}_{\alpha}(\mathbf{x}_f, t) + \frac{q_{\text{jump}}}{1+\sigma} - \frac{\varepsilon_D \phi_{\text{jump}}}{1+\sigma}. \tag{13b}$$

Furthermore, for the most simplified case of $\Delta = 0.5$ and $\sigma = 1$, Eqs. (13a) and (13b) reduce to:

$$g_{\bar{\alpha}}(\mathbf{x}_f, t+\delta t) = \hat{g}_{\bar{\alpha}}(\mathbf{x}_s, t) + q_{\text{jump}}/2 + \varepsilon_D \phi_{\text{jump}}/2, \tag{14a}$$

$$g_{\alpha}(\mathbf{x}_s, t+\delta t) = \hat{g}_{\alpha}(\mathbf{x}_f, t) + q_{\text{jump}}/2 - \varepsilon_D \phi_{\text{jump}}/2. \tag{14b}$$

For straight interfaces where $\Delta \neq 0.5$ and for curved interfaces, the complete interface conditions can be found in [9, 23].

It should be noted that second-order accurate boundary schemes can also be obtained using only the single lattice node next to the boundary, as demonstrated in [32], instead of using interpolations in Eqs. (9) and (10). However, such boundary schemes were constructed with complex coefficients that are related not only to geometry-related Δ, but also to the LB model-related relaxation coefficient. Interested readers are referred to [32] for details and such interface schemes are not discussed in this chapter.

4.2 Modified geometry-ignored interface schemes

The second-order interpolation based interface scheme developed in [23] has attracted much interest. In the last 5 years, there have been various modified interface schemes proposed with the objective of simplifying the original scheme, as it becomes complex and computationally expensive when applied to curved or irregular interfaces. The applicability of those modified schemes was demonstrated in those publications while their accuracy and convergence order have not been fully investigated. In this section, we present three groups of those modified schemes that do not account for the local interface geometry. Most of those schemes were formulated for conjugate heat transfer problems, and conjugate mass transfer can be similarly handled. Comparison of their numerical accuracy with the original interpolation based scheme will be presented in Section 5.1.

4.2.1 Group 1: sourcing term addition

In the first group, additional source terms were introduced to the lattice nodes in the domains next to the interface. For example, the following additional source was given in [34]:

$$S_{conj} = \frac{\partial}{\partial x_j}\left(\frac{1}{\rho c_p}\right) \cdot \left(-k\frac{\partial \phi}{\partial x_j} + \rho c_p u_j \phi\right) \tag{15}$$

In LBM, the total flux in the second bracket can be conveniently obtained from the moment of the nonequilibrium DFs [20, 21, 34]. In [34], the gradient of the heat capacity-related term in Eq. (15) was computed from a first-order one-sided finite-difference (FD) scheme: $\frac{\partial}{\partial x_j}\left(\frac{1}{\rho c_p}\right)_k = \left[\left(\frac{1}{\rho c_p}\right)_k - \left(\frac{1}{\rho c_p}\right)_{\text{avg}}\right] \Big/ 0.5\delta x$ with $\left(\rho c_p\right)_{\text{avg}} = \left[\left(\rho c_p\right)_k + \left(\rho c_p\right)_{k+1}\right]/2$.

It should be noted that the above introduction of the source term and the calculation of the heat capacity-related gradient cause two issues: first, for adjacent domains with distinct ρc_p values, such as fluid-solid interfaces, a discontinuity shows up and the gradient term cannot be resolved with FD schemes; second, with the simple first-order FD approximation, the LBM solution would preserve, at most, up to first-order accuracy. The first-order accuracy was demonstrated in [34], and it will be further discussed in Section 5 with a numerical test.

4.2.2 Group 2: enthalpy-based formulation

Another area of interest is found in enthalpic formulations for the LBE. With the definition of an "enthalpic term" $h^* = (\rho c_p)_0 \phi$, where $(\rho c_p)_0$ is a reference heat capacity, the governing CDE (2) becomes [27, 28]:

$$\frac{\partial h^*}{\partial t} + \nabla \cdot (\mathbf{u} h^*) = \nabla \cdot \left(\frac{k}{\rho c_p} \nabla h^* \right) - \frac{k}{(\rho c_p)_0} \nabla h^* \cdot \nabla \left(\frac{(\rho c_p)_0}{\rho c_p} \right) - \frac{h^* (\rho c_p)_0}{\rho c_p} \mathbf{u} \cdot \nabla \frac{\rho c_p}{(\rho c_p)_0} \tag{16}$$

Comparing Eq. (16) with Eq. (2), the last two terms in Eq. (16) need to be included in the source term implementation in LBM simulations. Clearly, those additional terms also have the heat capacity-related gradients and thus the discontinuity effect. In [36], the gradient was approximated from $\nabla_j f = \frac{1}{c_s{}^2 \delta t} \sum_\alpha w_\alpha f(\mathbf{x} + (\mathbf{e}_\alpha \cdot \mathbf{x})\delta t) e_{\alpha j}$, which reduces to a central FD schemes in the Cartesian grid.

4.2.3 Group 3: modified equilibrium distribution functions

In this group, modified equilibrium DFs were introduced. The heat capacity is typically involved in the modified equilibrium DFs, such as [37, 38]

$$g_\alpha^{\mathrm{eq}} = \begin{cases} \phi(\eta c_p - c_{p0}) + \omega_\alpha \phi c_p \left[c_{p0}/c_p + (\mathbf{e}_\alpha \cdot \mathbf{u})/c_s^2 \right], & \alpha \neq 0 \\ \omega_\alpha \phi c_p \left[c_{p0}/c_p + (\mathbf{e}_\alpha \cdot \mathbf{u})/c_s^2 \right], & \alpha = 0 \end{cases} \tag{17}$$

When using the MRT D2Q5 model the equilibrium moments are calculated to be

$$\mathbf{m}^{\mathrm{eq}} = \left[c_p \phi, \; u c_p \phi, \; v c_p \phi, \; 4 c_p \phi - \frac{10}{3} c_{p0} \phi, \; 0 \right]^T. \tag{18}$$

The temperature is solved from $\phi = \sum_\alpha g_\alpha / \eta c_p$. A key note should be made regarding the relationship between transport properties and the relaxation coefficient, τ, in LBM for Group 3. In all previous models, the thermal diffusivity is related to τ as $D = (\tau - 0.5) c_s^2 \delta t$. However, since a virtual heat capacity correction is employed in this modified equilibrium DF group, the thermal conductivity, rather than the diffusivity, is related to τ as $k = c_{p0}(\tau - 0.5) c_s^2 \delta t$.

5. Numerical accuracy of interface schemes

In order to verify the applicability of the different interface schemes to simulate conjugate heat and mass transfer and compare their accuracy, we consider two benchmark cases with analytical solutions: (i) 2D convection-diffusion in a channel with two-layered fluids, and (ii) 2D diffusion within a circular domain of two solids with interfacial jump conditions. The computational domains are depicted in **Figures 2** and **9**, respectively. Those two cases have been widely employed in [9, 23, 32–34] to verify the various interface schemes.

For straight interfaces, the following relative L2 norm errors are defined to check the numerical accuracy and convergence orders following [23]:

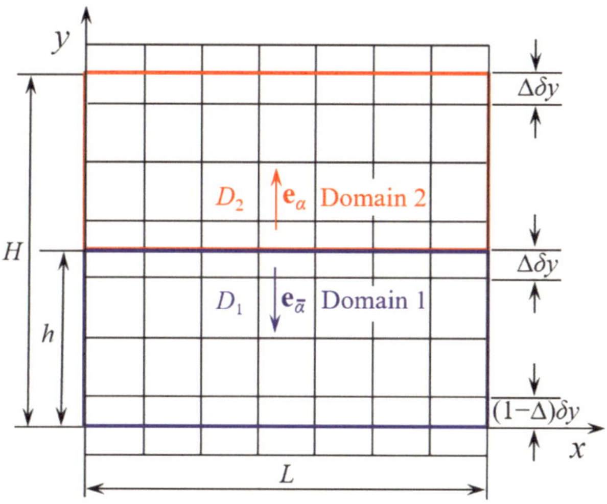

Figure 2.
Schematic layout of the lattice on a 2D channel containing two fluids in domain 1 (0 ≤ y ≤ h) and domain 2 (h ≤ y ≤ H). With permission from [23].

$$E_2 = \left[\sum_{x,y} (\phi_{\text{LBE}} - \phi_{\text{ex}})^2 / \sum_{x,y} \phi_{\text{ex}}^2 \right]^{1/2} \tag{19}$$

$$E_{2,\,\phi\text{int}} = \left[\sum_{x,\,y=h} \left(\phi_{1,2}|_{\text{LBE}} - \phi_{1,2}|_{\text{ex}}\right)^2 / \sum_{x,y=h} \left(\phi_{1,2}|_{\text{ex}}\right)^2 \right]^{1/2} \tag{20}$$

$$E_{2,\,\text{qint}} = \left[\sum_{x,y=h} \left(D_{1,2} \frac{\partial \phi_{1,2}}{\partial y}|_{\text{LBE}} - D_{1,2} \frac{\partial \phi_{1,2}}{\partial y}|_{\text{ex}} \right)^2 / \sum_{x,y=h} \left(D_{1,2} \frac{\partial \phi_{1,2}}{\partial y}|_{\text{ex}} \right)^2 \right]^{1/2} \tag{21}$$

where E_2 evaluates the overall error in all the interior lattice nodes in the two domains, $E_{2,\,\phi\text{int}}$ and $E_{2,\,\text{qint}}$ evaluate the respective relative errors of the interfacial ϕ value and its flux. For circular interfaces, $E_{2,\,\phi\text{int}}$ and $E_{2,\,\text{qint}}$ are evaluated at the interface nodes along the curved geometry. The computation of the interfacial quantities follows that in [21, 23].

5.1 Two-D convection-diffusion in a channel with two fluids

The computational domain is depicted in **Figure 2**. The two fluids are assumed immiscible and both have the same velocity $\mathbf{u} = (U, 0)$. The characteristic Péclet number, Pe, is defined as $Pe = UH/D_1$.

When considering isotropic diffusion, the governing CDE can be expressed as

$$\frac{\partial \phi_{1,2}}{\partial t} + U \frac{\partial \phi_{1,2}}{\partial x} = D_{1,2} \left(\frac{\partial^2 \phi_{1,2}}{\partial x^2} + \frac{\partial^2 \phi_{1,2}}{\partial y^2} \right) \tag{22}$$

We consider only the steady case with sinusoidal boundary conditions on the horizontal walls $\phi_1(x, y = 0) = \phi_2(x, y = H) = \cos(2\pi x/L)$, and periodic conditions in the x-direction. Taking into account the standard conjugate conditions: $\phi_1 = \phi_2$, $k_1 \partial\phi_1/\partial y = k_2 \partial\phi_2/\partial y$ at $y = h$, the analytical solution to Eq. (22) can be solved (see [23] for details).

The interpolation-based interface scheme in Section 4.1 and the modified schemes in Groups 1–3 in Section 4.2 are implemented with the D2Q5 MRT-LB model. For all cases presented, Pe = 20 is used with $H = 2\,h$.

Figures 3a and **b** show the scalar value and flux profiles at selected locations along the x-direction using a diffusivity ratio $D_{21} = D_2/D_1 = 2$ and a thermal conductivity ratio $k_{21} = k_2/k_1 = 3$ (consequently $\sigma = (\rho c_p)_2/(\rho c_p)_1 = 1.5$). Note that $\Delta = 0.5$ is used for all schemes and thus the geometry effect is not included. Other simulation parameters are: $\tau_1 = 0.65$, $H = 64$. Good agreement between LBM and analytical solutions is observed in **Figures 3a** and **b** for all interface schemes. As a further step, **Figures 4a** and **b** compares the interfacial scalar and flux values using the same parameters as for **Figure 3**. Noticeable discrepancy between numerical and analytical solutions is observed in **Figure 4a** and **b** when using the interface schemes of Group 1 and Group 2; while both Group 3 and the original interface scheme in [23] have good agreement. This is mainly due to the presence of the

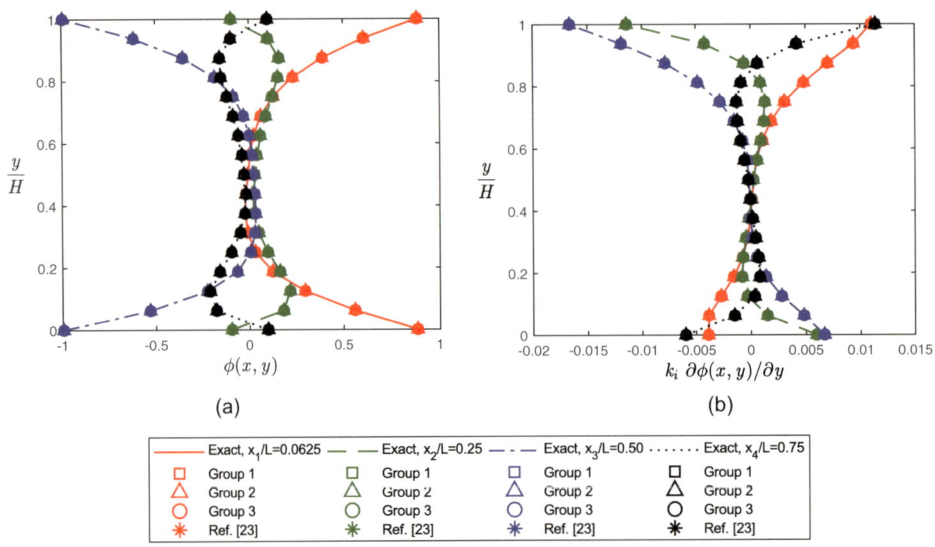

Figure 3.
Profiles of the (a) scalar variable, and (b) flux values at selected vertical lines in the channel.

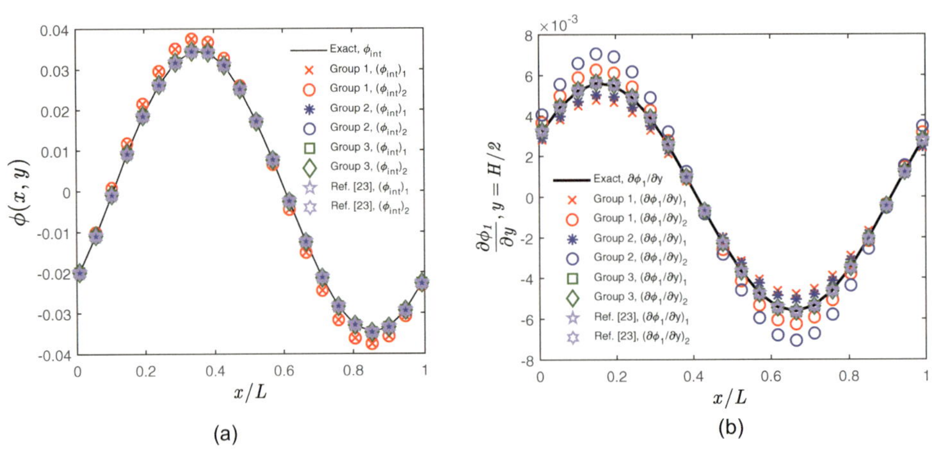

Figure 4.
Comparison of (a) interfacial scalar value, and (b) interfacial fluxes.

discontinuity in approximating the heat capacity gradient in Groups 1 and 2 when $\sigma = (\rho c_p)_2/(\rho c_p)_1 \neq 1$. This will be further demonstrated in the L_2 norm errors.

Figures 5–7 show the L_2 errors defined in Eqs. (19)–(21), respectively. The physical and geometric parameters are $D_{21} = 10$, $k_{21} = 100$, $\sigma = 10$, and $\Delta = 0.5$. Simulation parameters in LBM include $\tau_1 = 0.55$, $\tau_2 = 1.0$ for the original scheme [23] and Groups 1 and 2 with $(\tau_2\_0.5)/(\tau_1\_0.5) = D_{21}$, and $\tau_1 = 0.5025$, $\tau_2 = 0.75$ for Group 3 with $(\tau_2\_0.5)/(\tau_1\_0.5) = k_{21}$.

Figure 5 clearly shows that the original scheme in [23] and Group 3 are able to preserve the second-order accuracy in LBM. However, Groups 1 and 2 show only a linear convergence at low resolution, and it reduces further towards zeroth-order convergence at high resolution. Similar observations can be found in the errors for the interfacial scalar values in **Figure 6**. For the interfacial flux errors shown in **Figure 7**, Group 2 always exhibits zeroth-order accuracy, while Group 1 exhibits linear convergence in one domain and zeroth-order in the other. As previously mentioned, a discontinuity approximation is present in the development of interface schemes in Groups 1 and 2. This is the direct cause for the degradation of the order of accuracy and the much higher error magnitude in these two groups.

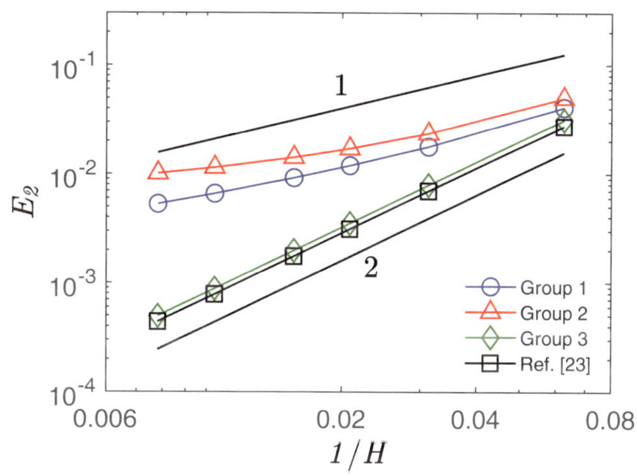

Figure 5.
Relative L2 norm error, E2, for the interior scalar versus the grid resolution, 1/H, for steady convection-diffusion in the channel at Δ = 0.5.

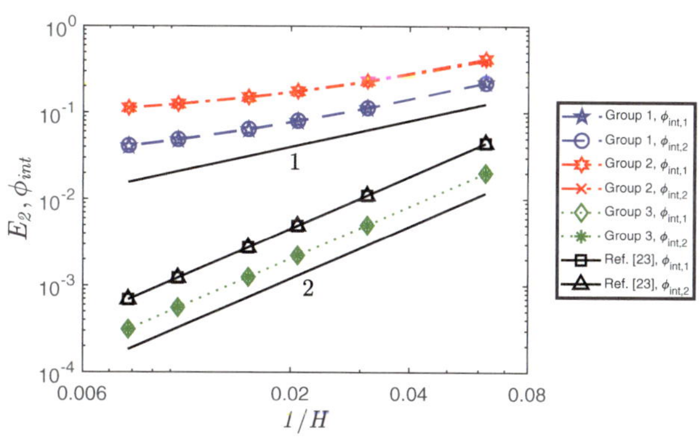

Figure 6.
Relative L2 norm error, E2, ϕint, for the interfacial scalar versus the grid resolution, 1/H, for steady convection-diffusion in the channel at Δ = 0.5.

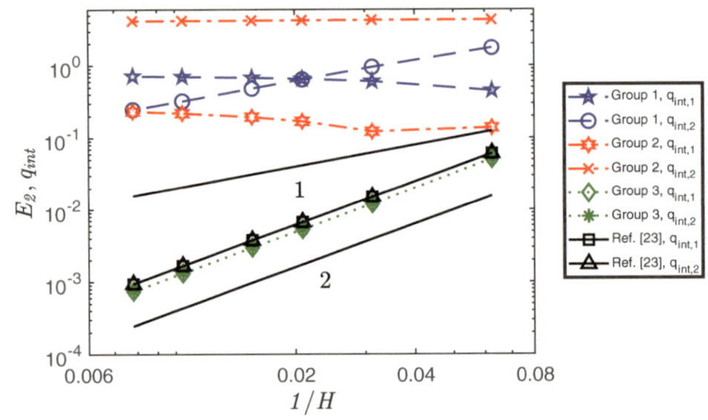

Figure 7.
Relative L2 norm error, E2, qint, for the interfacial flux versus the grid resolution, 1/H, for steady convection-diffusion in the channel at Δ = 0.5.

To further demonstrate the necessity of taking into account the local geometry to preserve second-order accuracy, **Figure 8** presents the L_2 norm errors for the interior field at different Δ values. The parameters used are $D_{21} = 5$, $k_{21} = 50$, $\sigma = 10$, $(\tau_1, \tau_2) = (0.515, 0.575)$ for the interpolation-based scheme [23] and Groups 1 and 2, and $(\tau_1, \tau_2) = (0.515, 1.25)$ for Group 3. For the interpolation-based scheme with an adjustable variable, $c_{d1} = -1$ is used.

It is clear in **Figure 8** that only the interpolation-based scheme considering the local geometry is able to preserve second-order accuracy at different Δ values. The error behavior for Groups 1 and 2 is similar to that in **Figure 5**, the near identical errors at high resolution for different Δ values confirm the error caused by the same discontinuity; moreover, while Group 3 presents second-order convergence for Δ = 0.5, the convergence order drops to first-order for Δ ≠ 0.5. This is similar to the behavior of the "half-lattice division" scheme discussed in [23]. It is evident that when the local interface geometry is not considered for cases Δ ≠ 0.5, the inherent second-order accuracy in LBM computation can be lost.

5.2 Two-D diffusion in a circular domain with jump conditions

This test is applied to demonstrate the applicability and accuracy of the interpolation-based interface scheme to simulate transport in complex geometry and with interfacial jump conditions. **Figure 9** schematically presents the lattice

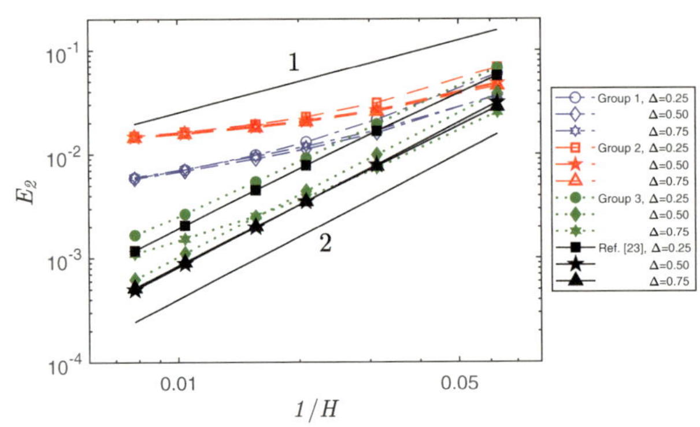

Figure 8.
Relative L2 norm error, E2, for the interior scalar versus the grid resolution, 1/H, for steady convection-diffusion in the channel at different Δ values.

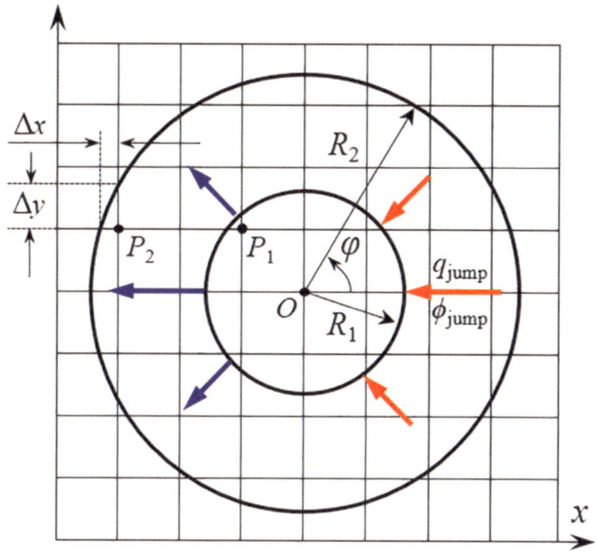

Figure 9.
Schematic layout of the computational domain for circular diffusion. With permission from [9].

layout and computational domain for 2D diffusion within a circular domain. On the outer circle with radius, R_2, a Dirichlet boundary condition is applied as $\phi_2(r = R_2) = \cos\varphi$. With the following conjugate conditions at the interface:

$$\phi_1 = \phi_2 + \phi_{jump} = \phi_2 + \phi_0 \cos\varphi \text{ at } r = R_1 \tag{23}$$

and

$$-D_1 \frac{\partial \phi_1}{\partial r} = -D_2 \frac{\partial \phi_2}{\partial r} + q_{jump} = -D_2 \frac{\partial \phi_2}{\partial r} + q_0 \cos\varphi \text{ at } r = R_1 \tag{24}$$

the analytical solution can be found [9]. For the numerical LBM computation, the parameters are $R_2/R_1 = 2$, $D_2/D_1 = 10$, $\tau_1 = 0.525$, and $\tau_2 = 0.75$. As previously mentioned, three schemes were presented in [21] for adjustable parameters of c_{d1},

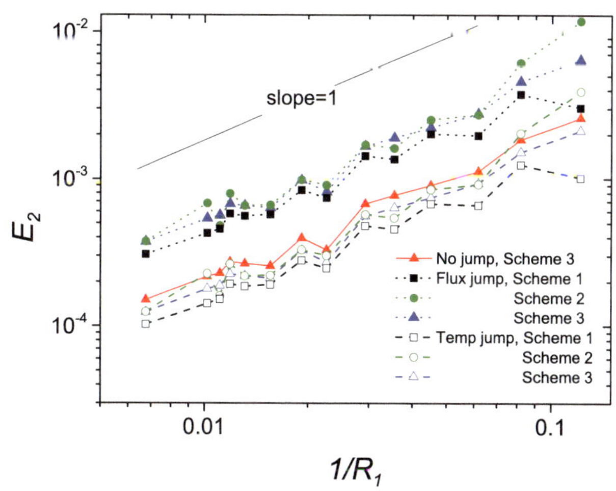

Figure 10.
Relative L2 norm error, E2, for the interior temperature versus the grid resolution for 2D diffusion in a circular domain. With permission from [9].

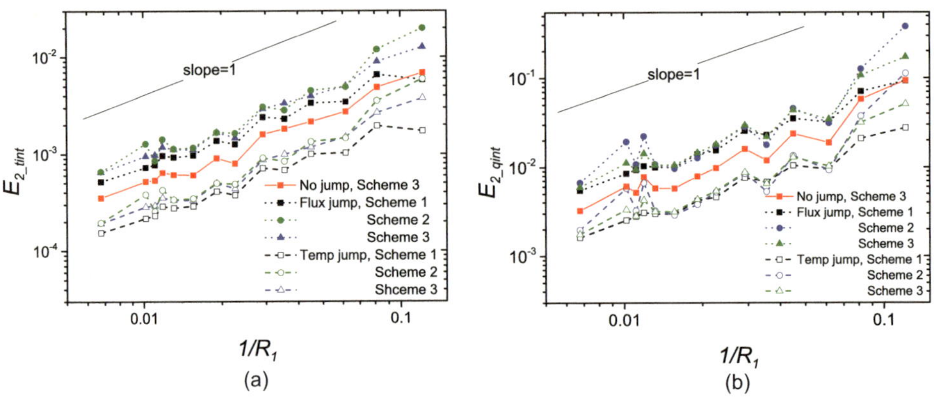

Figure 11.
Relative L2 norm errors (a) E2_tint for the interfacial temperature, and (b) E2_qint for the interfacial flux, versus the grid resolution for 2D diffusion in a circular domain. With permission from [9].

those three schemes in Eqs. (12a)–(12c) are also applied here. For the flux jump case, the jump conditions are set as $\phi_0 = 0$ and $q_0 = D_2/R_1$; and for the temperature jump case, those are $\phi_0 = 0.5$ and $q_0 = 0$.

Figures 10 and **11a, b** provide the respective relative L_2 norm errors for the interior temperature and the interfacial temperature and flux with the conjugate schemes implemented.

The results in **Figures 10** and **11** demonstrate the first-order accuracy at high resolution for all cases. It should also be noted that the temperature and flux jump conditions at the interface do not affect the order of convergence. The decrease of the convergence order from second-order in Section 5.1 to first-order in this test is due to the implementation of the Cartesian decomposition method [21] that was used to convert the normal fluxes into those in the discrete velocity directions. It is expected that the modified geometry-ignored interface schemes in Groups 1–3 would result in much higher error magnitude for curved interfaces, and Groups 1 and 2 would yield only zeroth-order convergence for all the three quantities of interest. This will be presented in future publications.

6. Conclusions

The chapter presents a brief review of the interface schemes within the scope of the lattice Boltzmann method (LBM) for conjugate transport between multiphases or different materials. Compared to the interface schemes developed to satisfy the macroscopic conjugate conditions using traditional CFD methods, the LBM method deals with the microscopic distribution functions (DFs); the physical conjugate conditions can be converted to those for the DFs, and they are satisfied in each time-marching step without iterations. In the last decade, a number of interface schemes have been proposed. The interpolation-based schemes [9, 23, 33] taking into account the local interfacial geometry are able to preserve the second-order accuracy in LBM for straight interfaces; while those "modified" geometry-ignored schemes [34–38] have at most first-order accuracy in general, and with the introduction of heat capacity-related discontinuity in those schemes (e.g., Groups 1 and 2), the order of accuracy becomes essentially zeroth order.

Furthermore, it is verified that when using the interpolation-based schemes, the interfacial jump conditions can be conveniently modeled with no effect on the order of accuracy of the LBM solutions.

Curved geometry also has a substantial effect and it reduces the order of accuracy of LBM solutions in modeling conjugate heat and mass transfer problems. In addition, the interpolation-based schemes would demand for a higher computational cost than those modified schemes. The readers are thus recommended to take into account both numerical accuracy and computational cost when selecting effective interface schemes for curved geometries.

Acknowledgements

LL acknowledges the support and start-up fund from the Department of Mechanical Engineering at Mississippi State University.

Author details

David Korba and Like Li*
Department of Mechanical Engineering, Mississippi State University, Mississippi, USA

*Address all correspondence to: likeli@me.msstate.edu

IntechOpen

© 2019 The Author(s). Licensee IntechOpen. This chapter is distributed under the terms of the Creative Commons Attribution License (http://creativecommons.org/licenses/by/3.0), which permits unrestricted use, distribution, and reproduction in any medium, provided the original work is properly cited.

References

[1] Roe B, Jaiman R, Haselbacher A, Geubelle PH. Combined interface boundary condition method for coupled thermal simulations. International Journal for Numerical Methods in Fluids. 2008;**57**:329

[2] Dorfman A, Renner Z. Conjugate problems in convective heat transfer: Review. Mathematical Problems in Engineering. 2009;**2009**:927350

[3] Errera MP, Chemin S. Optimal solutions of numerical interface conditions in fluid-structure thermal analysis. Journal of Computational Physics. 2013;**245**:431

[4] Li L, Mei R, Klausner JF, Hahn DW. Heat transfer between colliding surfaces and particles. ASME Journal of Heat Transfer. 2012;**134**:011301

[5] Zhang J, Yang C, Mao ZS. Unsteady conjugate mass transfer from a spherical drop in a simple extensional creeping flow. Chemical Engineering Science. 2012;**79**:29

[6] Wang J, Wang M, Li Z. A lattice Boltzmann algorithm for fluid-solid conjugate heat transfer. International Journal of Thermal Sciences. 2007;**46**: 228-234

[7] Wang M, Wang J, Pan N, Chen S. Mesoscopic predictions of the effective thermal conductivity for microscale random porous media. Physical Review E. 2007;**75**:036702

[8] Chen L, Kang Q, He YL, Tao WQ. Pore-scale simulation of coupled multiple physicochemical thermal processes in micro reactor for hydrogen production using lattice Boltzmann method. International Journal of Hydrogen Energy. 2012;**37**:13943

[9] Guo K, Li L, Xiao G, AuYeung N, Mei R. Lattice Boltzmann method for conjugate heat and mass transfer with interfacial jump conditions. International Journal of Heat and Mass Transfer. 2015;**88**:306-322

[10] Lu JH, Lei HY, Dai CS. Analysis of Henry's law and a unified lattice Boltzmann equation for conjugate mass transfer problem. Chemical Engineering Science. 2019;**199**:319-331

[11] Peskin CS. The immersed boundary method. Acta Numerica. 2002;**11**: 479-517

[12] Mittal R, Iaccarino G. Immersed boundary methods. Annual Review of Fluid Mechanics. 2005;**37**:239-261

[13] LeVeque RJ, Li ZL. The immersed interface method for elliptic equations with discontinuous coefficients and singular sources. SIAM Journal on Numerical Analysis. 1994;**31**:1019-1044

[14] Deng SZ, Ito K, Li ZL. Three-dimensional elliptic solvers for interface problems and applications. Journal of Computational Physics. 2003;**184**: 215-243

[15] Fedkiw RP, Aslam T, Merriman B, Osher S. A non-oscillatory Eulerian approach to interfaces in multimaterial flows (the ghost fluid method). Journal of Computational Physics. 1999;**152**: 457-492

[16] Liu XD, Fedkiw RP, Kang M. A boundary condition capturing method for Poisson's equation on irregular domains. Journal of Computational Physics. 2000;**160**:151-178

[17] Udaykumar HS, Mittal R, Rampunggoon P, Khanna A. A sharp interface Cartesian grid method for simulating flows with complex moving boundaries. Journal of Computational Physics. 2001;**174**:345-380

[18] Marella S, Krishnan S, Liu H, Udaykumar HS. Sharp interface Cartesian grid method I: An easily implemented technique for 3D moving boundary computations. Journal of Computational Physics. 2005;**210**:1-31

[19] Zhou YC, Zhao S, Feig M, Wei GW. High order matched interface and boundary method for elliptic equations with discontinuous coefficients and singular sources. Journal of Computational Physics. 2006;**213**:1-30

[20] Yoshida H, Nagaoka M. Multiple-relaxation-time lattice Boltzmann model for the convection and anisotropic diffusion equation. Journal of Computational Physics. 2010;**229**:7774

[21] Li L, Mei R, Klausner JF. Boundary conditions for thermal lattice Boltzmann equation method. Journal of Computational Physics. 2013;**237**:366

[22] Li L, Mei R, Klausner JF. Heat transfer evaluation on curved boundaries in thermal lattice Boltzmann equation method. ASME Journal of Heat Transfer. 2014;**136**:012403

[23] Li L, Chen C, Mei R, Klausner JF. Conjugate heat and mass transfer in the lattice Boltzmann equation method. Physical Review E. 2014;**89**:043308

[24] Krivovichev GV. Linear Bhatnagar-Gross-Krook equations for simulation of linear diffusion equation by lattice Boltzmann method. Applied Mathematics and Computation. 2018; **325**:102-119

[25] Perko J. A single-relaxation-time lattice Boltzmann model for anisotropic advection-diffusion equation based on the diffusion velocity flux formulation. Computational Geosciences. 2018;**22**: 1423-1432

[26] Zhang XX, Bengough AG, Crawford JW, Young IM. A lattice BGK model for advection and anisotropic dispersion equation. Advances in Water Resources. 2002;**25**:1-8

[27] Ginzburg I. Equilibrium-type and link-type lattice Boltzmann models for generic advection and anisotropic-dispersion equation. Advances in Water Resources. 2005;**28**:1171-1195

[28] Ginzburg I, d'Humières D, Kuzmin A. Optimal stability of advection-diffusion lattice Boltzmann models with two relaxation times for positive/negative equilibrium. Journal of Statistical Physics. 2010;**139**:1090-1143

[29] Chai Z, Shi B, Guo Z. A multiple-relaxation-time lattice Boltzmann model for general nonlinear anisotropic convection-diffusion equations. Journal of Scientific Computing. 2016;**69**: 355-390

[30] Chai Z, Zhao TS. Lattice Boltzmann model for the convection-diffusion equation. Physical Review E. 2013;**87**: 063309

[31] Shi B, Guo Z. Lattice Boltzmann model for nonlinear convection-diffusion equations. Physical Review E. 2009;**79**:016701

[32] Hu Z, Huang J, Yong WA. Lattice Boltzmann method for convection-diffusion equations with general interfacial conditions. Physical Review E. 2016;**93**:043320

[33] Mu Y-T, Gu Z-L, He P, Tao W-Q. Lattice Boltzmann method for conjugated heat and mass transfer with general interfacial conditions. Physical Review E. 2018;**98**:043309

[34] Karani H, Huber C. Lattice Boltzmann formulation for conjugate heat transfer in heterogeneous media. Physical Review E. 2015;**91**:023304

[35] Rihab H, Moudhaffar N, Sassi BN, Patrick P. Enthalpic lattice Boltzmann formulation for unsteady heat

conduction in heterogeneous media. International Journal of Heat and Mass Transfer. 2016;**100**:728

[36] Chen S, Yan YY, Gong W. A simple lattice Boltzmann model for conjugate heat transfer research. International Journal of Heat and Mass Transfer. 2017; **107**:862

[37] Gao D, Chen Z, Chen L, Zhang D. A modified lattice Boltzmann model for conjugate heat transfer in porous media. International Journal of Heat and Mass Transfer. 2017;**105**:673-683

[38] Chen S. Simulation of conjugate heat transfer between fluid-saturated porous media and solid wall. International Journal of Thermal Sciences. 2018;**124**:477-483

[39] Yu D, Mei R, Luo L-S, Shyy W. Viscous flow computations with the method of lattice Boltzmann equation. Progress in Aerospace Science. 2003;**39**: 329-367

[40] Li L, Mei R, Klausner JF. Lattice Boltzmann models for the convection-diffusion equation: D2Q5 vs. D2Q9. International Journal of Heat and Mass Transfer. 2017;**108**:41-62

Section 2

CFD Application in Chemical and Environmental Engineering

Chapter 5

Hydrodynamic Analysis on a Photocatalytic Reactor Using ANSYS Fluent®

Adolfo Ruiz-Soto, Diana Barraza-Jiménez, Abel Hurtado-Macias, Sandra Iliana Torres-Herrera, Carlos Omar Ríos-Orozco, Miriam López-Guzmán, Elva Marcela Coria-Quiñones, Raúl Armando Olvera-Corral and Manuel Alberto Flores-Hidalgo

Abstract

Solar technology includes a wide variety of developments in environmental applications that include photovoltaic cells and photocatalytic devices, among others. Sunlight usage as a clean energy source is highly desirable in technology applications. The main interest of this proposal is to carry on with hydrodynamic analysis in photocatalytic reactors applications where sunlight is used to activate a chemical reaction to degrade water pollutants and calculations are based in computational fluid dynamics (CFD) using ANSYS®. The different steps, geometric domain, preprocessing steps, setup, and postprocessing steps, are described to display an analysis of a numerical calculation during the design of a photocatalytic reactor using the commercial software ANSYS Fluent®. This work may help as a guide for chemical reactor design and includes a numerical solution of one case for a photocatalytic reactor during its design process. In addition, simplifications are explained which enable the designer to make an efficient process of the numerical calculation. Calculations and analysis are carried over in ANSYS Fluent® a powerful multi-physics program suite to develop photocatalytic reactors.

Keywords: hydrodynamics, CFD, ANSYS®, photocatalysis, chemical reactor

1. Introduction

Chemical reactor design is a complex task that requires multiple disciplines working together to obtain an efficient design. Among the different tools that can be used during the chemical reactor design development, computational fluid dynamics (CFD) is one that has attracted attention due to the interesting extent of simulation capabilities that latest code versions offer. In this chapter, the main focus relates to the first step that needs to be resolved during a CFD simulation, that is, fluid dynamic simulation.

Supported by brief theoretical fundamentals as design guidelines and software tools such as computer-aided design (CAD) and ANSYS Fluent®, a case of study for a photocatalytic reactor is presented in this chapter. The case of study focuses in the fluid dynamic simulation for practical reasons in regard to the extent needed to be explained further from this simulation. The chapter intends to capture the essence of the first step in the CFD simulation task on the reactor design development and be a general guide for other developments of similar reactors.

2. Chemical reactors

Chemical reactor design is a complex task that requires multiple disciplines to interact, so a final product may be achieved. Due to the high complexity involved, there are extensive literature covering the chemical reactor design. The present chapter intends to be a brief view to a case where a chemical reactor design intends to simulate the hydrodynamics of the reactor [1, 2].

Chemical reaction engineering (CRE) is mentioned in several sections assuming the reader has an idea of this area, but the case displayed in this chapter only simulates hydrodynamics without going further in the process to develop CRE. For more information related to CRE, the authors recommend to consult textbooks on this matter.

A chemical reactor may be defined as an equipment unit in a chemical process (plant) where chemical transformations (reactions) take place to generate a desirable product at a specified production rate, using a given chemistry. Usually, the performance of the chemical reactor plays a pivotal role in the operation and economics of the entire process since its operation affects most other units in the process (separation units, utilities, etc.), so an efficient reactor design will reflect deeply in the plant performance [2].

2.1 Chemical reactor brief background

Classifying chemical reactors is a difficult task due to the great variation available; in fact, reactor features may be as unlimited as the designer's imagination. In general, reactors may be classified by three main characteristics:

1. Mode of operation (e.g., batch, continuous, semi-batch).

2. Geometric configuration (e.g., tubular, agitated tank, radial flow).

3. Contacting patterns between phases (e.g., packed bed, fluidized bed, bubble column).

Another practical classification is based in reactor operations that may be based in the way their temperature (or heat transfer) is controlled. Three operational conditions are commonly used:

i. Isothermal operation—the same temperatures exist throughout the reactor.

ii. Adiabatic operation—no heat is transferred into or out of the reactor.

iii. Non-isothermal operation—the operation is neither isothermal nor adiabatic.

There are terms commonly used that are recommended to at least have an idea of what they mean, for example, batch reactors, semi-batch reactor, distillation reactor, continuous reactor (flow reactors), residence time, etc. These terms are explained in textbooks related to reaction engineering or chemical reactor design. Also, there are common configurations of continuous reactors that a designer should have at least an idea of them, such as tubular reactor, continuous stirred-tank reactor (CSTR), and cascade of CSTRs, among others. For multiphase reactions, the contacting patterns are used as a basis for classifying the reactors. Common configurations include packed-bed reactor, moving-bed reactor, fluidized-bed reactor, trickle-bed reactor, bubbling column reactor, and others (e.g., spray reactor, slurry reactor, membrane reactor, etc.) [1–4].

2.2 Photocatalytic reactors

Photocatalysis is based in the photocatalyst absorption of radiant energy (visible or UV), normally an oxide semiconductor, which accelerates the reaction rate. The photocatalysis main advantage relates to the use of solar energy as a clean primary source of energy in its processes. The difference between photocatalysis and catalysis is found at the method to activate the catalyst, because the former uses photonic energy for activation in substitution of thermal activation used in conventional catalysis [4].

The photocatalytic process starts when the semiconductor receives photonic energy from the light source with an energy equivalent or higher than band gap ($h\upsilon \geq E_g$) and a pair electron hole ($e-/h+$) with opposite charge is generated. Charges $e-/h+$ generated migrate to the semiconductor surface. The electron at the interphase may be transferred to an acceptor or oxidant molecule in such a way that the excited electron in the conduction band produces the reduction semi-reaction. In contrast, the electron transfer from a reducing molecule to fill a hole in the valence band produces the oxidation semi-reaction, which potentially will generate the hydroxyl radicals or other radical species considered responsible of the photocatalytic activity. The simultaneous oxidation reduction reactions occur at the catalyst surface, at the interphase between the excited solid and the fluid (liquid or gas). During the design process, electronic transfer process effectiveness needs to be considered taking into account that there is a competition with the electron–hole recombination because this last process dissipates in heat the energy absorbed by the photon.

Another consideration the designer needs to account for is the operation mode for the reactor. For example, it may be continuous with single step or in batch. If degradation occurs in a single step, then reactor size and flow rate need to be carefully defined to make sure the desired pollutant degradation is reached, and it may be possible to increase the number of reactors for that purpose. In discontinuous mode or batch mode, the fluid is stored in a tank and is recirculated continuously through the reactor until the desired pollutant concentration is reached. Batch mode operation is the most used option [1–3].

A photoreactor specifically with two features is considered more important; these features are (a) catalyst load and configuration and (b) light source type.

TiO_2 photoactivation requires light radiation with a wavelength of at least 384 nm and a maximum absorbance approximated to 340 nm. The required radiation may be generated by artificial illumination (lamps) or with sunlight illumination. Artificial light sources are in general multidirectional, and it is common to use reflectors to direct or focus the light emitted to the reactor. In contrast, sunlight is considered unidirectional. In consequence, reactor design depends on the light source since its geometry and dimensions are oriented to capture the most ideal way to capture available light [3, 4].

2.3 Chemical reactor design highlights

Due to the diversity of applications and numerous configurations of chemical reactors, a generic design procedure is impossible in order to describe reactor operations. It is necessary to identify the characteristics of the chemical reaction and the main features that the reactor should provide and from there define the next steps. Once these are identified, the appropriate physical and chemical concepts are applied to describe the selected reactor operation [1–5].

3. Computational fluid dynamics

Computational fluid dynamics (CFD) is the analysis of systems involving fluid flow, heat transfer, and associated phenomena such as chemical reactions by means of computer-based simulation. This methodology is very powerful and may be used in a wide range of industrial and nonindustrial application areas. For hydrodynamic simulations, CFD is the proper methodology that may be used during reactor design procedures.

3.1 CFD methodology

CFD codes may be considered a series of numerical algorithms that can resolve fluid flow problems. Commercial CFD packages include sophisticated user interfaces to facilitate the process of input problem parameters and to examine the results, and our interest within this chapter is the code ANSYS® [6, 7]. Most codes contain three main elements: (a) a preprocessor, (b) a solver, and (c) a postprocessor. There are external functions that can be coded as additional or complementary functions that can add up to the calculations and results; for ANSYS Fluent® these functions are known as user-defined functions (UDFs) [8]. Briefly, the function of each of these elements within a CFD code is mentioned [1–5].

3.1.1 Preprocessing

Preprocessing consists of the input of a flow problem to a CFD program by means of an interface and the subsequent transformation of this input into a data ready to use by the solver. The user activities at the preprocessing stage involve geometry definition (domain), grid generation (mesh), physical and chemical phenomena to be modeled, fluid properties, and boundary conditions.

3.1.2 Processing

There are three different schemes for numerical solution techniques: finite difference, finite element, and spectral methods. In this work only finite volume method will be considered, a special finite difference formulation that is central to the most well-established CFD codes including the program used in our case of study ANSYS Fluent®. The numerical algorithm to resolve the problem at hand consists of the following steps: governing equations, discretization, and solution of algebraic equations.

3.1.3 Postprocessing

In this step, results are available, and thanks to the visualization code tools, the data may be presented in different ways to facilitate the results analysis. The leading CFD packages are equipped with versatile data visualization tools. These may include domain geometry and grid display, vector plots, line and shaded contour plots, 2D and 3D surface plots, particle tracking, and view in perspective (translation, rotation, scaling, etc.), and more recently, animation has been included.

Most codes produce trusty alphanumeric output and have data export facilities for further manipulation external to the code which also represents a possibility to enhance the result analysis and solution presentation/explanation procedures.

3.2 Governing equations

The governing equations of fluid flow represent mathematical statements of the conservation laws of physics: (a) mass conservation, (b) momentum conservation, and (c) energy conservation.

The fluid will be subject to the theory of the continuum. For the analysis of fluid flows at macroscopic length scales (around 1 μm and larger), the molecular structure of matter and molecular motions may be ignored. The behavior of the fluid is described in terms of macroscopic properties, such as velocity, pressure, density and temperature, and their space and time derivatives. These may be thought of as averages over a suitable large number of molecules. We will save the mathematical development because it is beyond the scope of this chapter and recommend the reader to consult CFD textbooks for such purpose.

4. Hydrodynamic calculations

As mentioned in prior sections, the design of a chemical reactor involves several steps and disciplines, and simulation hydrodynamic calculations are the first to be carried on. Fluid kinematics deals with describing the motion of fluids without necessarily considering the forces and moments that cause the motion. In this section, we introduce fundamental kinematic concepts related to flowing fluids. We discuss briefly the material derivative and its role in transforming the conservation equations from the Lagrangian description of fluid flow (following a fluid particle) to the Eulerian description of fluid flow (pertaining to a flow field).

The kinematics of experiments under the Lagrangian description involves keeping track of the position vector of each object, x_A, x_B,. ..., and the velocity vector of each object, V_A, V_B, ..., as functions of time. When this method is applied to a flowing fluid, we call it the Lagrangian description of fluid motion. From a microscopic point of view, a fluid is composed of billions of molecules that are subject to continuous collisions into one another, somewhat like billiard balls, but the task of following even a subset of these molecules is quite difficult, even for our biggest computers.

A more common method of describing fluid flow is the Eulerian description of fluid motion. In the Eulerian description of fluid flow, a finite volume called a control volume is defined, through single piece of volume fluid flows in and out. We do not need to keep track of the position and velocity of a mass of fluid particles of fixed identity. Instead, we define field variable functions of space and time, within the control volume [3–5]. For example, the pressure field is a scalar field variable for general unsteady tridimensional fluid flow in Cartesian coordinates:

Pressure field:

$$\vec{P}=P\left(\vec{x},\vec{y},\vec{z},t\right) \tag{1}$$

We define the velocity field as a vector field variable in a similar fashion:

Velocity field:

$$\vec{V} = V\left(\vec{x},\vec{y},\vec{z},t\right) \tag{2}$$

Likewise, the acceleration field is also a vector field variable:

Acceleration field:

$$\vec{a} = a(\vec{x}, \vec{y}, \vec{z}, t) \tag{3}$$

Collectively, these (and other) field variables define the flow field. The velocity field of Eq. 2 is expanded in Cartesian coordinates (x, y, z) and (i, j, k) as:

$$V = (\vec{u}, \vec{v}, \vec{w}) = u(x,y,z,t)\hat{i} + v(x,y,z,t)\hat{j} + w(x,y,z,t)\hat{k} \tag{4}$$

We will leave for other texts the mathematical development of the Lagrangian and Eulerian descriptions since we are interested in a qualitative interpretation more than the mathematical concepts. In the Eulerian description, we do not really care what happens to individual fluid particles, but we are concerned with the pressure, velocity, acceleration, etc., of whichever fluid particle happens to be at the location of interest at the time of interest. The Eulerian description is often more convenient for fluid mechanic applications. Furthermore, experimental measurements are generally more suited to the Eulerian description [3–5].

4.1 First hydrodynamics

In this work, the intention is to resolve and analyze one of the most important steps during the chemical reactor design process, in particular, for a case related to a photocatalytic reactor. This section establishes the importance of hydrodynamics in chemical reactor design for water remediation focusing in photocatalytic reactors. The fluid dynamics is a part of the reactor design process that among the simulation activities should be resolved in the first place. Fluid dynamic results can be complemented by chemical reaction simulation, chemical species, radiation from light source effects, etc.

4.2 Proposed model to incorporate CFD calculations in chemical reactor design

In this type of reactors, the phenomena occurring in the device may be explained by four different processes: (i) fluid mechanics, (ii) heat transfer (radiation), (iii) mass transport, and (iv) chemical reaction. These different processes may be considered the steps to follow or a design methodology. It works better if used as a recirculated cyclic process because at some point some experimental parameters will be needed. For example, a kinetic intrinsic model for the chemical degradation of the selected pollutant(s) is needed as input. The kinetic parameters are obtained from experimentation and will be used as input data in the CFD modeling. Another advantage obtained from CFD is the possibility to build new scenarios using these input data, for example, the use of a bigger size or more complex reactor under different flow dynamics or optical conditions.

A photocatalytic reactor used a light source or sunlight as energy source to trigger the chemical reaction. Modeling facilitates the designer to run the chemical reactor under different scenarios that may be later compared with results obtained with experimentation when a physical device is available. There are different chemical reactor methodologies or guidelines; in **Figure 1** the reader may find a general methodology proposed for the specific case of photocatalytic reactor design.

In the design of a chemical reactor, knowledge of the chemical kinetics process involved is needed. Kinetic parameters should follow a mechanistic model, so they may be used during the simulation beyond the operational limits used to obtain them. The reaction modeling is an area under development due to the difficulty of accurate methodologies to obtain appropriate kinetic parameters that may be used during the

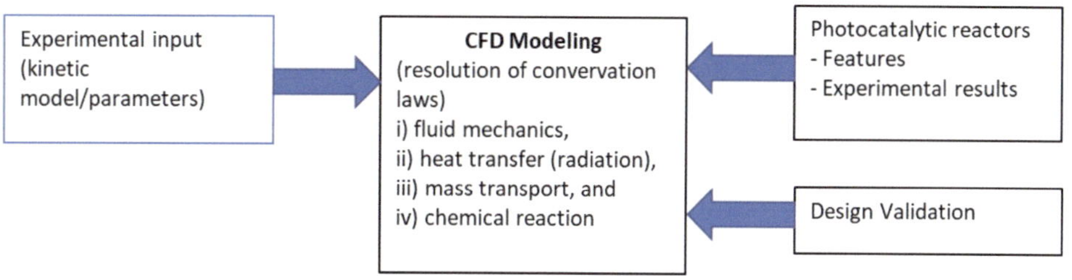

Figure 1.
Proposed methodology for photocatalytic reactor design.

simulation. For each specific reactor type, the recommendation is to consult the latest literature to define the guidelines in order to define the kinetic parameters [3–5].

4.3 CFD calculations

As mentioned in prior sections, this chapter will deal with hydrodynamic calculations related to a photocatalytic reactor. While any intelligent, computer-literate person can run a CFD code, the results obtained may not be physically correct. The case proposed within this chapter has been included as an academic example only. Therefore, the goal of this chapter is to present guidelines or the basic steps for a hydrodynamic simulation.

The examples presented here have been obtained with the commercial computational fluid dynamic code ANSYS Fluent® [6, 7]. Other CFD codes would yield similar but not identical results. Sample CFD solutions are shown for incompressible laminar flow, without heat transfer and chemical reaction.

5. Case of study

In this case, an academic example of a photocatalytic reactor design process is included. More specifically, the intention is to develop the hydrodynamic simulation of the photocatalytic reactor. The data that can be obtained from hydrodynamic simulation permits a detailed definition of the flow motion features, for example, fluid speed in the reaction zone, fluid pressure distribution, and the use of these data as input for further calculations such as distribution of energy from light source, photocatalytic reaction, etc. To achieve these calculations, literature and operational basic requirements were studied in order to define fundamental criteria to assign basic dimensions to the reactor geometry.

With the definition of basic dimensions, it is possible to start using CAD tools to build 3D models that will provide the reactor geometric features that will be needed during CFD simulations. Creo 4.0 and SolidWorks 2016 were used to develop the 3D models of the proposed reactor. For hydrodynamic calculations ANSYS Fluent® versions 18.2 and 19 were used [6, 7].

5.1 Initial calculations

To be able to grow an understanding of the basic reactor design criteria, it was needed to consult the bibliography in different functional areas. The design intent is to build a photochemical reactor of an appropriate size to support laboratory capabilities for testing of nanoparticles after those are synthetized by our research group. It is important for this design to simplify as much as possible the reactor operation.

The basic dimensions defined for the reaction zone container are 300 x 40 x 25 mm. From these dimensions everything else was defined until a 3D model was ready for each component of the reactor including a file with the part assembly that contains the whole reactor. Also, the basic dimensions allow fundamental initial calculations such as volume, inlet/outlet diameter, Reynolds number, etc. Some of the initial assumptions made for this reactor relate to the shape of the reaction zone container which was defined as cuboid (a rectangular hexahedron or a polyhedron bounded by six quadrilateral faces) as will be displayed in the next pages. Within this reaction zone, polluted water will be subject to a chemical reaction to degrade the pollutant into harmless components. So, the fluid chosen is water with a pollutant in low concentration.

Another criterion that needs to be covered in this initial part is the pollutant that will be considered during the design procedures. For this case hydrogen peroxide (H_2O_2) at very low concentrations (20 mg/L) was selected, so for hydrogen calculations the effects of the pollutant may be ignored, and the fluid may be considered as water. Inlet velocity is considered completely axial to the inlet face, and this face is considered exposed to the atmospheric pressure (P_{abs} = 1 atm). All the walls in the domain are considered steady nonslip conditions. The analysis was performed in steady state (nondependent of time) and in laminar regime considering that Reynolds number can be calculated with the next equation:

$$Re = \frac{V \rho \, Lc}{\mu} \tag{5}$$

V = velocity in m/s.
ρ = density in kg/m3.
L_c = characteristic length in m.
μ = dynamic viscosity in Pa-s.

For the geometry employed in the reactor, characteristic length may be calculated with the next equation (for rectangular ducts):

$$L_c = \frac{4 \times A}{I} \tag{6}$$

So, we can calculate for a velocity with the next value: V = 0.025 m/s:

$$A = 0.25\,m \times 0.003\,m = 75 \times 10^{-6} m^2$$

$$P = (2 \times 0.025\,m) + (2 \times 0.003\,m) = 0.056\,m$$

$$L_c = \frac{4 \times (75 \times 10^{-6}\,m^2)}{0.056\,m} = 0.00535\,m$$

$$Re = \frac{0.025\,{}^{m}\!/\!{}_{s} \times 998.2\,{}^{kg}\!/\!{}_{m^3} \times 0.00535\,m}{0.001003\,{}^{kg}\!/\!{}_{m-s}} = 133.10$$

Then, the values for Re may be calculated for different inlet velocities (**Table 1**).

For the hydrodynamic simulation, the inferior face to work as the reaction surface as well as the inlet and outlet flow face was defined. The liquid selected will be water, and the properties are defined as explained in prior pages. The properties that will be used for water are:

Density = 998.2 kg/m^3.
Viscosity = 0.001003 kg/(m-s).
Inlet velocity = 0.05 m/s.
Temperature = 288.16 K = 15.01 C.

5.2 Tridimensional models using CAD tools

With the initial parameters mainly in the reaction zone, the CAD 3D models were developed to continue further building the rest of the components. After designing each one of the components, the assembly of the whole reactor was built, and engineering prints were also generated to complete the CAD development. In this chapter the reaction zone will be the main focus for calculations. The rest of the assembly will be displayed to complement the reactor context. Engineering prints will be mentioned only, but the information within the prints which is intended to manufacture the components falls out of the scope of this chapter. The assembly of the reactor design is displayed in **Figure 2**.

Once the geometry definition of the reactor is completed, then it is needed to define the domain where the hydrodynamic calculations will be performed. The reaction zone was defined from the beginning and is considered the central part of the reactor. From the reaction zone, the volume that will be used for hydrodynamic calculations is extracted using CAD software and ANSYS Fluent® tools. The geometry of the reaction zone is shown in **Figure 3**. This domain is the central part of

Item	Inlet velocity (m/s)	Reynolds number	Residence time in the reaction zone (s)
1	0.05	266.2	5.00
2	0.10	532.4	2.50
3	0.15	798.6	1.33
4	0.20	1064.8	1.25
5	0.25	1331.0	1.00

Table 1.
Effects of cell count for an experiment under V = 0.05 m/s during the mesh analysis.

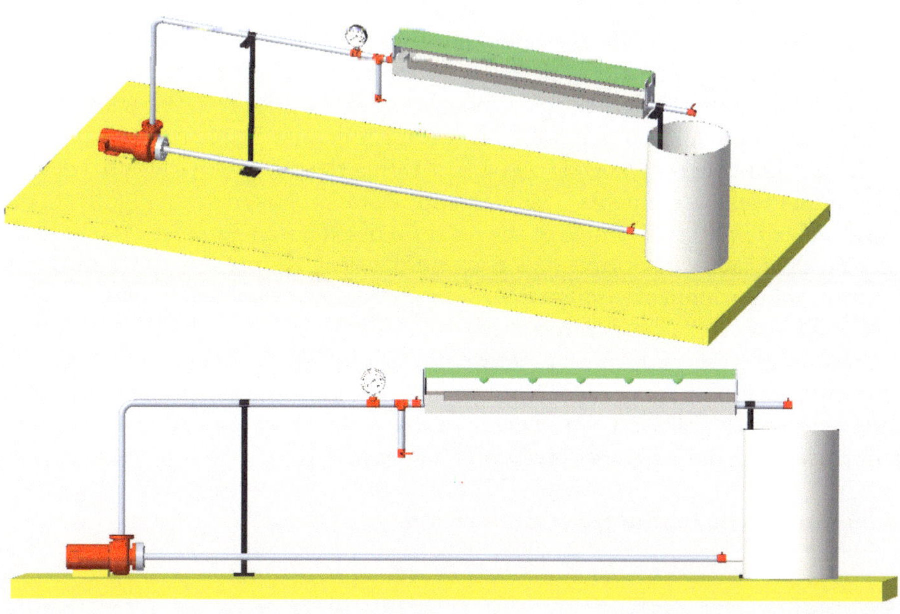

Figure 2.
Photocatalytic reactor proposed. (a) Reactor assembly isometric view and (b) reactor assembly side view.

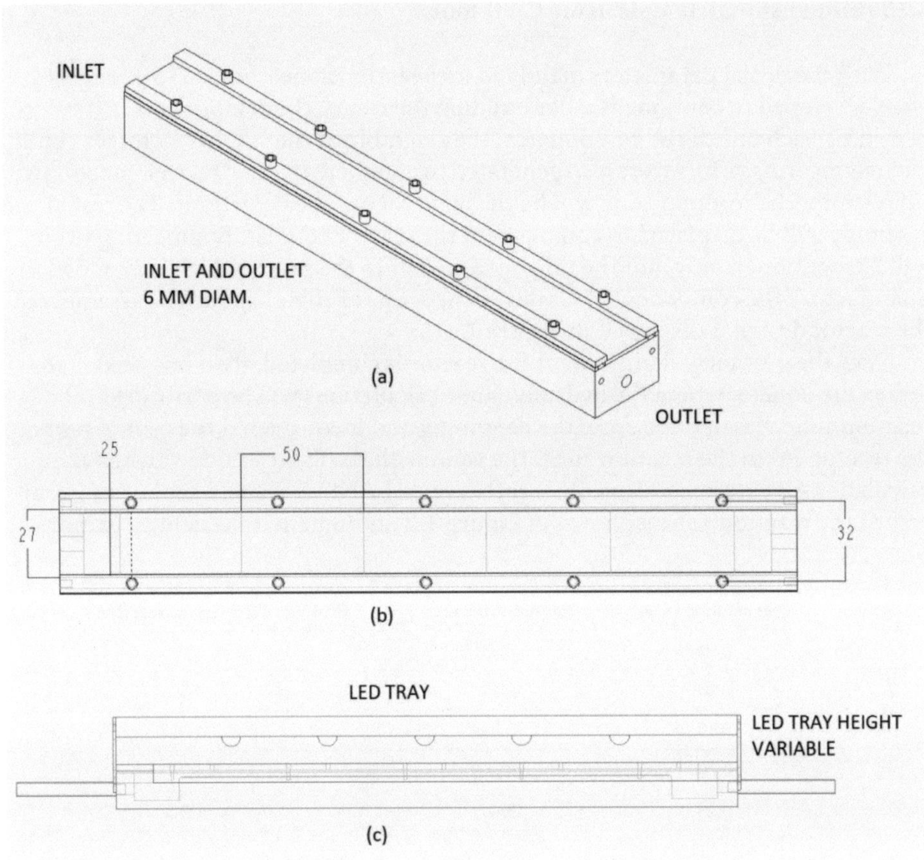

Figure 3.
Reaction zone, (a) isometric view of reaction zone, (b) top view of reaction zone, and (c) lateral view of reaction zone including LEDs.

CFD calculations because the methodology used is finite volume. For that purpose, the domain will be processed using the software by first discretizing the domain in small control volumes where the governing equations will be resolved with the help of numerical methods to obtain valuable results.

Finite volume methodology is based in dividing the domain in a number of control volumes (cells or elements), the elements should not overlap among them, and the variable of interest is located at the centroid of each element. If the nodes in the border wall of each consecutive region are unidentical, the mesh is nonconformed. The walls should connect through their interphase, and calculated flows through these walls may be assigned from a mesh to the other.

Conformed mesh is the most precise connectivity between regions. Nonconformed mesh may reduce the complexity of the meshing process but would increase the error, at least the local error. In this work, there is an effort to obtain the conformed mesh during the discretization process.

5.3 Photocatalytic reactor general operation

The reactor operation is simple; the design intent is having an inlet circular port with a diameter of 6 mm and arrives into a rectangular cavity which is filled until reaching the reaction zone level. Then the fluid circulates through the reaction zone under a laminar flow regime. The fluid reaches the outlet and finally goes out through a 6-mm-diameter outlet port. The reaction zone has the dimensions of 3 mm x 25 mm x 250 mm, and from there the calculated volume is 18,750 mm3. With these dimensions the reactor complies with one of its purposes since it is intended for laboratory testing. **Figure 4** shows an image of the domain.

The reactor has a cover made of a transparent material that allows light to pass through; in this case we selected Pyrex glass with dimensions of 40 x 300 mm, and it will be fastened to the main block using ten fasteners and two fixing devices made of metal with dimensions of 7.5 x 300 mm.

The light source will not be analyzed in this chapter, but initial estimations will be mentioned in this section. The distance to locate the light source from the reaction zone is adjustable. For the light source, it is intended to use 10 lamps which will be UV LED devices with 365 nm mounted in a bench that will have dimensions of 25 x 300 mm. LEDs are distributed along the reaction zone in 250 mm of the total length. Light source selection is based in some of the advantages this device offer such as low electric consumption, long life cycle, and easy control of light intensity using low-cost electronic devices.

The inlet and outlet have a 6 mm diameter with a threaded connection. The centerline of the inlet is located at a height of 18 mm with respect to the reactor base. The outlet is located at a height of 10 mm with respect to the reactor base with the intention of helping out the fluid.

5.4 Mesh details

An initial mesh of *m* (0.75 mm) hexahedral-type cubic cells was used considering the geometry is simple enough, applying a meshing method known as hexa-dominant within the native options within ANSYS® Mesh. The initial meshing process generated 52,615 cells. This is the first step in an iterative process where the cell count increases in order to find the optimal cell count where the effect of the number of cells in the calculation is minimized. This is done comparing the iterative residual values in a simple calculation; when the difference between one experiment and the subsequent is minimum, then the cell count may be considered irrelevant, and this may be considered the optimum cell count for the problem.

The option facing was used to standardize the cell size and its orthogonality (these are native options within ANSYS® Mesh). During the mesh analysis,

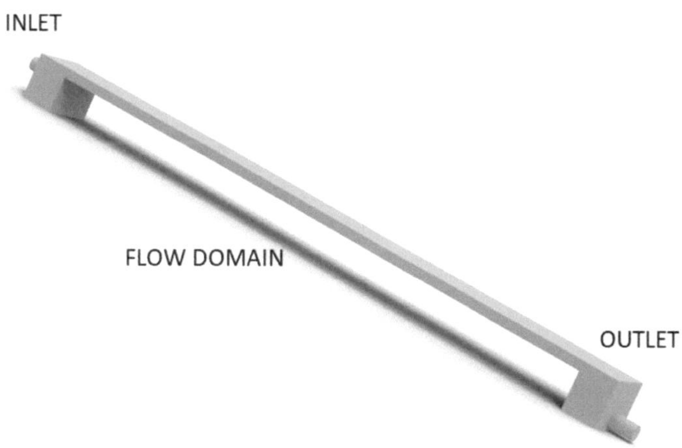

Figure 4.
Definition of reaction zone for CFD simulations.

AANSYS® version 19.2 was also used. For this analysis, the cell count was increased until reaching 500,000 cells. The simulation was performed in the reaction zone which has a simple geometry that facilitates all the meshing process and in general reduces the problem complexity. The results of the mesh analysis are displayed in **Tables 2** and **3**, and the best choices for the meshing process are marked *.

CFD will resolve governing equations for mass, momentum, and energy. The discretization process is also known as the meshing process. In this process ANSYS Fluent® has improved a lot, and since the geometry was simple for this reactor, meshing was resolved easily. There are two meshing types: structured mesh and nonstructured mesh. Examples of structured and nonstructured cases are shown in **Figure 5**.

In the former the mesh is identified by a triple index (i, j, k), in three dimensions. The cell borders form continuous lines of meshing with the adjacent cells which help a lot in the subsequent steps. In nonstructured meshing, cells and nodes do not have an exact match and cannot identify easily between neighboring cells.

Meshing has been regarded as the most difficult process in a CFD simulation. The latest versions of ANSYS® incorporate new tools that have enhanced the program capabilities and make the job easier for the users. Among the different tools incorporated, the known tools have been improved. An important factor to consider while meshing the domain is the cell shape. For this case, the geometry is a regular shape, so the cell shape was resolved easily. Conformed mesh is the desired status for the mesh because the nodes and boundaries will work smoothly if the individual cells are conformed.

In this way, the user needs to consider the interphase interaction, for example, interaction between the walls and the fluid and others. Discretized domain is shown in **Figure 6**.

Different scenarios were calculated with variants in the mesh, so the results can be compared in search for the optimum mesh as shown in **Tables 2–4**. Results such as velocity, pressure, and mainly the residuals were used to reach a conclusion on which mesh and how many cells are the best choice for this work.

Item	Mesh size (m)	Adjustments used	Cells generated
1	7.5 e-4	54,615	1e-3, 1e-4, 1e-5 y 1e-6
2	5.00E-04	200000*	1e-3, 1e-4, 1e-5 y 1e-6
3	4.00E-04	385020*	1e-3, 1e-4, 1e-5 y 1e-6

* *Optimized value for elements quantity.*

Table 2.
Effects of cell count for an experiment under V = 0.05 m/s during the mesh analysis.

Inlet velocity (m/s)	Maximum pressure absolute (Pa)	Likeliness between values	Outlet pressure at central zone (Pa)	Likeliness between values
0.1	27.906	91.6	11.310	92.6
0.1	27.906	91.6	11.310	92.6
0.1	29.802	97.8	11.265	92.2
0.1	29.804	97.8	11.265	92.2
0.1	30.481	100.0	12.218	100.0
0.1	30.482	100.0	12.218	100.0

Table 3.
Effects of cell count for an experiment under V = 0.05 m/s during the mesh analysis.

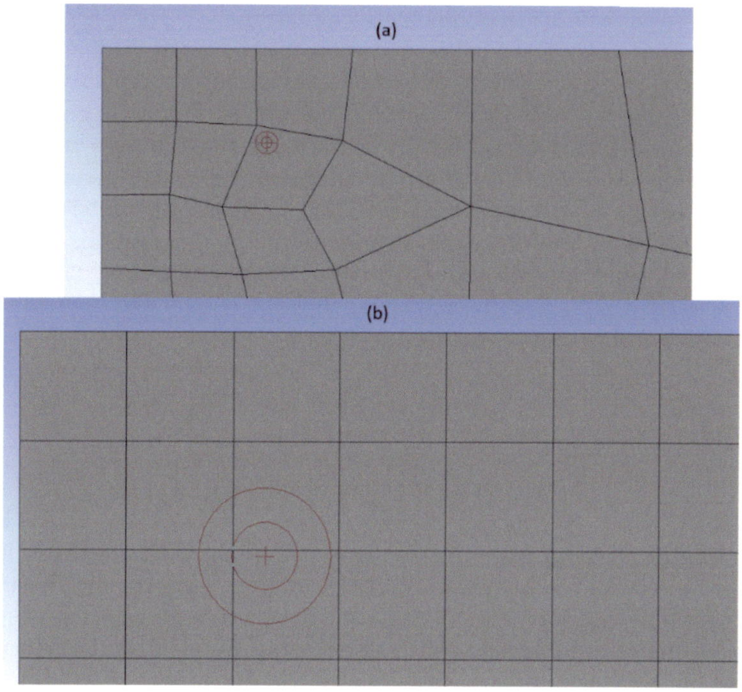

Figure 5.
Mesh examples: (a) non-structured mesh and (b) structured mesh.

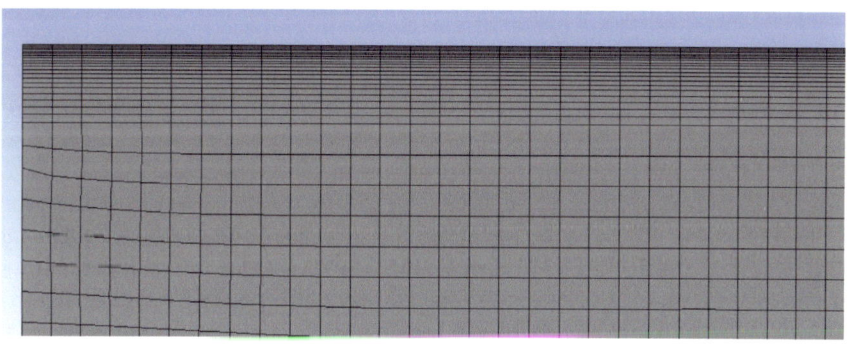

Figure 6.
Discretized domain using tools incorporated in ANSYS fluent® version 19.2.

Inlet velocity (m/s)	Maximum pressure absolute (Pa)	Outlet pressure at central zone	Minimum pressure absolute (Pa)	Mesh variant	Inlet velocity (m/s)	Outlet velocity (m/s)	Residuals
0.1	27.906	11.310	0.245	1	0.103	0.151	1.00E-05
0.1	27.906	11.310	0.245	1	0.103	0.151	1.00E-06
0.1	29.802	11.265	0.007	2	0.100	0.160	1.00E-05
0.1	30.481	12.218	0.004	3	0.101	0.154	1.00E-05
0.1	29.804	11.265	0.007	2	0.100	0.160	1.00E-06
0.1	30.482	12.218	0.004	3	0.101	0.154	1.00E-06

Table 4.
Effects of cell count for an experiment under V = 0.05 m/s during the mesh analysis.

5.5 Processing and resolution of governing equations

As mentioned in prior sections, flow dynamics is the first step during the design development of a photocatalytic reactor. A good velocity field with an appropriate distribution will be important for the photocatalyst distribution and to avoid possible nonuniform zones. Since the estimated operational parameters indicate Reynolds will oscillate between 266.2 and 1331.0, calculations will be carried on under laminar flow regime. Since the regime and the fluid conditions may be considered homogeneous, calculations can be performed in steady state and assume one can proceed on to resolve the continuity equation (Eq. 7) and the tridimensional classical equation of Navier–Stokes (Eq. 8):

$$\nabla(\rho\overline{v}) = 0 \tag{7}$$

$$\nabla\cdot(\rho\overline{v}\overline{v}) = (-\nabla P) + (\nabla\cdot\overline{\overline{\tau}}) + \left(\rho\vec{g}\right) \tag{8}$$

where ∇ is the delta operator and ρ, $\overline{v}$, P, $\overline{\overline{\tau}}$, and are the fluid density, velocity vector, pressure, stress tensor, and gravity acceleration, respectively. These equations and Newton's viscosity law as a constitutive equation to relate the stress tensor with the continuous fluid motion will enable the user to calculate the velocity field for this reactor. Since we are interested only in fluid dynamics, we will skip energy equation. The resolution of these equations keeps representing one of the most complicated problems analytically and numerically in fluid mechanics for complex geometries. An analytical solution is unavailable, but a reasonably accurate solution may be reached numerically using methodologies such as finite volume as it was described briefly in prior sections.

In this case the resolution algorithm semi-implicit, linked equations (SIMPLE) was used. In this algorithm an initial value for pressure from there calculates the velocity and verifies that the outlet flow is identical to the inlet value and increases or decreases the input according to the best option. The program will resolve numerically for each of the finite volumes created during the discretization process or in other words will resolve the equations in the centroid for each cell in the mesh. This step may be considered a numerical solution for the discrete volumes but need an integration to display global results. The integration is done by producing an algebraic version of the differential equations, and the solution for this version is reached with a matrix. The processing means the software will resolve the governing equations according to the inputs provided. The inputs include the geometry, the mesh, fluid properties, boundary conditions, the algorithm to be used, and other details that the software will require if following its natural sequence as coded by ANSYS Fluent®.

5.6 Result analysis and postprocessing

When the processing step completes its respective work cycle, the governing equations are resolved for the centroid of each cell, and the integration process was also completed for global results.

From these results the program offers a module to create views of the domain geometry that facilitate the analysis by the users or the presentation of the results to a specific audience. In this problem, the results obtained relate to the reactor zone hydrodynamics. A useful tool generates contours of the domain for the desired parameters such as velocity, pressure, drag force, residuals, mass imbalances, etc.

The analysis of pressure for the system and in specific regions of the reaction zone may be an interesting contour to start with the analysis procedures. **Figure 7** shows the results obtained for pressure within the domain by generating an image at the inlet and outlet. In **Figure 8**, one may see another interesting vector graphic to display the velocity profile at the inlet and outlet considering an inlet velocity of 0.1 m/s. The maximum pressure calculated at the inlet is 29.8 Pa, while the minimum pressure at the outlet is 0.0742 Pa. Then, with this data the designer has an idea of the pump dimensions considering the inlet pressure is needed, so the flow can pass through the reaction zone under the selected conditions. Pressure losses due to the interaction of the fluid with pipes and other reactor components in its way to the inlet in the reaction zone will not be included within this chapter but can easily be resolved by the designer.

Reaction zone pressure in addition with pressure losses will provide the minimum value needed to dimension the appropriate pump for a correct operation of this reactor. An interesting tool that helps to know the exact value in a specific location within the domain may be completed with the help of the probe tool. With this tool the software is capable of calculating this parameter on a specific location, for example, if the user would like to know the average velocity, the probe tool should indicate the specific region and the desired parameter, and, in this case, the average

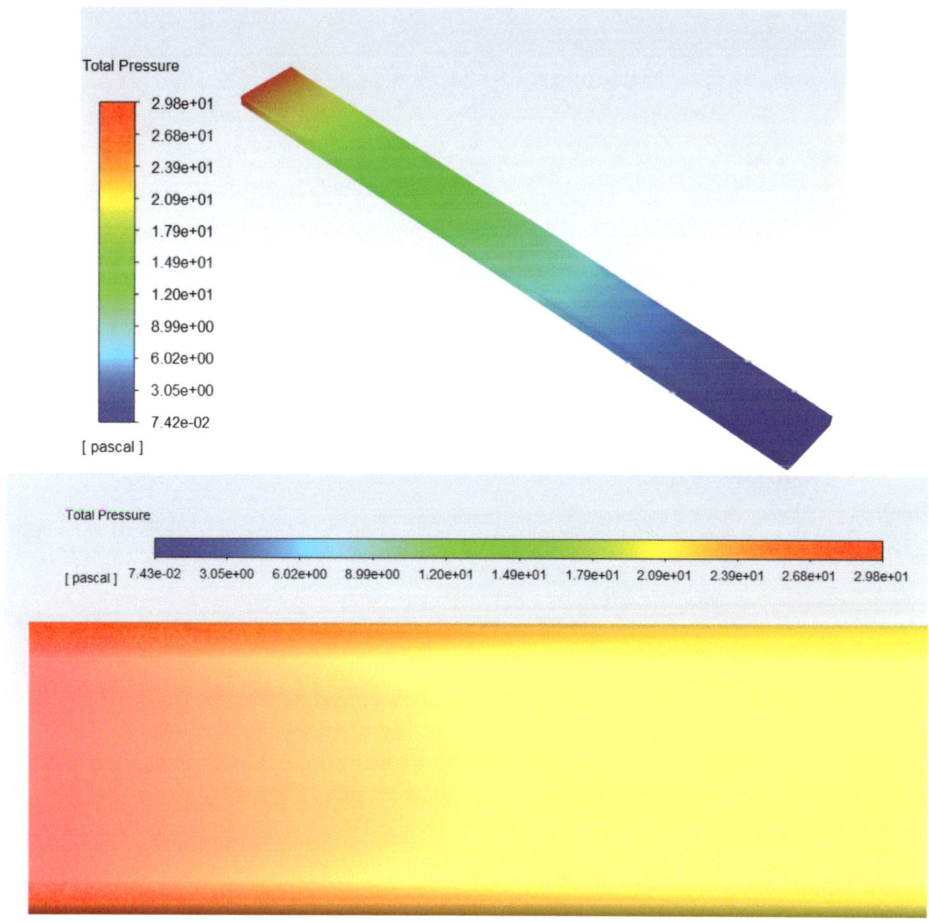

Figure 7.
Postprocessing contour for pressure distribution in the reaction zone in (a) isometric view and (b) lateral view.

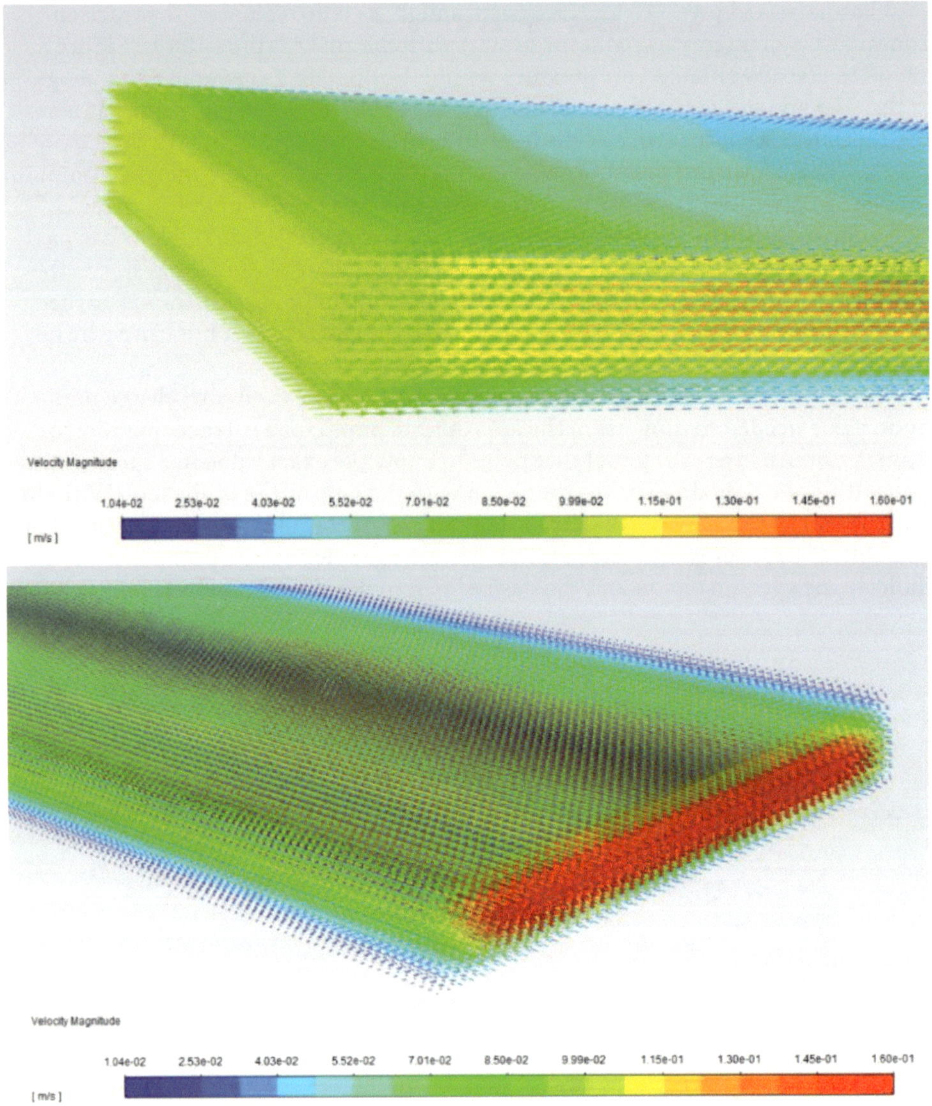

Figure 8.
(a) The graph showing velocity vectors at the inlet in the reaction zone considering an inlet velocity of 0.1 m/s and (b) velocity profile at the outlet considering the case of an inlet fluid velocity of 0.1 m/s.

velocity at central region is 0.1000459 m/s. In this measurement it is interesting to observe that the expected value was 0.1 m/s for the inlet speed at the central region and the software calculated 0.1000459 m/s instead. The difference is very small, so a conclusion that may be reached is that the calculation accuracy is acceptable for the case of the velocity in this location. The minimum pressure obtained was 0.0742 Pa, for a velocity of 0.1 m/s in the reaction zone. **Figure 9** shows the velocity contour.

ANSYS® in its latest version offers a whole module to carry on with the postprocessing procedures. This module facilitates a lot the designer work and allows the user to generate a great deal of graphics, charts, animations, etc. which will be of great help for the presentation of results. In this example the graphics generated so far will be considered enough to illustrate the analysis done but represent only a few of the potential analysis tools that can be obtained.

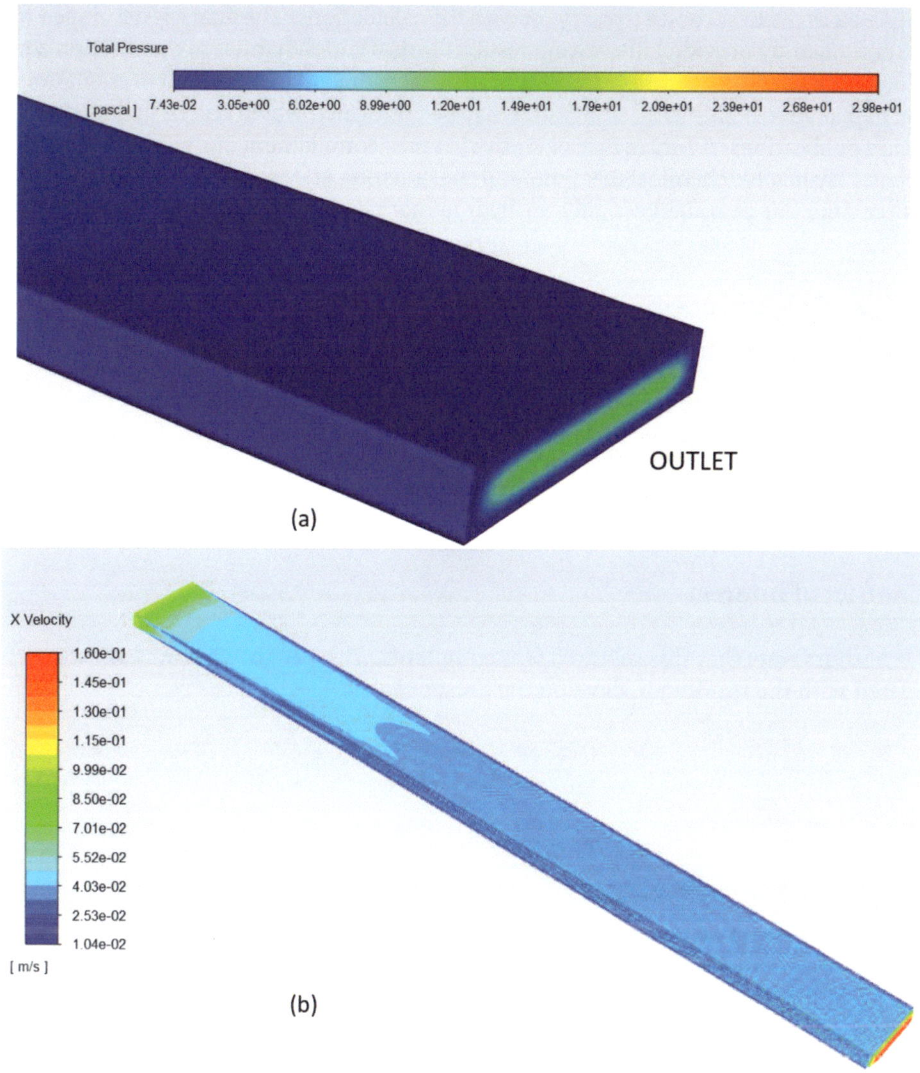

Figure 9.
(a) Contour graphic for pressure obtained at the outlet of the reaction zone. (b) Contour graphic for velocity in the reaction zone with an inlet velocity of 0.1 m/s.

6. Conclusions

In conclusion, CFD codes are becoming common tools used in chemical reactor design development. The simulation possibilities are quite interesting and include resolving the governing equations for fluid dynamics but also may include chemical species and multiphase, among other, additional possibilities for the simulation of the chemical reactor operation under different scenarios. CAD tools complement the CFD code because the geometry is the central piece of information required to carry on with CFD simulation. Discretization process required a formidable amount of efforts in the past, in particular for complex geometries, but this problem has decreased in the latest versions of CFD codes. In particular, ANSYS® has incorporated a series of tools at different stages of the simulation process that facilitate in a great deal in the meshing process.

In the case of study, a simple geometry was used for the reaction zone of the chemical reactor proposed for a laboratory scale, and a mesh with 385,020 elements was proposed as the best choice to carry on with the calculations. The analysis developed for the case of study provided interesting results for the fluid dynamics in the reaction zone which were used to generate graphics and images that facilitate the result presentation and explanation. Additional simulation may be performed which will be included in future publications. A future case of study that may complement our work presented herein may involve chemical reactions and the radiation effects caused by the light source since this case analyzes only the hydrodynamics of a photochemical reactor.

Acknowledgements

This work was financed by CONACyT (Mexican Science and Technology National Council) through the 2015 CONACyT SEP-CB (Basic Science-Public Education Ministry) project fund 258553/CONACyT/CB-2015-2101. Thanks to the Scientific Computational Laboratory at FCQ-UJED for the computational resources. Thanks to the Academic Group UJED-CA-129 for the valuable discussions.

Conflict of interest

Authors state that this research was completed without any conflicts of interest related with the funding to develop the present work.

Author details

Adolfo Ruiz-Soto[1], Diana Barraza-Jiménez[1], Abel Hurtado-Macias[2], Sandra Iliana Torres-Herrera[3], Carlos Omar Ríos-Orozco[4], Miriam López-Guzmán[5], Elva Marcela Coria-Quiñones[1,4], Raúl Armando Olvera-Corral[1] and Manuel Alberto Flores-Hidalgo[1*]

1 Faculty of Chemical Science, Juarez University of Durango State, Durango, México

2 Research Center for Advanced Materials (CIMAV), Campus Chihuahua, Chihuahua, Mexico

3 Faculty of Forestry Science, Juarez University of Durango State, Durango, México

4 TecNM/Durango Institute of Technology, Durango, Mexico

5 Research Center for Advanced Materials (CIMAV), Campus Durango, Durango, Mexico

*Address all correspondence to: maflores.hidalgo02@gmail.com

IntechOpen

© 2019 The Author(s). Licensee IntechOpen. This chapter is distributed under the terms of the Creative Commons Attribution License (http://creativecommons.org/licenses/by/3.0), which permits unrestricted use, distribution, and reproduction in any medium, provided the original work is properly cited.

References

[1] Versteeg HK, Malalasekera W. An Introduction to Computational Fluid Dynamics. 2nd ed. Prentice Hall: Pearson; 2007. pp. 242-247

[2] Mann U. Principles of Chemical Reactor Design, Analysis and Design. 2nd ed. Hoboken, New Jersey, USA; John Wiley & Sons, Inc; 2009. pp. 103-303

[3] Cimbala JM, Cengel YA. Fluid Mechanics Fundamentals and Applications. 3rd ed. New York, NY, USA: McGraw-Hill; 2014

[4] Merino CC. Aplicación de técnicas de fluido-dinámica computacional al modelado y simulación de reactores fotocatalíticos a partir de parámetros cinéticos intrínsecos [Doctoral Thesis]. Universidad Rey Juan Carlos; 2017

[5] Krasberg N, Hohmann L, Bieringer T, Bramsiepe C, Kockmann N. Selection of technical reactor equipment for modular, continuous small-scale. Plants Processes. 2014;**2**:265-292. DOI: 10.3390/pr2010265

[6] ANSYS Fluent®. Customization Manual. USA: Published for licensees' users by ANSYS® INC; 2017

[7] ANSYS Fluent® Academic Research CFD, Release 18.2, 19.1

[8] Barraza-Jiménez D, Ruiz-Soto A, Torres-Herrera SI, Coria-Quiñones EM, Olvera-Corral RA, Romero-Soto DJ, et al. Radiative Transference Equation Algorithm as an ANSYS® User-Defined Function for Solar Technology Applications. IntechOpen; 2019. DOI: 10.5772/intechopen.88767. Available from: https://www.intechopen.com/online-first/radiative-transference-equation-algorithm-as-an-ansys-user-defined-function-for-solar-technology-app

Chapter 6

Computational Fluid Dynamics of Mixing Performance in Microchannel

Siti Nor Azreen Ahmad Termizi
and Syamsul Rizal Abd Shukor

Abstract

In microchannel, fluid viscous effect becomes dominant, and the micro-flow typically falls in laminar regime. Mixing of fluid in the absence of turbulence is a slow molecular process as it is solely dependent on diffusion. Fast and complete mixing of relevant fluids is of crucial importance in many chemical engineering processes, thus computational fluid dynamics simulation on mixing in microchannel is the main topic in this chapter. The simulation was based on laminar flow and convective diffusion equation model. The factors affecting the mixing performance in microchannel was further simulated. The finding provides some insight of transport phenomena on mixing in microchannel.

Keywords: mixing, microchannel, laminar, diffusion, simulation

1. The microreactors and microchannel

Microreactor is more commonly known in the field of process intensification and microsystems technology that has attracted significant interest in several years. The channel of microreactor is known as microchannel due to the micro size, while under microreactor group, there are micro mixers which are designed for mixing purpose. Numerous plausible advantages of microreactors for the pharmaceutical and fine chemical industries have been realized, thanks to their excellent capability for mixing and for thermal exchanges which increase yields and selectivity of reactions [1–4].

Microreactors have two major advantages with respect to smaller physical size and the increase in numbers of units. Benefits from reduction of physical size became more apparent in chemical engineering aspects. The difference of physical properties between microreactors and conventional reactor such as temperature, concentration, density, or diffusional flux increase with decreasing of linear dimension [5, 6]. Consequently, the driving forces for heat transfer, mass transport increase when using the microreactors. Besides, a significant reduction in volume for microreactor as compared to conventional reactors lead to smaller hold up that increase process safety and improves selectivity due to shorter residence time [7, 8].

IntechOpen

2. Fluidic and mixing environment

The smooth and constant fluid motion represents the laminar flow, whereas the vortices and flow fluctuation are properties of turbulent flow. These two types of flow are determined by using Reynolds number. Reynolds number (Re) measured the relative importance of viscous force and inertial forces. The Re is defined as

$$\mathrm{Re} = \frac{\rho v D_\mathrm{h}}{\mu} \tag{1}$$

where ρ and μ are the fluid density (kg/m^3) and viscosity (kg/m s), respectively; v is the velocity of the fluid (m/s) and D_h is the hydraulic diameter of the channel (m). Due to specific microstructuring technology employed to build microreactor, the channels of the microreactor have rectangular or trapezoidal cross section [9]. Thus, the hydraulic diameter D_h has to be properly defined. The hydraulic diameter of rectangular shapes is defined as [10]:

$$D_h = \frac{2wh}{(w+h)} \tag{2}$$

where w is the width and h is the height of the microchannel.

On the other hand, mixing is a physical process with a goal of achieving a uniform distribution of different components in a mixture, usually within a short period of time [7]. Conventionally, at a macroscale level, the decrease in the mixing path and increase in the contact surface are achieved by a turbulent flow. The segregation of the fluid into small domain occurred by the help of vortices and flow fluctuation [11].

The fluid entity is constantly subdivided into thinner layers by an induced circular motion of fluid compartments, the so-called eddies, and subsequent breaking into fragments. In a laminar regime, a similar breaking of fluid compartments cannot occur due to the high viscous forces. Instead, the fluid entity has to be continuously split and recombined, forming regularly sized fluid embodiments [7].

Due to small dimension of microreactor, the fluid in microreactor is considered as microfluidic. The mixing in microfluidic is achieved and improvised by the decrease of mixing path and increase of surface area. The designed microreactor that highlights reduction of mixing path increases the effect of diffusion and advection on the mixing [11].

On top of that, mixing characterization is important to show how mixing performances in certain mixing process are described. Mixing performance of microreactor can be measured by evaluating the mixing quality as be done numerously in literatures [12–16]. A common definition of the mixing quality is based on Danckwerts' intensity of segregation [17] and is defined by

$$I_s = \frac{\sigma^2}{\sigma^2_{\max}} \tag{3}$$

where $\sigma^2_{\max}$ is the maximum possible variance (which is 0.5 for symmetrical boundary condition) and σ^2 is defined by

$$\sigma^2 = \frac{1}{|V|}\int_V (c - \bar{c})^2 dV \tag{4}$$

and also can be written as

$$\sigma^2 = \frac{1}{N}\sum (c - \bar{c})^2 \tag{5}$$

where $\bar{c}$ denotes the mean value of the concentration field c and N as the sampling point inside the cross section. Then, a measure for the intensity of mixing or the so-called mixing intensity can be deduced as

$$I_M = 1 - \sqrt{I_s} = 1 - \frac{\sigma}{\sigma_{\max}} \tag{6}$$

Since I_s is normalized, the quantity I_s reaches a value of 0 for a completely segregated system and a value of 1 for the homogeneously mixed case.

3. Simulation of mixing in microchannel

3.1 Geometric and meshing

Geometry is defined as the computational domain of the flow region where the governing equation of fluid flow, mass, and energy is applied with its boundary condition. The computational domain is different from physical domain as the physical domain is the real physical flow that might include the wall, etc. [19]. The geometry may result from measurement of the existing configuration or may come with a design study.

In this work, the geometry is chosen, aspired by the actual geometry domain (physical domain) which is depicted in **Figure 1** of the standard slit interdigital micro mixer (SSIMM) mixing element. The mixing element of the micro mixer consists of the corrugated wall of microchannels and discharge slit. Simulation of the complete geometry of this mixing element required large number of degree of freedom to be solved. This can only be achieved by large computational resources. However, due to the limitation of the computational resource, simplification of the geometry is preferred and required. Thus, to simplify the computational work, only the middle part of the mixing element structure domain was taken to represent the overall mixing element as shown in **Figure 2**.

The middle part was chosen based on strategies of the macro model approach of computational fluid dynamics in [9] that partitioned the reactor domain prior to simulation. It was noticed that the mixing element of the SSIMM has trapezoidal shape with two bifurcations and parallel microchannel that served as flow guide to avoid maldistribution of the fluid stream. A fluid maldistribution would induce unequal residence time in different channels, with undesired consequences for the product distribution in the micromixer [9]. However, this is not considered in this study as only the middle part of the mixing element is taken as computational fluid domain.

There are two inlets and an outlet assigned to this model geometry as in **Figure 2**. An additional geometry domain with a straight-line microchannel was built for the purpose of comparisons with the SSIMM mixing element. The geometry has the same dimension as the simplified corrugated microchannel domain. On the other hand, meshing is the process of generating mesh or grid cell overlaying the whole domain geometry. In CFD, the domain is required to be subdivided into a number of smaller, non-overlapping subdomains in order to solve the flow physics within the domain. COMSOL Multiphysics software chosen as the simulation platform in this work provides two option types of meshing that can be used by the user which are physics-controlled meshing and user-controlled meshing. Physics-controlled meshing sequence will build the mesh for the domain which is adapted to

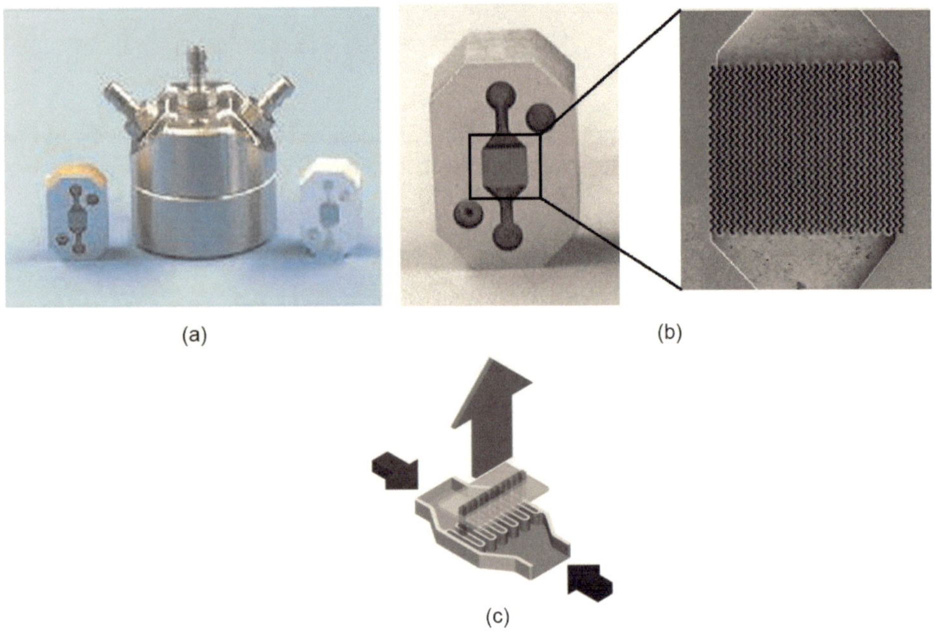

Figure 1.
(a) The photograph of SSIMM and (b) the mixing element; (c) the flow principle of SSIMM [18].

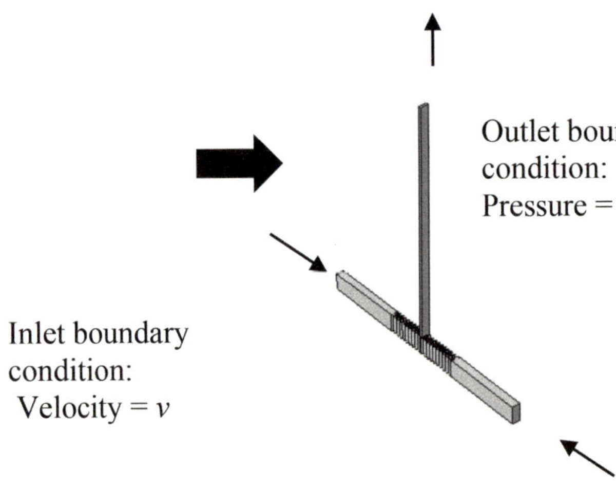

Figure 2.
Schematic diagram of the simulation system.

the physics setting of the model, while the user-controlled meshing builds the mesh based on the user input of size, element type, etc. [19, 20]. **Figures 3** and **4** are the meshed geometry domains with different types of the mesh element can be seen. The mesh element of corrugated microchannel has a high number at the shape of the corrugated section, while the straight microchannel has a high number at the entrance of the discharge slit.

3.2 Governing equation

Laminar flow interface is used to model and simulate fluid mechanics for laminar and incompressible fluids by using Navier-Stokes equation. Since the fluid flow is laminar flow in the microreactor, this interface is suitable to be implemented in this simulation work. The Navier-Stokes equation for incompressible flow is given as [20]:

$$\rho \frac{\partial v}{\partial t} + \rho(v \bullet \nabla)v = \nabla \bullet \left[-pI + \mu\left(\nabla v + (\nabla v)^T\right)\right] + F \\ \rho \nabla \bullet v = 0 \tag{7}$$

where v is velocity vector (SI unit: m/s); p is the pressure (SI unit: Pa); ρ is the density (SI unit: kg/m^3); F is the volume force vector (SI unit: N/m^3); μ is the dynamic viscosity (SI unit: Pa.s); and T is the absolute temperature (SI: K).

The density and the viscosity data are those of water ($\rho = 1 \times 10^3$ kg/m^3 and $\mu = 1 \times 10^{-3}$ Pa s).

The driving force for the fluid to flow through the mixing slot to the outlet is the applied inlet velocity boundary conditions on the inputs while the pressure boundary condition is assumed to be equal to zero. Meanwhile, the chamber wall is assumed to have a nonslip boundary condition. Mixing is obtained by diffusion of various species in the fluid. The species are diluted in the water, thus having material properties like water. The transfer equation is then taken as the convection-diffusion equation with a reaction term as shown below [20]:

$$\frac{\partial c}{\partial t} + \nu \bullet \nabla c = \nabla \bullet (D \nabla c) + R \tag{8}$$

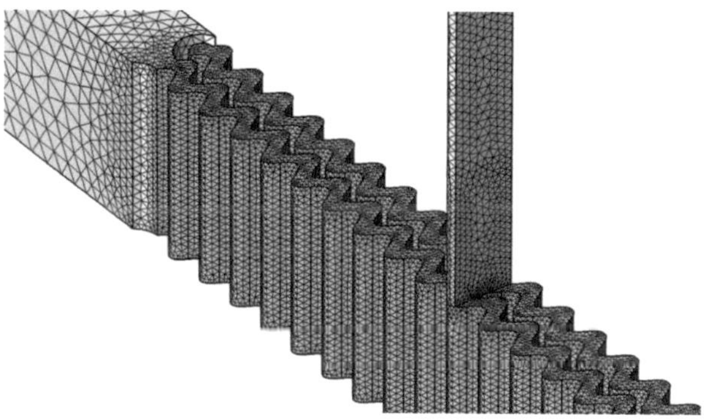

Figure 3.
The meshing for corrugated microchannel.

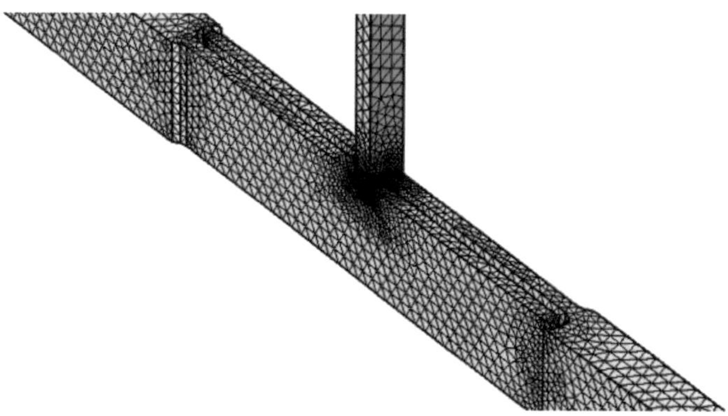

Figure 4.
The meshing for straight microchannel.

where c is the concentration of the species (SI unit: mol/m^3); D is the diffusion coefficient (SI unit: m^2/s); R is a reaction rate expression for the species (SI unit: mol/(m^{3}s)); and v is velocity vector (SI unit: m/s).

In this model, $R = 0$, because there is no reaction occurred. The species is introduced at different concentration from the range of 0–1 mol/m^3 where one species is at a concentration of 1 mol/m^3 on one of the input boundaries, while the other is at zero concentration. At the output boundary, the substance flows through the boundary by convection [21].

3.3 Velocity and concentration profile visualization

As mentioned earlier, there are two physics interface models that were solved in this work which are laminar flow (LF) and transport of diluted species (TDS). The LF interface model considers the fluid flow of the system with inlet velocity ranging from 1 to 10,000 μm/s chosen as input parameter. The LF interface model is solved independently. However, the TDS interface model is solved by obtaining a data of velocity field from the solution of the LF interface model. This is the reason why both physics interface models are used together. **Figures 5** and **6** show the velocity profile from the top view (XY view) of the geometric configuration comprised of corrugated and straight microchannel, respectively. The color gradient shows the maximum velocity of the microchannel at the middle of the channel which can be

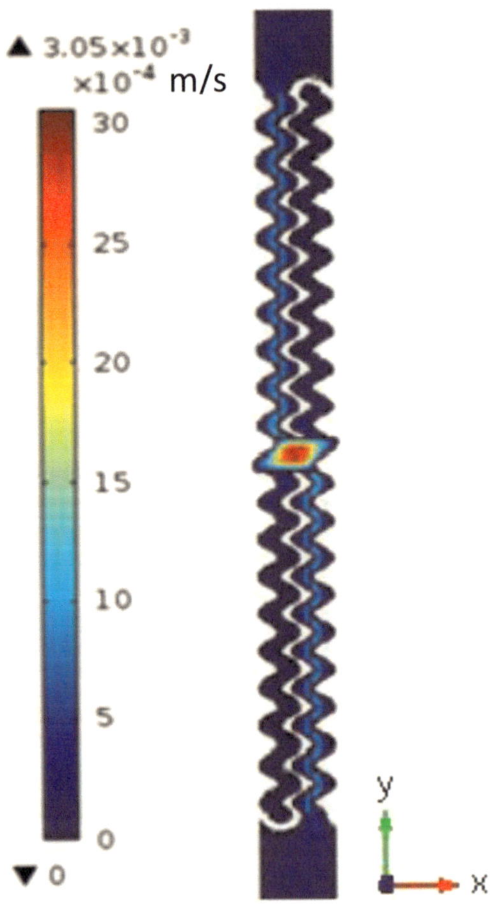

Figure 5.
Velocity profile of corrugated microchannel from XY-axis view.

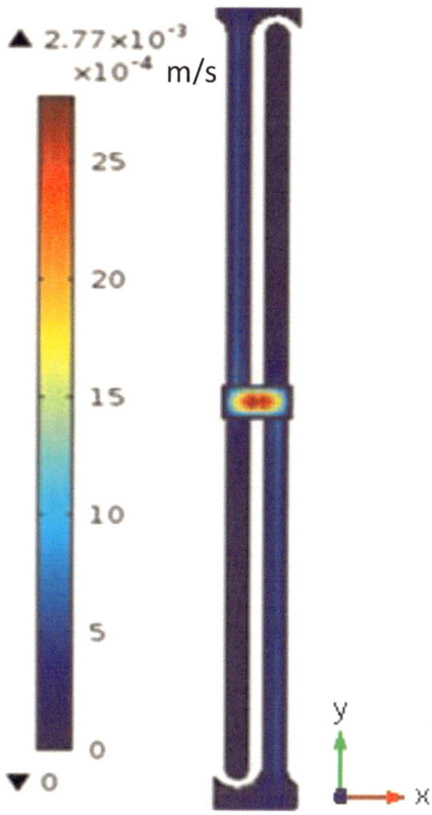

Figure 6.
Velocity profile of straight microchannel from XY-axis view.

seen in red color, while the blue color represents the low velocity value which is at the domain wall. This phenomenon indicates laminar parabolic flow where the velocity varies parabolically across the discharge slit with the maximum velocity at the center.

Visualization of mixing process in this work can be seen clearly by the plotted concentration profile of the species in which the different color gradients represent the species before and after the mixing process. In particular, the unmixed species is represented by blue and red colors, and the green color represented the mixed one.

Figures 7 and **8** shows the concentration profile of the corrugated and straight microchannel respectively for all the inlet velocities studied in this work. Both geometric domains have similar dimension of length and width but different configuration of microchannel. The mixing starts when the fluids with different concentrations denoted as blue and red enter the discharge slit. A clear separation of the

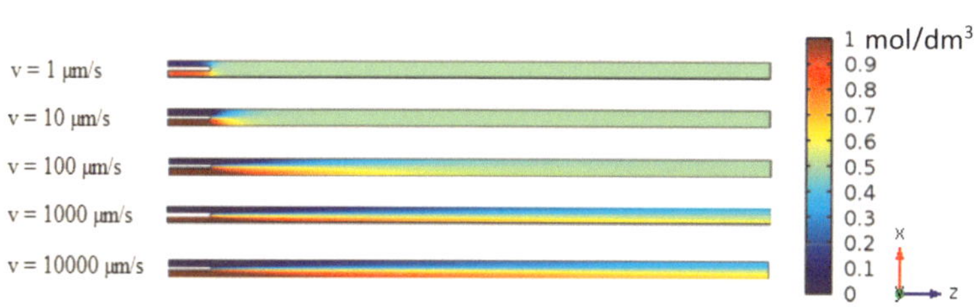

Figure 7.
Concentration profile of corrugated microchannel for various inlet velocities.

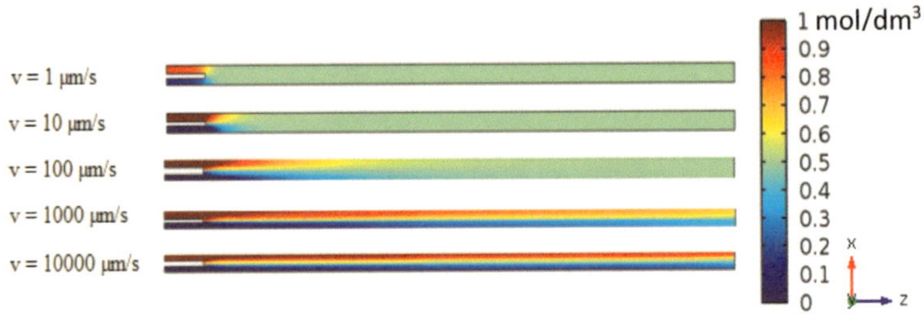

Figure 8.
Concentration profile of straight microchannel for various inlet velocity.

concentration is observed at the entrance, but this diminishes toward the end of the discharge slit for inlet velocity equal to 100 µm/s. Thus the mixing process is completed. For inlet velocity lower than 100 µm/s, the mixing completely occurred instantaneously as the fluids enter the discharge slit. For inlet velocity higher than 100 µm/s, the mixing is not complete as distinct color can be seen from the entrance until the end of discharge slit.

In short, complete mixing occurred at low inlet velocity, and the mixing is incomplete at higher inlet velocity of 100 µm/s for both configurations of microchannel.

3.4 Mixing intensity evaluation

As mentioned in previous section, the Danckwerts segregation intensity or the so-called mixing intensity is defined with the mean square deviation of the concentration profile of the component i in a cross section of the discharge slit. The segregation intensity can be transformed to a value between 0 (completely segregated) and 1 (completely mixed) [22].

In this work, to determine the mixing quality with respect to discharge slit length, the value of mixing intensity is evaluated at every 100 µm of discharge slit position starting from 300 µm where the fluid starts to mix until 4300 µm which is the end of discharge slit. The mixing intensity value against the discharge slit position for both corrugated and straight microchannels at inlet velocity of 10,000 µm/s is plotted in **Figure 9**. The mixing intensity of corrugated microchannel is higher than the mixing intensity of straight microchannel.

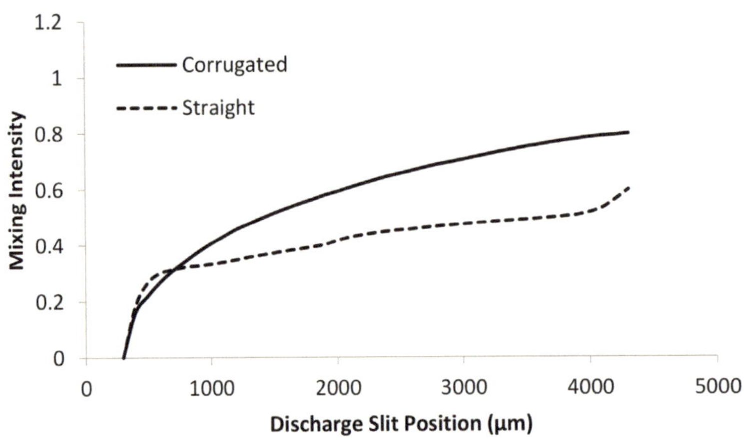

Figure 9.
Comparison of mixing intensity between geometric configuration at inlet velocity of 10,000 µm/s.

As compared to concentration profiles, the mixing intensity profile gave information of mixing quality with respect to discharge slit position. The discharge slit position can represent the mixing length. These mean that complete mixing occurred at different mixing length. The mixing profile shows the difference of mixing intensity profile among the microchannel configurations. The corrugated microchannel has gradually increased the profile of mixing intensity from the entrance toward the end of the discharge slit. This might be due to the corrugated shape of the microchannel which serves to form multi-lamination of fluid that gives even distribution of concentration which then results in a smooth mixing intensity profile. This might prove that the concept of multi-lamination of fluid as the purpose of microreactor is designed in such way.

4. Conclusions

This chapter discussed a study of mixing simulation in microchannel. An analysis is carried out to investigate the effect of the changes of inlet velocity toward mixing intensity over the two different microchannel configurations. The simulation results show the visualization of velocity and concentration profiles along the microchannel. A laminar parabolic flow of velocity profile is observed for two microchannel configurations simulated. The concentration profile gave visualization on the mixing process that occurred in the microchannel. Evaluation of the mixing intensity value represents the mixing performance of the geometry structure. It also gave information on the mixing length requirement to achieve complete mixing. The microchannel needs longer discharge slit to achieve complete mixing if high inlet velocity is used. The result showed that inlet velocity has significant effects on the mixing performance which is represented by the mixing intensity in this study. The higher the inlet velocity, the lower the mixing quality. Careful observation on the mixing intensity profiles among geometry configurations shows different trends of mixing intensity between the corrugated and straight microchannels.

Acknowledgements

I would like to thank everybody who was important to the successful realization of this chapter.

Appendices and nomenclature

c	concentration (mol/dm^3)
D	diffusion coefficient (m^2/s)
D_h	hydraulic diameter (m)
F	volume force vector (N/m^3)
p	pressure (Pa)
L	length (m)
w	width (m)
h	height (m)
R	reaction ($mol/(m^3s)$)
T	absolute temperature (K)
v	velocity (m/s)
μ	dynamic viscosity (Pa.s)

ρ	density (kg/m^3)
V	volume (m^3)
λ	molecular diameter (m)
D	characteristic dimension (m)
σ	variance
σ_{max}	maximum variance
N	number of sampling point
I_S	intensity of segregation
I_M	mixing intensity
Re	Reynolds number

Author details

Siti Nor Azreen Ahmad Termizi[1*] and Syamsul Rizal Abd Shukor[2]

1 Department of Chemical Engineering Technology, Faculty of Engineering Technology, Universiti Malaysia Perlis, Padang Besar, Perlis, Malaysia

2 School of Chemical Engineering, Universiti Sains Malaysia, Nibong Tebal, Pulau Pinang, Malaysia

*Address all correspondence to: sitinorazreen@unimap.edu.my

IntechOpen

© 2019 The Author(s). Licensee IntechOpen. This chapter is distributed under the terms of the Creative Commons Attribution License (http://creativecommons.org/licenses/by/3.0), which permits unrestricted use, distribution, and reproduction in any medium, provided the original work is properly cited.

References

[1] Choe J, Kwon Y, Kim Y, Song H-S, Song KH. Micromixer as a continuous flow reactor for the synthesis of a pharmaceutical intermediate. Korean Journal of Chemical Engineering. 2003; **20**(2):268-272

[2] Haverkamp V, Ehrfeld W, Gebauer K, Hessel V, Lowe H, Richter T, et al. The potential of micromixers for contacting of disperse liquid phase. Fresenius' Journal of Analytical Chemistry. 1999;**364**:617-624

[3] Lomel S, Falk L, Commenge JM, Houzelot JL, Ramdani K. The microreactor: A systematic and efficient tool for the transition from batch to continuous process? Chemical Engineering Research and Design [Internet]. 2006;**84**(5):363-369

[4] Song KH, Kwon Y, Choe J. Microreaction technology in practice. Proceedings of the 4th Asia-Pacific Chemical Reaction Engineering Symposium, Gyeongju, Korea. 2006; 435-436

[5] Ponce-Ortega JM, Al-Thubaiti MM, El-Halwagi MM. Process intensification: New understanding and systematic approach. Chemical Engineering and Processing - Process Intensification [Internet]. 2012;**53**(2010):63-75

[6] Tsouris C, Porcelli JV. Process intensification—Has its time finally come? Chemical Engineering Progress. 2003;**99**(10):50-55

[7] Ehrfeld W, Hessel V, Löwe H. Microreactors New Technology for Modern Chemistry. Weinheim: WILEY-VCH Verlag GmbH; 2000. 288 p

[8] Moulijn JA, Stankiewicz A, Grievink J, Górak A. Process intensification and process systems engineering: A friendly symbiosis. Computers & Chemical Engineering [Internet]. 2008 Jan;**32**(1–2):3-11

[9] Keil FJ, editor. Modeling of Process Intensification. Weinheim: WILEY-VCH Verlag GmbH & Co. KGaA; 2007. 422p

[10] Sharp KV, Adrian RJ, Santiago JG, Molho JI. The MEMS Handbook, Chapter 10: Liquid flows in microchannels. In: Gad-el-Hak M, editor. 2nd edition. Boca Raton: Taylor & Francis Group; 2002. 53 p

[11] Capretto L, Cheng W, Martyn Hill XZ. Micromixing within microfluidic devices. Topics in Current Chemistry. 2011;**304**:27-68

[12] Engler M, Kockmann N, Kiefer T, Woias P. Numerical and experimental investigations on liquid mixing in static micromixers. Chemical Engineering Journal. 2003;**101**:315-322

[13] Kockmann N, Kiefer T, Engler M, Woias P. Convective mixing and chemical reactions in microchannels with high flow rates. Sensors and Actuators B: Chemical [Internet]. 2006; **117**(2):495-508

[14] Bothe D, Stemich C, Warnecke H-J. Fluid mixing in a T-shaped micro-mixer. Chemical Engineering Science [Internet]. 2006;**61**(9):2950-2958

[15] Zhendong L, Yangcheng L, Jiawei W, Guangsheng L. Mixing characterization and scaling-up analysis of asymmetrical T-shaped micromixer: Experiment and CFD simulation. Chemical Engineering Journal [Internet]. 2012;**181–182**:597-606

[16] Jain M, Nandakumar K, Jain M, Nandakumar K. Novel index for micromixing characterization and comparative analysis. Biomicrofluidics. 2010;**4**:031101

[17] Danckwerts PV. The definition and measurement of some characteristics of

mixtures. Applied Scientific Research, Section A. 1952;**3**(4):279-296

[18] Ehrfeld W, Golbig K, Hessel V, Lowe H, Richter T. Characterization of mixing in micromixers by a test reaction: Single mixing units and mixer arrays. Industrial and Engineering Chemistry Research. 1999;**38**:1075-1082

[19] COMSOL. COMSOL Multiphysics 4.3a Reference Guide. 2012. p. 1-750

[20] COMSOL. COMSOL Multiphysics 4.3a. User's Guide. 2012. p. 1-1292

[21] Baccar N, Kieffer R, Charcosset C. Characterization of mixing in a hollow fiber membrane contactor by the iodide–iodate method: Numerical simulations and experiments. Chemical Engineering Journal [Internet]. 2009; **148**(2–3):517-524

[22] Kockmann N. Transport Phenomena in Micro Process Engineering. Berlin/Heidelberg/New York: Springer; 2008. 365p

Chapter 7

Numerical Modeling of Nanotechnology-Boosted Chemical Enhanced Oil Recovery Methods

Pablo D. Druetta

Abstract

Since it was theorized more than 50 years ago, nanotechnology has become the perfect boost for existing old technologies. The unique properties exhibited by materials at these scales have a potential to improve the performance of mature oil fields along with enhanced oil recovery (EOR) processes. Regarding polymer flooding, the influence of the (macro) molecules' architecture on the fluid properties has been lately stressed. This chapter presents the numerical simulation of the combination of both agents in a single, combined recovery process. The presence of the nanoparticles affects the rheological behavior and the rock's wettability, increasing the organic phase mobility. Undesirable effects such as (nano) particle aggregation and sedimentation are also considered. The polymer's architecture has a major influence on the recovery process, improving the rheological and viscoelastic properties. On the other hand, although nanoparticles improve the viscosity as well, its main mechanism is their adsorption onto the rock and wettability modification. This chapter shows the importance of a good polymer characterization for EOR, the potential of nanoparticles acting as a boost of traditional EOR processes, and the vital role CFD techniques play to assess the potential of these agents and the optimization of the recovery strategies.

Keywords: enhanced oil recovery, polymer, nanotechnology, reservoir simulation, nanofluids

1. Introduction

The so-called era of discovery and exploitation of the denominated "easy oil" is since some years reaching to an end [1–5]. The exploitation of a conventional oil field can be mainly divided in three stages, which depend on the physical mechanisms acting during the oil recovery [6–8]. The first step consists in taking advantage of the natural-driven mechanisms present in the oil field, without the injection of fluids or specific agents. This stage, known as primary recovery, finished when the pressure in the reservoir or the amount of oil produced is no longer enough to render a profitable production. Subsequently, water or gas starts being injected with a dual goal: repressurize the rock formation and sweep the remaining oil toward the

IntechOpen

producer wells (**Figure 1**). Both primary and secondary recoveries account for approximately a 50% of the original oil in place (OOIP). During the last 50 years, oil companies began applying more advanced processes after secondary recovery, which are known as enhanced oil recovery (EOR) or tertiary processes, which involve the injection of different fluids in order to modify the physical properties of the different fluids and/or rock formation. This renders an increase in the oil production and lifetime of conventional oil sources, which is necessary while newer and greener energy sources are developed and optimized (**Figure 2**).

Among EOR techniques, chemical EOR comprises the addition of certain agents (e.g., polymers, surfactants, alkali) and presents a great potential in low- and medium-viscosity mature oil fields to increase their productivity. Recently, polymers were also applied in high-viscosity reservoirs showing promising results when specific geological conditions are met. Nevertheless, there are certain problems associated with these products which hinder their efficiency and limit their applicability. Thus, during the last 20 years, researchers have begun making use of the nanotechnology in order to boost the efficiency or EOR processes, based on the novel properties of materials exhibited at these scales. The combination of chemical EOR agents with different nanomaterials has shown increased recovery efficiencies in rock formations which had otherwise reached their operational limit [10, 11].

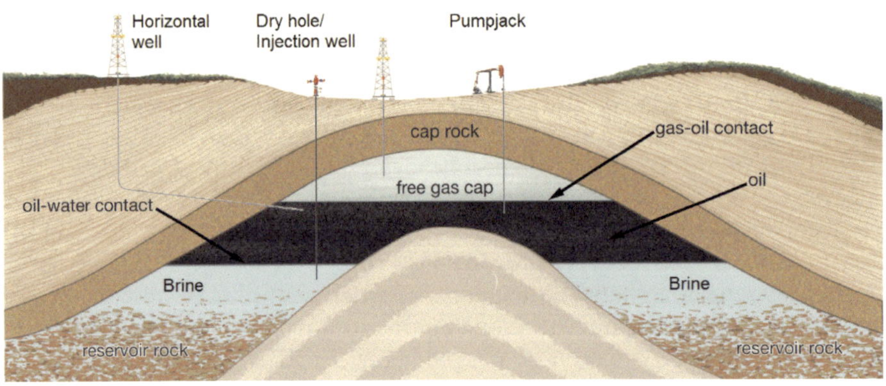

Figure 1.
Anticlinal type petroleum trap [9].

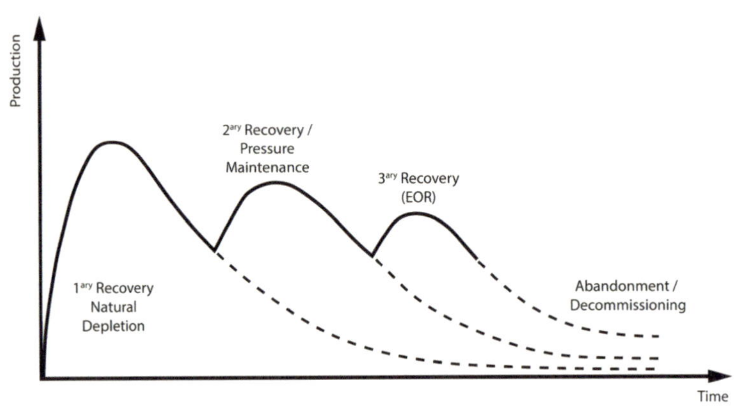

Figure 2.
Schematic representation as a function of time of the different oil recovery stages and their respective productivities [12].

However, the potential advantages of the application of nanomaterials in oil recovery should also be carefully considered from an environmental point of view. Some concerns have been raised since the same features that make nanotechnology so attractive to oil recovery processes might also have a negative impact on the environment and human health. These include the potential long-term side effects associated with medical applications as well as with the biodegradability of nanomaterials being used [13–19]. Even though during the last years many nanomaterials have been inserted into the market, the amount of information over their impact on the environment is minimal. For instance, there is almost no information about the associated risks in the manufacturing, usage, and final disposal of nanomaterials [20]. Focusing on EOR techniques, it has been shown that a percentage of the nanoparticles will remain underground and deposited in the rock formation, remaining for many years, and this could cause the contamination of groundwater sources. Thus, one of the desirable properties of these nanoparticles should be high durability and recyclability in a cost-effective process to decrease their impact on the environment [20].

An important point to mention is that the field test of these new techniques involves a significant use of resources and time in order to assess their efficiency. Therefore, for the last 40 years, scientists started using computers to perform this assessment and save considerable time and physical resources, which is known as reservoir simulation. This is a branch of computational fluid dynamics (CFD) which involves solving the balance equations present in porous media, which renders a number of coupled, highly nonlinear systems of equations dealing with temporal and spatial variations of pressure and mass concentrations. Different numerical and physical techniques have been applied during the years in order to simulate different recovery processes as well as to increase the numerical accuracy of the simulation.

The objective of this chapter is to present the potential of nanomaterials as a boost of traditional EOR techniques, focusing on the development of numerical models for reservoir simulation, especially in the combination of chemical EOR agents with nanoparticles, studying their advantages and synergy in order to increase the productivity of conventional oil sources. Reservoir simulation consists broadly speaking of three parts: geological, fluid, and well models which describe the main parts of the extraction system [21–24]. The accurate mathematical representation of the whole system is still a topic in which further research is needed. The risks associated with the uncertainties in the numerical model might lead to the failure of exploration and production (E&P) projects. This rendered a multiphase, multicomponent model, considering all the effects that chemical agents and nanoparticles provoke both in the fluid and rock formation. Regarding the nanoparticles, this includes the aggregation, retention, rheology, and changes in permeability and porosity. The chemical agent studied in this chapter is a polymer, which is modeled considering the influence of the (macro)molecules' architecture on the fluid properties. The salt present in the reservoir is also considered in the water phase. This model rendered a novel simulator, combining the benefits of nanotechnology with chemical EOR processes.

2. Model description

2.1 Physical model

The study of this combined model of polymer-boosted EOR flooding is presented in a 2D domain, based on a well configuration used in the oil industry. Indeed, the five-spot scheme consists in a square domain, in which the injection well is placed in the center and four production wells in each of the vertices (**Figure 3a**). From this, a mathematic simplification consists in dividing this domain using its symmetry, which is known as quarter five-spot (**Figure 3b**).

The physical model (Ω) represents an oil field of known geometric and physical properties, i.e., absolute permeabilities (K) and porosity (ϕ) and a rock compressibility (c_f). Furthermore, the fluid flow is considered isothermal and incompressible. Darcy's law is valid, and the gravitational forces are negligible compared to the viscous ones [25–27]. Thus, numerically speaking, the domain is discretized in a number of $n_x \times n_y$ blocks. The grid size is chosen based on the analysis of the representative elementary volume (REV), which is determined by the minimum size in which the rock properties remain approximately the same.

The recovery process consists in a two-phase (aqueous and organic), multicomponent (water, salt, polymer nanoparticles, and petroleum) flow. These components may be also mixtures of a number of pure ones, e.g., petroleum is a mixture of many hydrocarbons, water contains dissolved minerals (other than salt itself), and finally polymer is composed of different molecules of different lengths and architectures [27]. The polymer properties are determined by the average molecular weight, which in this model is assumed to be identical for all the molecules, which renders a polydispersity index (*PDI*) equal to unity. The nanoparticles affect the water phase and the rheology, using a function of their concentration and size. With respect to this, aggregation mechanisms present in the system tend to increment the nanoparticles' average size (**Figure 4**) [28–31].

The mathematical description of the system is represented by a number of strongly nonlinear partial differential equations complemented by a set of algebraic relationships describing the physical properties of fluids and rock, which are aggregation of nanoparticles, degradation of polymer molecules, interfacial tension, residual phase saturations, relative permeabilities, rock wettability, phase viscosities, capillary pressure, adsorption and retention of both polymer and nanoparticles onto the formation, inaccessible pore volume (IAPV), disproportionate permeability reduction (DPR), nanoparticles-polymer interactions, and dispersion.

2.2 Mathematical model

To study the flow of multiphase, multicomponent system in porous media, the mass, momentum, and energy balance equations are applied. Therefore, the

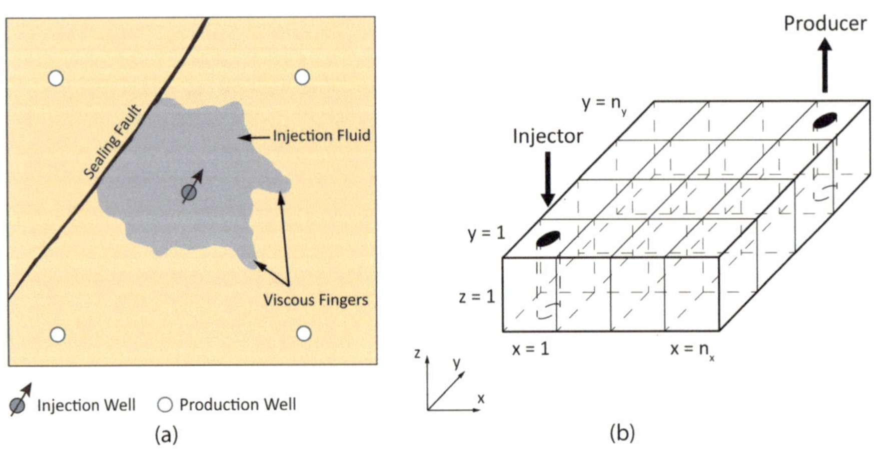

Figure 3.
Five-spot scheme (a) indicating the well's location and the possible presence of faults and its simplification to the quarter five-spot used during this chapter (b) [12, 32].

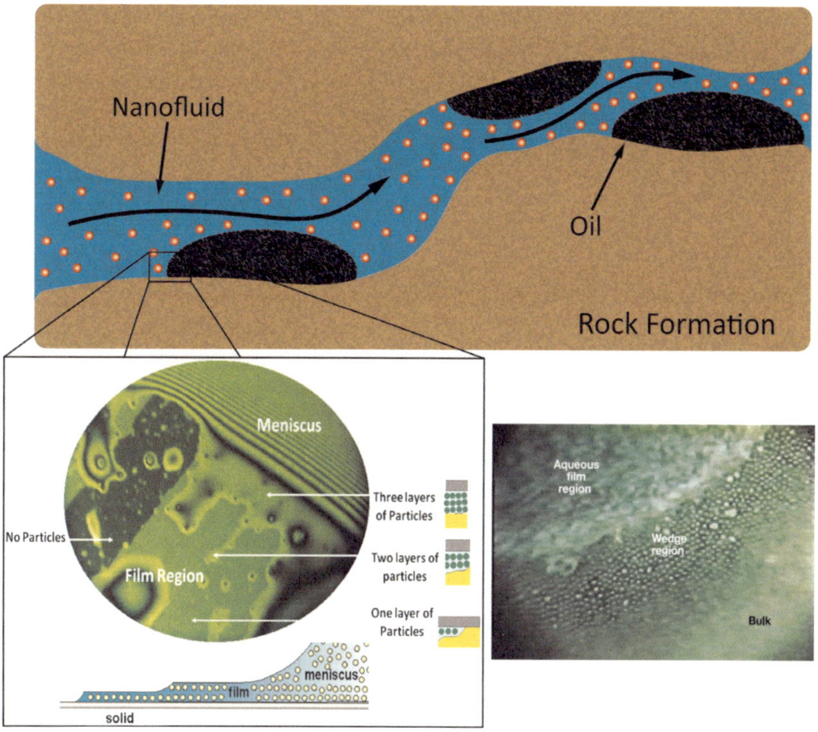

Figure 4.
Schematic representation as a function of time of the different oil recovery stages and their respective productivities [33].

equations used to describe the process are Darcy's equation applied to each phase and the mass conservation valid for each component [33]. Since in most of chemical EOR processes the exchange of energy (i.e., temperature changes) is not significant, the energy balance equation is not considered in the simulation. The compositional approach is chosen because of its versatility to model the different physical properties according to the components' concentrations. These equations are then applied on a REV of the porous medium. Considering Darcy's equation for each phase first,

$$\vec{u}^j = -\underline{\underline{K}} \cdot \frac{k_r^j}{\mu^j} \cdot \vec{\nabla} p^j; \quad j = o, a \tag{1}$$

$$\frac{\partial(\phi z_i)}{\partial t} + \nabla \cdot \sum_j V_i^j \cdot \vec{u}^j - \nabla \cdot \sum_j \underline{\underline{D_i^j}} \cdot \nabla \cdot V_i^j = -\frac{\partial(\phi A d_i)}{\partial t} + q_i; \quad i = p, np, w, s, pol \tag{2}$$

$$\underline{\underline{D_i^j}} = dm_i^j \cdot \phi \cdot S^j \cdot \delta_{ij} + \|\vec{u}^j\| \cdot \left[\frac{dl^j}{\|\vec{u}^j\|^2} \cdot \begin{vmatrix} \left(u_x^j\right)^2 & u_x^j \cdot u_y^j \\ u_y^j \cdot u_x^j & \left(u_y^j\right)^2 \end{vmatrix} + dt^j \cdot \begin{vmatrix} 1 - \frac{\left(u_x^j\right)^2}{\|\vec{u}^j\|^2} & -\frac{u_x^j \cdot u_y^j}{\|\vec{u}^j\|^2} \\ -\frac{u_y^j \cdot u_x^j}{\|\vec{u}^j\|^2} & 1 - \frac{\left(u_y^j\right)^2}{\|\vec{u}^j\|^2} \end{vmatrix} \right] \tag{3}$$

$$\phi c_r \frac{\partial p^a}{\partial t} + \vec{\nabla} \cdot (\lambda \cdot \nabla p^a) = \frac{\partial}{\partial t}\left(\phi \cdot \sum_i A d_i\right) - \vec{\nabla} \cdot (\lambda^o \cdot \nabla p_c) + q_t \tag{4}$$

As in every compositional simulation, the number of equations generated by the conservation laws is not sufficient to mathematically determine the system. Thus, a number of auxiliary relationships are needed in order to find the solution, which for a five-component, two-phase system this renders a total of $N_{comp} \cdot (N_{phases} - 1) = 5$ equations, determined by system's phase behavior [24, 34, 35].

2.2.1 Discretization of the partial differential equations

The above system of equations are discretized and then solved in this chapter by a finite difference method, which is one of the most well-known techniques in CFD. The first equation is the aqueous phase pressure (Eq. (4)), which is implicitly discretized using a centered scheme for the pressure terms and a second-order Taylor approximation for the time derivatives. This scheme chosen for the simulator is often used in systems with derived second-order with coefficients that are not constants. Besides the Darcy momentum equation, the discretization of the total and aqueous velocities is also done using a centered difference scheme. Therefore, Eqs. (1) and (4) are discretized as

$$
\begin{aligned}
&c_r\left(\phi+\frac{\Delta t}{2}\frac{\partial\phi}{\partial t}\right)_{m,n}^{<n+1>,[k]}\left(\frac{p_{m,n}^{a,<n+1>}-p_{m,n}^{a,<n>}}{\Delta t}\right)^{[k+1]}+\frac{\lambda_{x,m+12,n}^{<n+1>,[k]}}{\Delta x^2}\cdot\left(p_{m+1,n}^{a}-p_{m,n}^{a}\right)^{<n+1>,[k+1]}-\ldots\\
&-\frac{\lambda_{x,m-12,n}^{<n+1>,[k]}}{\Delta x^2}\cdot\left(p_{m,n}^{a}-p_{m-1,n}^{a}\right)^{<n+1>,[k+1]}+\frac{\lambda_{y,m+12,n}^{<n+1>,[k]}}{\Delta y^2}\cdot\left(p_{m,n+1}^{a}-p_{m,n}^{a}\right)^{<n+1>,[k+1]}-\ldots\\
&-\frac{\lambda_{y,m-12,n}^{<n+1>,[k]}}{\Delta y^2}\cdot\left(p_{m,n}^{a}-p_{m,n-1}^{a}\right)^{<n+1>,[k+1]}=\left(\phi+\frac{\Delta t}{2}\frac{\partial\phi}{\partial t}\right)_{m,n}^{<n+1>,[k]}\left(\frac{Ad_{m,n}^{<n+1>}-Ad_{m,n}^{<n>}}{\Delta t}\right)^{[k+1]}+\ldots\\
&+\left(Ad+\frac{\Delta t}{2}\frac{\partial Ad}{\partial t}\right)_{m,n}^{<n+1>,[k]}\left(\frac{\phi_{m,n}^{<n+1>}-\phi_{m,n}^{<n>}}{\Delta t}\right)^{[k+1]}+\frac{\lambda_{x,m+12,n}^{o,<n+1>,[k]}}{\Delta x^2}\cdot\left(p_{m+1,n}^{c}-p_{m,n}^{c}\right)^{<n+1>,[k+1]}-\ldots\\
&-\frac{\lambda_{x,m-12,n}^{o,<n+1>,[k]}}{\Delta x^2}\cdot\left(p_{m,n}^{c}-p_{m-1,n}^{c}\right)^{<n+1>,[k+1]}+\frac{\lambda_{y,m+12,n}^{o,<n+1>,[k]}}{\Delta y^2}\cdot\left(p_{m,n+1}^{c}-p_{m,n}^{c}\right)^{<n+1>,[k+1]}-\ldots\\
&-\frac{\lambda_{y,m-12,n}^{o,<n+1>,[k]}}{\Delta y^2}\cdot\left(p_{m,n}^{c}-p_{m,n-1}^{c}\right)^{<n+1>,[k+1]}+q_{m,n}^{t,<n+1>,[k]}
\end{aligned}
\tag{5}
$$

$$
\begin{aligned}
&\vec{u}_{m,n}^{<n+1>,[k+1]}=\left[-\frac{\lambda_{x,m,n}^{[k]}}{2\cdot\Delta x}\cdot\left(p_{m+1,n}^{a}-p_{m-1,n}^{a}\right)^{[k+1]}-\ldots\right.\\
&\left.-\frac{\lambda_{x,m,n}^{o,[k]}}{2\cdot\Delta x}\cdot\left(p_{c,m+1,n}^{a}-p_{c,m-1,n}^{a}\right)^{[k+1]}\right]^{<n+1>}\cdot\hat{i}+\left[-\frac{\lambda_{y,m,n}^{[k]}}{2\cdot\Delta y}\cdot\left(p_{m,n+1}^{a}-p_{m,n-1}^{a}\right)^{[k+1]}-\ldots\right.\\
&\left.-\frac{\lambda_{y,m,n}^{o,[k]}}{2\cdot\Delta y}\cdot\left(p_{c,m,n+1}^{a}-p_{c,m,n-1}^{a}\right)^{[k+1]}\right]^{<n+1>}\cdot\hat{j}
\end{aligned}
\tag{6}
$$

$$
\begin{aligned}
&\vec{u}_{m,n}^{a,<n+1>,[k+1]}=\left[-\frac{\lambda_{x,m,n}^{a,[k]}}{2\cdot\Delta x}\cdot\left(p_{m+1,n}^{a}-p_{m-1,n}^{a}\right)^{[k+1]}\right]^{<n+1>}\cdot\hat{i}+\ldots\\
&+\left[-\frac{\lambda_{y,m,n}^{a,[k]}}{2\cdot\Delta y}\cdot\left(p_{m,n+1}^{a}-p_{m,n-1}^{a}\right)^{[k+1]}\right]^{<n+1>}\cdot\hat{j}
\end{aligned}
\tag{7}
$$

where m, n represent the cells of the numerical domain $(x, y) = (m \cdot \Delta x, n \cdot \Delta y)$, respectively, $<n>$ is the temporal step (time $= <n> \cdot \Delta t$), and $[k], \forall k \in \mathbb{N}^+$, is the iteration number within each time-step. Finally, mass conservation equation is discretized using a second-order approach. Eq. (2) is the typical advection-diffusion PDE used to describe the mass transport in porous media. The advective terms are of hyperbolic nature, and first-order numerical schemes cause an artificial diffusion in the solution. There are different approaches in order to overcome this, and one of them is the use of higher-order schemes. The proposed simulator in this chapter uses a full second-order explicit discretization scheme in time and space, based on total variation diminishing (TVD) and flux-limiting techniques. This increases the numerical accuracy of the simulator as well as decreases the influence of numerical diffusion and dispersion. Diffusive terms are discretized using a centered second-order scheme. The second order in time is achieved using a Taylor expansion up to the second order [12]. The flux-limiting techniques require to establish a functional relationship between the gradient of the volumetric concentration and the limiting function ψ. Different second-order methods have been studied, with the functions utilized in the development of this simulator presented in **Table 1** (together with the standard upwind method). These functions depend on the ratio of the concentrations' consecutive gradients in the numerical grid $\left(r_{x,i} = \left(V_{i,m,n}^{j,[k]} - V_{i,m-1,n}^{j,[k]}\right) / \left(V_{i,m+1,n}^{j,[k]} - V_{i,m,n}^{j,[k]}\right)\right)$.

Thus, the discretized mass conservation equation is

$$
\begin{aligned}
\frac{C_1}{\Delta t} z_i^{<n+1>} = {} & C_2 z_i^{<n>} + \frac{C_3}{\Delta x} \cdot \sum_j F_{LIM,x}^{j,<n+1>,[k+1]} \left(u_{x,m,n}^{j,[k+1]} \cdot V_{i,m,n}^{j,[k]} - u_{x,m-1,n}^{j,[k+1]} \cdot V_{i,m-1,n}^{j,[k]}\right)^{<n+1>} + \dots \\
& + \frac{C_3}{\Delta y} \cdot \sum_j F_{LIM,y}^{j,<n+1>,[k+1]} \left(u_{y,m,n}^{j,[k+1]} \cdot V_{i,m,n}^{j,[k]} - u_{y,m,n-1}^{j,[k+1]} \cdot V_{i,m,n-1}^{j,[k]}\right)^{<n+1>} + \dots \\
& + \frac{1}{\Delta x^2} \cdot \sum_j \left[\left(S^j \phi dm_i^j\right)_{m+1/2,n} \cdot \left(V_{i,m+1,n}^j - V_{i,m,n}^j\right) - \left(S^j \phi dm_i^j\right)_{m-1/2,n} \cdot \left(V_{i,m,n}^j - V_{i,m-1,n}^j\right)\right]^{<n+1>,[k]} + \dots \\
& + \frac{1}{\Delta y^2} \cdot \sum_j \left[\left(S^j \phi dm_i^j\right)_{m,n+1/2} \cdot \left(V_{i,m,n+1}^j - V_{i,m,n}^j\right) - \left(S^j \phi dm_i^j\right)_{m,n-1/2} \cdot \left(V_{i,m,n}^j - V_{i,m,n-1}^j\right)\right]^{<n+1>,[k]} - \dots \\
& - \frac{1}{\Delta t}\left(\phi + \Delta t \frac{\partial \phi}{\partial t}\right)_{m,n}^{<n+1>,[k+1]} \cdot \left(Ad_i^{<n+1>} - Ad_i^{<n>}\right)_{m,n}^{[k]} - \frac{Ad_{i,m,n}^{<n+1>,[k]}}{\Delta t} \cdot \left(\phi^{<n+1>} - \phi^{<n>}\right)_{m,n}^{[k]} + \dots \\
& + q_{i,m,n}^{<n+1>,[k+1]} + \sum_j \frac{u_{x,m,n}^{j,[k+1]} \Delta t}{2\phi_{m,n}^{<n+1>,[k+1]}} \cdot \left(\frac{\partial Ad}{\partial x}\frac{\partial \phi}{\partial t} + Ad \frac{\partial^2 \phi}{\partial t \partial x} + \frac{\partial \phi}{\partial x}\frac{\partial Ad}{\partial t}\right)_{i,m,n}^{<n+1>,[k]} + \dots \\
& + \sum_j \frac{u_{y,m,n}^{j,[k+1]} \Delta t}{2\phi_{m,n}^{<n+1>,[k+1]}} \cdot \left(\frac{\partial Ad}{\partial y}\frac{\partial \phi}{\partial t} + Ad \frac{\partial^2 \phi}{\partial t \partial y} + \frac{\partial \phi}{\partial y}\frac{\partial Ad}{\partial t}\right)_{i,m,n}^{<n+1>,[k]} - \dots \\
& - \sum_j \left(\frac{u_{x,m,n}^{j,[k+1]} \Delta t \, \partial^2 Ad}{2 \, \partial t \partial x} + \frac{u_{y,m,n}^{j,[k+1]} \Delta t \, \partial^2 Ad}{2 \, \partial t \partial y}\right)_{i,m,n}^{<n+1>,[k]} + \dots \\
& + \sum_j \frac{\Delta t}{2\phi_{m,n}^{<n+1>,[k+1]}} \left[\left(u_{x,m,n}^{j,[k+1]}\right)^2 \frac{\partial^2 V_{i,m,n}^{j,[k]}}{\partial x^2} + \left(u_{y,m,n}^{j,[k+1]}\right)^2 \frac{\partial^2 V_{i,m,n}^{j,[k]}}{\partial y^2} + 2u_{x,m,n}^{j,[k+1]} u_{y,m,n}^{j,[k+1]} \frac{\partial^2 V_{i,m,n}^{j,[k]}}{\partial x \partial y}\right]^{<n+1>}
\end{aligned} \tag{8}
$$

Flux limiter	Function
Upwind	0
Superbee	$\max\left[0, \min(2r, 1), \min(r, 2)\right]$
Minmod	$\max\left[0, \min(r, 1)\right]$
MUSCL	$\max\left[0, \min\left(2r, \frac{1+r}{2}, 2\right)\right]$

Table 1.
General parameters used for the simulations.

and the coefficients $C_{1,2,3}$ are calculated as follows:

$$C_1 = \left(\phi_{m,n} + \Delta t\frac{\partial\phi}{\partial t} - \frac{u_{tx,m,n}\Delta t}{2\phi_{m,n}}\frac{\partial\phi}{\partial x} - \frac{u_{ty,m,n}\Delta t}{2\phi_{m,n}}\frac{\partial\phi}{\partial y}\right)^{<n+1>,[k+1]}$$

$$C_2 = \left(\frac{\phi_{m,n}}{\Delta t} - \frac{u_{tx,m,n}}{2\phi_{m,n}}\frac{\partial\phi}{\partial x} - \frac{u_{ty,m,n}}{2\phi_{m,n}}\frac{\partial\phi}{\partial y} + \frac{u_{tx,m,n}\Delta t}{2\phi_{m,n}}\frac{\partial^2\phi}{\partial t\partial x} + \frac{u_{ty,m,n}\Delta t}{2\phi_{m,n}}\frac{\partial^2\phi}{\partial t\partial y}\right)^{<n+1>,[k+1]}$$

$$C_3 = \left(1 - \frac{\Delta t}{2\phi_{m,n}}\frac{\partial\phi}{\partial t}\right)^{<n+1>,[k+1]} \tag{9}$$

2.2.2 Boundary conditions

At the beginning of the EOR process, the oil saturation in the reservoir is assumed to be equal to values after the application of primary schemes or the saturation after a waterflooding which reached the economical threshold at the producing well. Therefore, there is no chemical components present, and the initial pressure is constant throughout the reservoir. Thus:

$$t = 0 \quad ; \quad \forall(x, y) \in \Omega: \quad z_c = 0 \quad ; \quad z_p = S^{orH} \quad ; \quad p^a = p_i \tag{10}$$

The combined EOR process begins with the injection for a certain period of time of a polymer and/or nanoparticles at a constant concentration. After this period, the chemical slug is followed by a water-bank in order to sweep the remaining oil. As the boundary conditions in the domain, a "no flow" is assumed on the oil field contour (Γ), since it is assumed that the porous rock is surrounded by an impermeable rock layer. Regarding the advection mechanisms, this is satisfied imposing a zero mobility on the boundaries. As far as the diffusive mechanisms are concerned, Fick's law is applied rendering

$$\textbf{Injecting well} \Rightarrow \begin{cases} 0 \le t \le t_{in} : z_c = z_{in} \\ t > t_{in} : z_{in} = z_w, z_c = 0 \end{cases} \tag{11}$$

$$\textbf{Boundaries} \Rightarrow \lambda^j_{m,n} = 0 \wedge \frac{\partial z_i}{\partial n_\Gamma} = 0 \quad ; \quad i = p, c \quad ; \quad \forall t \wedge \forall(m,n) \in \Gamma \tag{12}$$

3. Physical properties

The properties for a polymer flooding are well described in the literature [12]. These are considered in this model, but during this chapter only the novel approaches will be discussed and presented, including the polymer architecture and the nanoparticles' modifications. Hence, the following phenomena are not included in the scope of this chapter: residual saturations, disproportionate permeability reduction (DPR), inaccessible pore volume (IAPV), polymer degradation, and capillary pressure.

3.1 Chemical component partition

As mentioned in the previous section, the conservation laws must be complemented with a number of extra relationships to mathematically find the solution of the system of equations. These come from what is known as the system's *phase behavior*. This is considered essential in order to understand how the components distribute among the phases. In this model it is assumed that both oil and water remain in the organic and aqueous phases, respectively. Moreover, polymer and salt remain only in the aqueous phase. As far as the phase behavior of the nanoparticles is concerned, a similar concept to the one commonly used for surfactants is employed in order to calculate how the (nano)particles distribute between the phases. Depending on the particles' wettability, namely: hydrophobic and lipophilic (HLPN); neutral wet (NWPN); or lipophobic and hydrophilic polysilicon nanoparticles (LHPN), these will be present in the organic, both, or the aqueous phase, respectively. Thus, three dimensionless relationships are used to account for this partition:

$$\text{Solubilization coefficient} = L^a_{pnp} = \frac{V^a_p}{V^a_{np}} \tag{13}$$

$$\text{Swelling coefficient} = L^o_{wnp} = \frac{V^o_w}{V^o_{np}} \tag{14}$$

$$\text{Partition coefficient} = k_{np} = \frac{V^o_{np}}{V^a_{np}} \tag{15}$$

Since in this model petroleum and water remain in their respective phases, the swelling and solubilization coefficients are zero $\left(L^a_{pnp} = L^o_{wnp} = 0\right)$. The value of the nanoparticle's partition coefficient is then determined by the physical characteristics of the nanomaterial used in the flooding, namely, HLPN (for $k_{np} \gg 1$), LHPN (for $k_{np} \approx 0$), and NWPN (for $k_{np} \approx 1$). Finally, the polymer and salt are assumed to remain completely in the aqueous phase, hence $V^o_s = V^o_{pol} = 0$.

3.2 Phase viscosities

The main functionality of the polymer addition is to increase the rheological and viscoelastic properties of the aqueous phase, improving the recovery efficiency and avoiding the *water fingering* phenomenon [6, 26, 27]. Although the nanoparticles also affect the viscosity, its influence is not as important as with the polymer. Several correlations have been proposed to account for the influence of nanoparticles on the viscosity, since Einstein's original work, used for low concentrations, to expanded studies considering, among others, the size and type of the nanoparticles, temperature, and the characteristics of the carrier fluid [36–40]. In order to consider these components in the calculation, a stepwise procedure is adopted. Since the polymer's architecture is also considered in this model, it is necessary to determine the viscosity ratio between a linear polymer used as reference and a (hyper)branched one which has the same total molecular weight [41–46]. Eq. (16) is used in the proposed model to calculate this relationship (**Figure 5**).

$$g_{viscosity} = \frac{1}{(1+\rho\cdot f)^3}\left[1+2f\cdot\rho+\left(2f+f^2\right)\rho^2+\left(3f^2-2f\right)\rho^3\right] \tag{16}$$

where f is the polymer's number of arms and ρ is the relationship between molecular weights of arms and backbone. The number of arms depends on the polymer used, with some authors reporting polymers for EOR up to 17 arms [47]. Subsequently, the aqueous phase viscosity at zero shear rate can be calculated using the Mark-Houwink relationship based on the polymer's molecular weight:

$$\mu_{0sr} = \mu_w \left\{ 1 + \left[k_1(g_v, f) \cdot [\eta] \cdot V_{pol}^{a} + k_2(g_v, f) \cdot [\eta] \cdot V_{pol}^{a^2} + k_3(g_v, f) \cdot [\eta] \cdot V_{pol}^{a^3} \right] C_{SEP}^{Sp} \right\} \tag{17}$$

The influence of nanoparticles is considered using a relationship for the nanofluid relative viscosity, expressed as a function of the particles' diameter, which is a function of time (see Section 3.3), the carrier fluid molecules diameter, and the particles' volumetric concentration [48–51]:

$$\mu_{np} = \mu_{cf} \cdot \left[1 + \left(2.5 + \eta_1 e^{-Dia_{np} d_w}\right) \cdot V_{np}^{a} + \left(6.2 + \eta_2 e^{-Dia_{np} d_w}\right) \cdot V_{np}^{a^2} \right] \tag{18}$$

where $\eta_{1,2}$ are correlation constants. It is important to mention that in the present model, the particular effects of associating polymers are not considered [40, 52, 53]. It is therefore recommended that future studies should be performed in order to establish a relationship between the rheological properties and the presence of both nanoparticles and these types of polymers.

3.3 Aggregation of nanoparticles

The aggregation of nanoparticles is a well-known and studied phenomenon, reported by many experiments which show that nanoparticles may aggregate until a critical size is reached, followed by their sedimentation [54–58]. This can be defined as the formation of clusters by (nano)particles or when small clusters aggregate to form bigger ones due to the forces present in the system, which are explained, for instance, by the Derjaguin-Landau-Verwey-Overbeek (DLVO) theory (**Figure 6**). As the flooding process evolves in the rock formation, the gradual increase in the clusters' size may provoke the sedimentation, affecting negatively the aqueous phase properties regarding the oil recovery efficiency. On the other hand, these clusters may also be split into smaller ones if their size reaches a critical limit (**Figure 7**) [59–65].

The mathematical modeling of this phenomenon was done in this chapter using a simple ODE with exponential growing, using the particle size and an aggregation constant to take into account the change in the particles' diameter as a function of time. This size is calculated for every time-step of the simulation since this affects also other related phenomena during the flooding process (see Section 3.5).

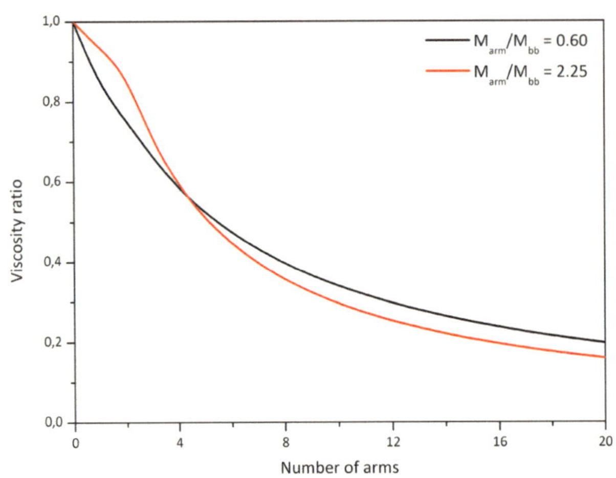

Figure 5.
Viscosity ratio $g_{viscosity}$ as a function of the numbers of arms and the molecular weight relationship [12].

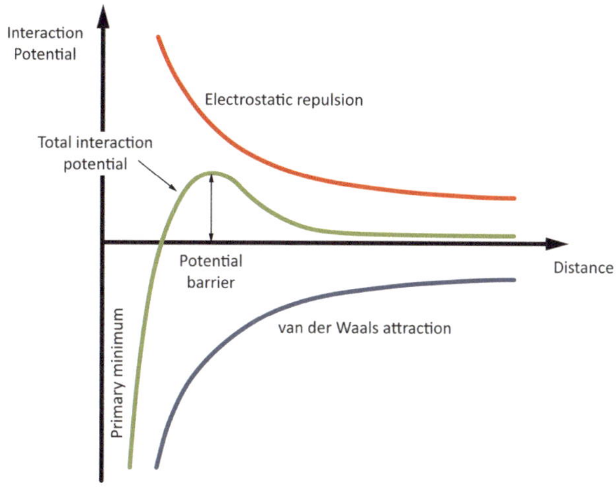

Figure 6.
Different forces at the molecular scale as a function of the particles' distance [65].

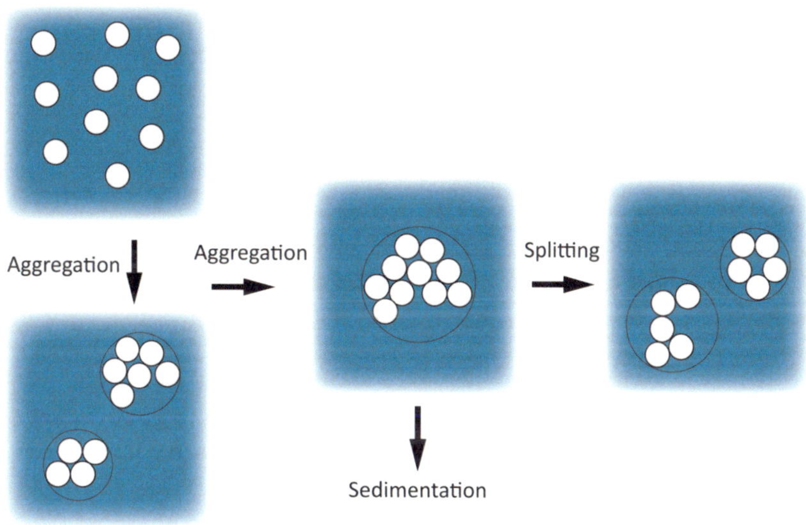

Figure 7.
Scheme of the aggregation mechanisms of nanoparticles and their subsequent splitting [33].

Furthermore, a maximum possible size for the clusters is also assumed, above which it splits into two identical, smaller clusters [59]:

$$\frac{dDia_{np}}{dt} = K_{aggreg} \cdot Dia_{np} \tag{19}$$

3.4 Diffusion of nanoparticles

The diffusion process can modify significantly the particles' transport and therefore their influence on the recovery process. In this chapter the influence of the polymers' molecules in this process will be also considered, which has not been reported previously [66]. In the proposed model in this chapter, the starting point is the Brownian diffusion model calculated by the Einstein-Stokes equation (Eq. (20)), which is valid for low concentrations, and it is therefore corrected for different values of the particles' volumetric concentration (Eq. (21)):

$$D^{a}_{np} = \frac{k_B T}{3\mu_{cf}\pi Dia_{np} f_{corr}} \tag{20}$$

$$f_{corr} = \left(1 - V^{a}_{np}\right)^{-6.55} \tag{21}$$

Then, this diffusion coefficient must be corrected in order to take into account the presence of the polymer molecules and their characteristics (architecture, composition, radius of gyration, and molecular weight). The radius of gyration depends on its structure, its chemical composition, and molecular weight. The calculation for a linear molecule is presented in Eq. (22) [67]:

$$R_{g,linear} = b_{rg} \cdot \left(\frac{M_{w,bb} + f \cdot M_{w,arm}}{M_{w,mon}}\right)^{0.588} \tag{22}$$

where b_{rg} is the segment length and $M_{w,mon}$ is the monomer's molecular weight. Using this value, the radius of gyration of the branched polymer can be calculated using a similar procedure to the one used in the viscosity calculation, expressing this relationship as a function of the number of arms (**Figure 8**):

$$g_{rg} = \frac{3f - 2}{f^2} \tag{23}$$

The nanoparticles' diffusion coefficient is then calculated using a stepwise function, expressed as a function of the particles' size, radius of gyration, and overlapping concentration, based on the study made by Flory [68–70]:

$$\phi^* = \frac{3(M_{w,bb} + f \cdot M_{w,arm})}{4\pi\rho_{pol} N_{Av} R_g^3} \tag{24}$$

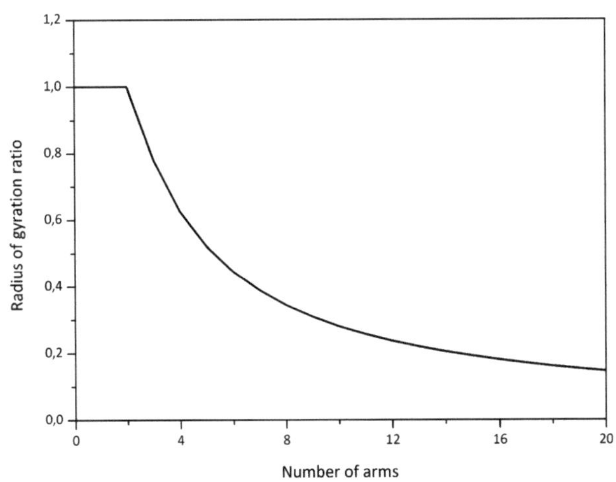

Figure 8.
Radius of gyration ratio g_{rg} as a function of the numbers of arms and the molecular weight relationship [12].

$$\chi = R_{g,pol}\left(\frac{V^a_{pol}}{\phi^*}\right)^{-0.76} \tag{25}$$

$$D^a_{NPp} \Rightarrow \begin{cases} D^a_{NP}\cdot e^{\alpha_D\left(\frac{Dia_{NP}}{2\chi}\right)^{\delta_D}} & \textbf{if} \quad \left(\frac{Dia_{NP}}{2}\right) \le R_g \\ D^a_{NP}\cdot e^{\alpha_D\left(\frac{R_g}{\chi}\right)^{\delta_D}} & \textbf{if} \quad \left(\frac{Dia_{NP}}{2}\right) \ge R_g \end{cases} \tag{26}$$

where ρ_{pol} is the polymer's density, N_{Av} is the Avogadro number, and α_D and δ_D are empirical constants.

3.5 Retention and adsorption

The adsorption takes place when nanoparticles or polymer molecules are deposited onto the surface of the rock formation. This phenomenon causes a loss of the EOR agents in the porous media, decreasing the efficiency and eventually making the whole process unprofitable, in case of high adsorption rates [48]. Similar to what is reported in combined chemical EOR processes, the presence of two or more agents will cause a competitive adsorption process, modifying their deposition rates, since the polymer molecules will cover part of the rock's surface, rendering a smaller available area for the adsorption of nanoparticles. This is modeled using two factors, one affecting the polymer's adsorption, and the other the nanoparticles, which are function of the injection scheme. This also alters the porous media properties (i.e., porosity and permeability) if the particles' size is larger or of the same order than the pore size or if large volumes of particles accumulate [71–75]. The loss is quantified in Eq. (4) and it can be calculated as [66, 72].

$$Ad_{np} = \left(F_{SP}\cdot v_1 + v_2 + v_3\right) \tag{27}$$

where v_1 is the volume of nanoparticles available on the pore surfaces, v_2 is the volume of nanoparticles entrapped in the pores of the porous medium due to plugging and bridging, and v_3 is the release rate from pore walls by colloidal forces, considering the salinity and the nanoparticles' possible charge [66]. The first term is usually expressed as a function of the critical velocity for surface deposition. Below this, only the retention phenomenon occurs, and above this value a combination retention and entrainment takes place:

$$v_1^{<n>} \Rightarrow \begin{cases} \alpha_1\cdot u^j\cdot V^a_{np} & \textbf{if} \quad u^j \le u^a_{crit} \\ \alpha_1\cdot u^j\cdot V^a_{np} - \alpha_2\cdot v_1^{<n-1>}\cdot\left(u^j - u^j_{crit}\right) & \textbf{if} \quad u^j \ge u^a_{crit} \\ v_1^{<0>} = 0 \quad \forall \quad \Omega \end{cases} \tag{28}$$

where α_1 is the coefficient for surface retention and α_2 is the coefficient for entrainment. The critical velocity can be expressed as a function of the particles'/clusters' size, which is a function of time (**Figure 9**) [71, 72, 76]. It is important to note here that this velocity is also a function of the particles' shape, which can vary considerable in the case of clusters:

$$u^j_{crit}\left[{}^{m}\!/_{day}\right] = 0.00992736\cdot Dia_{NP}[nm] + 0.0009936 \tag{29}$$

Regarding the term v_2, a formulation similar to the retention term in v_1 is adopted, expressed according to the following stepwise function:

$$v_2^{<n>} \Rightarrow \begin{cases} \alpha_3 \cdot u^j \cdot V_{np}^a \\ v_2^{<0>} = 0 \quad \forall \quad \Omega \end{cases} \tag{30}$$

where α_3 is the pore blocking empirical constant. Lastly, v_3 is calculated based on the salinity, expressed as the TDS in the rock formation [66]:

$$v_3^{<n>} \Rightarrow \begin{cases} -\alpha_4 \cdot v_1^{<n>} \cdot (C_{sc} - V_s^a)2mm \\ v_3^{<0>} = 0 \quad \forall \quad \Omega \end{cases} \tag{31}$$

where α_4 is the rate of colloidally induced mobilization. Eq. (31) implies that the colloidal release of particles from any phase j is limited by a critical salinity C_{sc}, which is a function of the nanoparticles' type and the mineralogy of the rock formation.

3.6 Change in absolute porosity and porosity

The processes studied in the previous sections alter, to a lesser or greater extent, the physical properties of the rock formation, i.e., porosity and absolute permeability. The nanoparticles' sedimentation, adsorption, and retention (by blocking and bridging) affect the EOR process and are one of the main causes of formation damage. The porosity decreases with the nanoparticles' deposition caused, among others, by aggregation mechanisms. Ju [72, 77–79] developed a numerical model to calculate the instantaneous porosity, which is independent of the alteration by compressibility:

$$\phi_{i,j}^{<n>} = \phi_{i,j}^{<n-1>} - \sum_{i=1}^{3}\left(v_i^{<n>} - v_i^{<n-1>}\right) \tag{32}$$

The absolute permeability is also affected by these processes, and its variation is related to the parameters calculated in the previous section, according to Eq. (33):

$$K_{i,j}^{<n>} = K_{i,j}^{<0>}\left[(1 - f_K)k_f + f_K \frac{\phi_{i,j}^{<n>}}{\phi_{i,j}^{<0>}}\right]^{n_K} \tag{33}$$

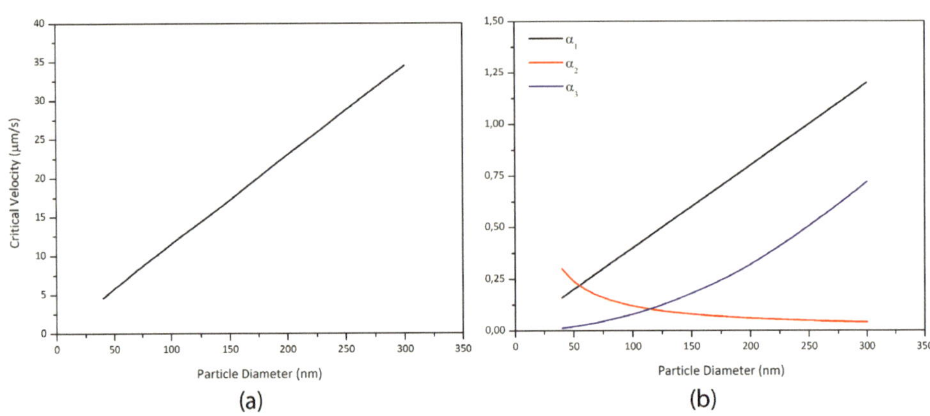

Figure 9.
Critical velocity (a) and retention parameters (b) as a function of the particle size [33].

where $K_{i,j}^{<0>}$ is the initial permeability, k_f is a constant for fluid seepage allowed by plugged pores, and n is an experimental coefficient which varies from 2.5 to 3.5. The term f_K is the fraction of original cross-sectional area or unplugged pores open to the flow:

$$f_K = 1 - \sum_j \gamma_{fi} \cdot v_2^j \tag{34}$$

where γ_{fi} is the coefficient of flow efficiency for the nanoparticles. The last part consists in describing how the nanoparticles also affect the rock's wettability by the retention/adsorption processes described previously in this chapter and, hence, the relative permeabilities. Firstly, how much area of the rock is covered by nanoparticles and how much area is available should be calculated. This is based on experimental observations which evidenced a functionality of the type of nanoparticles, their shape and size, the rock formation, its origin and initial porosity, permeability, and wettability. It is assumed for this calculation that the nanoparticles are spherical and touching each other [13]. Thus, the specific area of the particles is

$$a_{np} = \frac{Area}{Volume} = \frac{6}{Dia_{np}} \tag{35}$$

Therefore, with the total amount of nanoparticles adsorbed or entrapped and the specific area, the total surface of the porous medium covered by the particles is calculated according to Eq. (36):

$$a_{t,i,j} = \frac{6 \cdot \beta_{np}}{Dia_{np}} (v_1 + v_2 + v_3)_{i,j}^{<n>} \tag{36}$$

where β_{np} is the surface area coefficient. To calculate the total specific area, the Kozeny-Carman equation is used. This correlation originally evidenced a problem in its applicability to all rock formations and, moreover, not valid for complex poral geometries [80, 81]. Thus, Carman modified the expression adding a variable to Kozeny's equation, known as hydraulic tortuosity [82, 83]:

$$S_v = \frac{Ko \cdot \phi^3}{T^2 \cdot \sqrt{K_x^2 + K_y^2}} \tag{37}$$

where S_v is the total specific area, Ko is the Kozeny constant, and T is the tortuosity. This new variable was initially thought to be independent of the rock properties, but later it was proven to be a function of the porosity, with a minimum of 1 when the latter tended to unity, and then increasing monotonically as the porosity decreases (**Figure 10**). There are several relationships between tortuosity and porosity, for example, Eq. (38) based on the work presented by Matyka et al. [84]:

$$T = 1 - 0.77 \cdot \ln \phi \tag{38}$$

Using the results from previous equations, the fraction of the rock covered by the nanoparticles can be calculated, hence the wettability alteration (**Figure 11**). The reference condition is the relative permeability curves without nanoparticles in the system ($a_{t,i,j} = 0$). On the contrary, when the entire surface of the rock is covered by the particles, the other extreme case is obtained $a_{t,i,j}/S_v \geq 1$, and then the

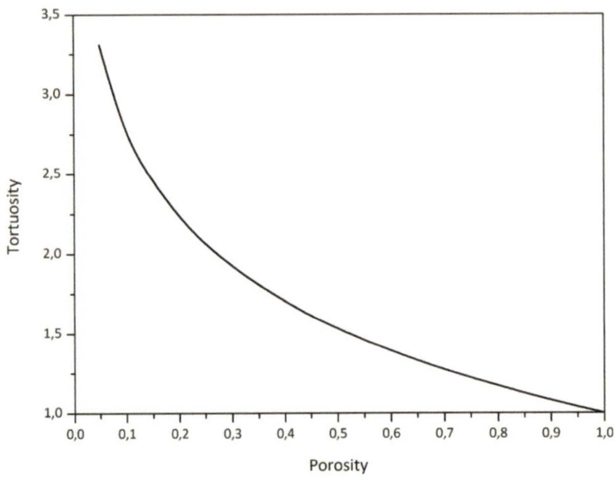

Figure 10.
Tortuosity-porosity relationship [12].

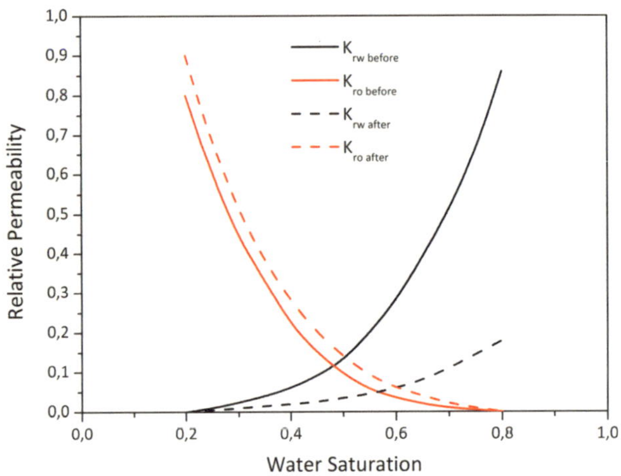

Figure 11.
Relative permeability modification based on the adsorption of nanoparticles [12].

maximum wettability change is reached, obtaining the new relative permeabilities $k_r^{a^*}$ and $k_r^{o^*}$. Between these two conditions means a partial coverage of the rock's surface ($0 < a_{t,i,j}/S_v < 1$), adopting a linear interpolation between the extreme cases to calculate the relative permeability modification (Eq. (39)) [71, 72, 78, 79, 85]:

$$k_{r,NP}^{a,o} = \min\left(1, k_r^{a,o} \cdot \left[1 + \frac{a_{t,i,j}}{S_v}(\theta_{a,o} - 1)\right]\right) \tag{39}$$

where $\theta_{a,o}$ represent the maximum achievable wettability modifications due to the nanoparticles' adsorption ($\theta_{a,o} = k_r^{a,o^*}/k_r^{a,o}$).

4. Results

The objective of this chapter, besides the development of a new CFD model for the simulation of a nanoparticle/polymer flooding process, is to present the potential advantages of combined techniques in EOR as well as the benefits of using CFD tools in order to optimize the recovery strategies in oil fields. Thus, a series of simulations were performed in a 2D oil field with a random permeability field (**Figure 12**), starting with a standard waterflooding process to be used as benchmark; followed by a traditional EOR polymer flooding using commercial, linear polymers to increase the viscosity of the sweeping phase; and finally a combined flooding with nanoparticles + branched polymers.

The recovery process was simulated during a period of 3000 days for the three different mechanisms. The results of both the recovery factor and the fractional flow as a function of time are presented in **Figure 13**.

At the beginning of the process, all the techniques render the same results, but the influence of the EOR agents is noticeable once the oil slug reaches the producing well. In the case of the linear polymer, there is a reduction of the oil in the field due to the increase in the viscosity. Furthermore, when the combined process with nanoparticles is used, these modify not only the rheology and viscoelastic properties of the polymer solution but also the wettability of the rock formation, allowing a

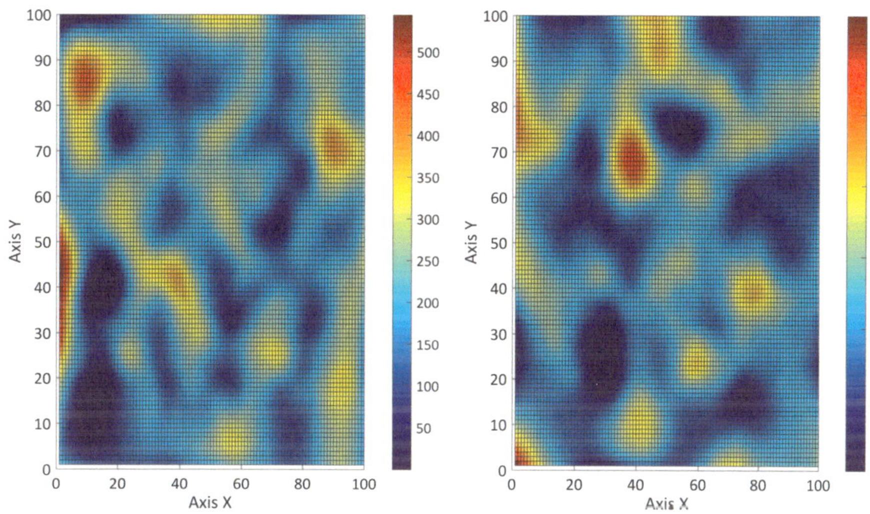

Figure 12.
Absolute permeability fields in the X (left) and Y (right) directions for the refined mesh, expressed in mD [12].

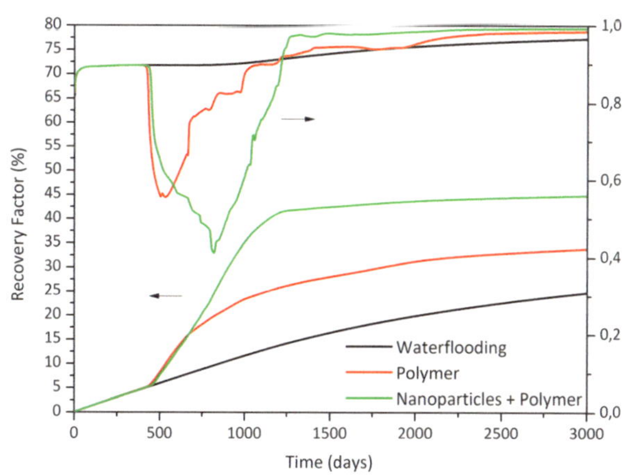

Figure 13.
Oil recovery and fractional flow as a function of time for the reference cases and the nanoparticles and polymer scheme in the refined mesh [12].

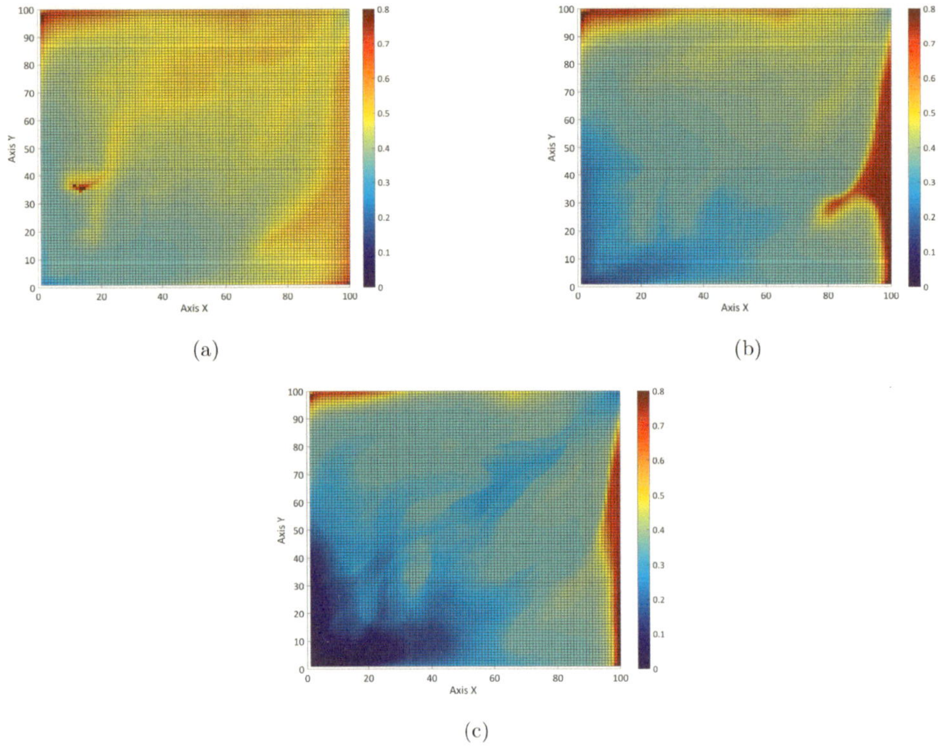

Figure 14.
Oil saturation after 3000 days for the waterflooding (a), linear polymer (b), and the nanoparticle and polymer (c) EOR flooding schemes [12].

further recovery of oil. This can be appreciated in **Figures 14** and **15**. The final oil saturation in the field decreases significantly when the nanoparticles are used, especially alongside the diagonal line connecting both wells, where the velocities reach the highest values.

The influence of the nanoparticles is more evident in the areas with low permeabilities (**Figures 14** and **15**) in which both waterflooding and linear polymer could not desaturate completely. On the other hand, the polymer + nanoparticle flooding modified the wettability of the formation, increasing the mobility of the oil phase, rendering lower residual saturations.

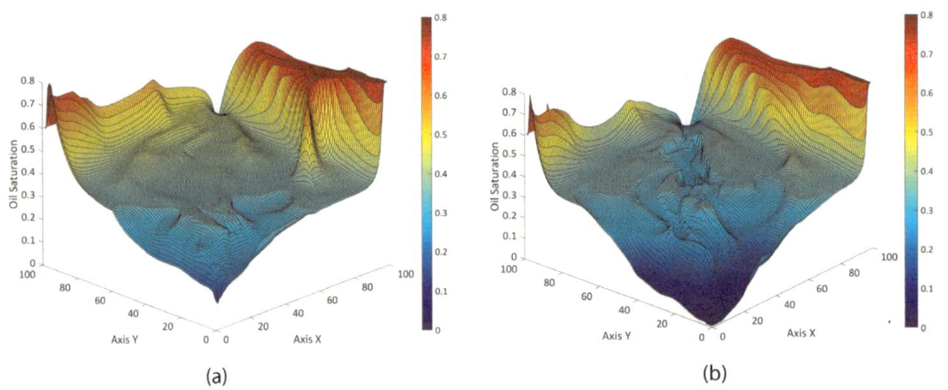

Figure 15.
Oil saturation after 3000 days for a linear polymer flooding (a) and a combined nanoparticles + polymer flooding (b) [12].

Numerically speaking, the solution of a chemical EOR process implies solving a set of highly nonlinear and coupled balance equations. In the proposed simulator, an iterative approach for each time-step of the simulation has been proposed. This rendered a pseudo-implicit scheme, guaranteeing the numerical stability of the system. This behavior can be proved using a matricial stability analysis of the system [12].

5. Conclusions

The goal of this chapter was to present the current strategies in oil recovery and how traditional techniques can be boosted by means of the nanotechnology, introducing the development of a reservoir simulation using CFD techniques. There is a need of optimizing the production of conventional oil sources while more sustainable energy resources are developed and a smooth transition between these can be carried out. One of the techniques used to evaluate the performance of these methods is reservoir simulation, a branch of engineering that emerged in recent years, used to justify and analyze the execution of E&P investments. Among EOR processes, chemical agents show a great potential in different oil fields, being mostly used in low- and medium-viscosity fields. A way to improve their efficiency is to use the nanotechnology in order boost the advantages of these chemical agents.

A novel mathematical model of porous media flow for a combined EOR/nanotechnology process is presented during this chapter, using a (hyper)branched polymer with several possible architectures, coupled with nanoparticles of different wettabilities. The mathematical model is represented by the momentum (Darcy) and mass conservation laws, using a compositional approach due to its versatility to model multiphase, multicomponent systems. There are several physical phenomena present in EOR flooding, and the combination of chemicals and nanoparticles affects some of them, studied in this chapter, presenting a set of formulas to implement these in a reservoir simulator. The polymer architecture is key factor in the oil recovery, with branched (e.g., comb/star) polymers yielding better recovery factors than linear ones. On the other hand, nanoparticle flooding increases the oil recovered by altering the rock wettability, allowing the organic phase to flow more easily. Thus, the synergy between both agents presents a great potential for its application in field tests.

All in all, nanotechnology-enhanced chemical EOR flooding could represent a novel and improved technique, considering the advantages and synergy of the agents being injected. Nanotechnology represents a breakthrough in EOR processes, and it is a perfect example of how well-developed, standard techniques can be enhanced by using the advantages of materials exhibited at the nanoscale.

Nomenclature

Ad	component adsorption $[1/\text{day}]$
c_r	rock compressibility $[1/\text{Pa}]$
$\underline{\underline{D}}$	dispersion tensor
f	number of arms (polymer)
K	absolute permeability $[\text{mD}]$
k_r	relative permeability
p	reservoir pressure $[\text{Pa}]$
p_{wf}	bottom-hole pressure $[\text{Pa}]$
q	flowrate $[\text{m}^3/\text{day}]$

r_w	well radius [m]
S	phase saturation
s	well skin factor
u	Darcy velocity [m/day]
v	nanoparticle adsorption
V	volumetric concentration
z	overall concentration

Greek letters

Γ	domain boundary
λ	phase mobility
μ	absolute viscosity [Pa s]
σ	interfacial tension [mN/m]
ϕ	formation porosity
Ω	reservoir domain

Superscripts

a	aqueous phase
c	capillary
H	water-oil system (no chemical)
j	phase
$<n>$	time-step
o	oil phase
r	residual

Subscripts

i	component
in	injection
m, n	spatial grid blocks
np	nanoparticle component
p	hydrocarbon component
pol	polymer component
s	salt component
t	total
w	water component

Author details

Pablo D. Druetta
Department of Chemical Engineering, ENTEG, University of Groningen, Groningen, The Netherlands

*Address all correspondence to: p.d.druetta@rug.nl

IntechOpen

© 2019 The Author(s). Licensee IntechOpen. This chapter is distributed under the terms of the Creative Commons Attribution License (http://creativecommons.org/licenses/by/3.0), which permits unrestricted use, distribution, and reproduction in any medium, provided the original work is properly cited.

References

[1] Owen NA, Inderwildi OR, King DA. The status of conventional world oil reserves-hype or cause for concern? Energy Policy. 2010;**38**:4743-4749

[2] Maugeri L. Oil: The next revolution discussion paper. In: Tech. Rep. 2012-10. Belfer Center for Science and International Affairs; 2012

[3] Maggio G, Cacciola G. When will oil, natural gas, and coal peak? Fuel. 2012; **98**:111-123

[4] Chapman I. The end of peak oil? Why this topic is still relevant despite recent denials. Energy Policy. 2014;**64**: 93-101

[5] Hughes L, Rudolph J. Future world oil production: Growth, plateau, or peak? Current Opinion in Environmental Sustainability. 2011;**3**:225-234

[6] Dake LP. Fundamentals of Reservoir Engineering. Amsterdam, the Netherlands: Elsevier; 1978. ISBN: 0-444-41830-X

[7] Donaldson EC, Chilingarian GV, Yen TF. Enhanced Oil Recovery, I: Fundamentals and Analyses. Amsterdam, The Netherlands: Elsevier Science; 1985. ISBN: 978-0-08086-872-1

[8] Satter A, Iqbal GM, Buchwalter JL. Practical Enhanced Reservoir Engineering. Tulsa, USA: PennWell Books; 2008. ISBN: 978-1-59370-056-0

[9] Encyclopædia Britannica Petroleum Trap. 2012. Available from: https://www.britannica.com/science/petroleum-trap

[10] Asrilhant B. A program for excellence in the management of exploration and production projects. In: Offshore Technology Conference. Houston, USA: Society of Petroleum Engineers; 2005

[11] Suslick SB, Schiozer D, Rodriguez MR. Uncertainty and risk analysis in petroleum exploration and production. Terrae. 2009;**6**:30-41

[12] Druetta P. Numerical simulation of chemical EOR processes [Ph.D. thesis]. Groningen, The Netherlands: University of Groningen; 2018

[13] Raffa P, Druetta P. Chemical Enhanced Oil Recovery: Advances in Polymer Flooding and Nanotechnology. Berlin, Germany: De Gruyter; 2019. ISBN: 978-3-11-064024-3

[14] The Royal Society & The Royal Academy of Engineering Nanoscience and Nanotechnologies: Opportunities and Uncertainties. London, UK: The Royal Society; 2004

[15] Holsapple M et al. Research strategies for safety evaluation of nanomaterials, part II: Toxicological and safety evaluation of nanomaterials, current challenges and data needs. Toxicological Sciences. 2005;**88**:12-17

[16] Balshaw D, Philbert M, Suk W. Research strategies for safety evaluation of nanomaterials, part III: Nanoscale technologies for assessing risk and improving public health. Toxicological Sciences. 2005;**88**:298-306

[17] Tsuji J et al. Research strategies for safety evaluation of nanomaterials, part IV: Risk assessment of nanoparticles. Toxicological Sciences. 2006;**89**:42-50

[18] Borm P et al. Research strategies for safety evaluation of nanomaterials, part V: Role of dissolution in biological fate and effects of nanoscale particles. Toxicological Sciences. 2006;**90**:23-32

[19] Powers K et al. Research strategies for safety evaluation of nanomaterials. part VI: Characterization of nanoscale

particles for toxicological evaluation. Toxicological Sciences. 2006;**90**:296-303

[20] Hashemi R, Nassar NN, Almao PP. Nanoparticle technology for heavy oil in-situ upgrading and recovery enhancement: Opportunities and challenges. Applied Energy. 2014;**133**: 374-387

[21] Lie K-A, Mallison BT. Mathematical models for oil reservoir simulation. In: Encyclopedia of Applied and Computational Mathematics. Vol. 1–8. Berlin, Germany: Springer-Verlag Berlin Heidelberg; 2013

[22] Fanchi JR. Principles of Applied Reservoir Simulation. Burlington, USA: Gulf Professional Publishing; 2005. ISBN: 978-0-08046-045-1

[23] Aziz K, Aziz K, Settari A. Petroleum Reservoir Simulation. Amsterdam, The Netherlands: Springer; 1979. ISBN: 978-0-85334-787-3

[24] Chen Z, Huan G, Ma Y. Computational Methods for Multiphase Flows in Porous Media. Philadelphia, USA: Society for Industrial and Applied Mathematics; 2006

[25] Bidner MS, Savioli GB. On the numerical modeling for surfactant flooding of oil reservoirs. Mecanica Computacional. 2002;**XXI**:566-585

[26] Lake LW. Enhanced Oil Recovery. Englewood Cliffs, USA: Prentice-Hall Inc.; 1989. ISBN: 0-13-281601-6

[27] Sheng J. Modern Chemical Enhanced Oil Recovery. Amsterdam, The Netherlands: Elsevier; 2011. ISBN: 978-1-85617-745-0

[28] Wasan D, Nikolov A. Spreading of nanofluids on solids. Nature. 2003;**423**: 156-159

[29] Kondiparty K, Nikolov A, Wu S, Wasan D. Wetting and spreading of nanofluids on solid surfaces driven by the structural disjoining pressure: Statics analysis and experiments. Langmuir. 2011;**27**:3324-3335

[30] Wasan D, Nikolov A, Kondiparty K. The wetting and spreading of nanofluids on solids: Role of the structural disjoining pressure. Current Opinion in Colloid & Interface Science. 2011;**16**: 344-349

[31] Nikolov A, Kondiparty K, Wasan D. Nanoparticle self-structuring in a nanofluid film spreading on a solid surface. Langmuir. 2010;**26**:7665-7670

[32] Al-Mjeni R et al. Has the time come for EOR? Oilfield Review. 2010;**2011**:4

[33] Druetta P, Picchioni F. Polymer and nanoparticles flooding as a new method for enhanced oil recovery. Journal of Petroleum Science and Engineering. 2019;**177**:479-495

[34] Barrett R et al. Templates for the Solution of Linear Systems: Building Blocks for Iterative Methods. Philadelphia, USA: Society for Industrial and Applied Mathematics; 1994

[35] Kamalyar K, Kharrat R, Nikbakht M. Numerical aspects of the convection-dispersion equation. Petroleum Science and Technology. 2014;**32**:1729-1762

[36] Duan F, Kwek D, Crivoi A. Viscosity affected by nanoparticle aggregation in Al_2O_3-water nanofluids. Nanoscale Research Letters. 2011;**6**:248

[37] Khandavalli S, Rothstein JP. Extensional rheology of shear-thickening fumed silica nanoparticles dispersed in an aqueous polyethylene oxide solution. Journal of Rheology. 2014;**58**:411-431

[38] Meyer JP, Adio SA, Sharifpur M, Nwosu PN. The viscosity of nanofluids: A review of the theoretical, empirical,

and numerical models. Heat Transfer Engineering. 2016;**37**:387-421

[39] Mishra P, Mukherjee S, Nayak S, Panda A. A brief review on viscosity of nanofluids. International Nano Letters. 2014;**4**:109-120

[40] Saito Y, Hirose Y, Otsubo Y. Size effect on the rheological behavior of nanoparticle suspensions in associating polymer solutions. Colloid and Polymer Science. 2012;**290**:251-259

[41] Graessley W, Masuda T, Roovers J, Hadjichristidis N. Rheological properties of linear and branched polyisoprene. Macromolecules. 1976;**9**:127-141

[42] Berry G. Thermodynamic and conformational properties of polystyrene.3. dilute solution studies on branched polymers. Journal of Polymer Science Part A-2-Polymer Physics. 1971; **9**:687

[43] Graessley WW. Effect of long branches on the flow properties of polymers. Accounts of Chemical Research. 1977;**10**:332-339

[44] Phillies G. Dynamics of polymers in concentrated solutions—The universal scaling equation derived. Macromolecules. 1987;**20**:558-564

[45] Shanbhag S. Analytical rheology of polymer melts: State of the art. ISRN Materials Science. 2012;**2012**:ID732176

[46] Berry G. Thermodynamic and conformational properties of polystyrene. 2. Intrinsic viscosity studies on dilute solutions of linear polystyrenes. Journal of Chemical Physics. 1967;**46**:1338

[47] Wever DAZ, Polgar LM, Stuart MCA, Picchioni F, Broekhuis AA. Polymer molecular architecture as a tool for controlling the rheological properties of aqueous polyacrylamide solutions for enhanced oil recovery. Industrial & Engineering Chemistry Research. 2013; **52**:16993-17005

[48] Delshad M, Pope G, Sepehrnoori K. UTCHEM version 9.0 Technical Documentation. The University of Texas at Austin, Austin, USA: Center for Petroleum and Geosystems Engineering; 2000

[49] Litchfield DW, Baird DG. The rheology of high aspect ratio nano-particle filled liquids. Rheology Reviews. 2006;**2006**:1

[50] Maurya NK, Mandal A. Studies on behavior of suspension of silica nanoparticle in aqueous polyacrylamide solution for application in enhanced oil recovery. Petroleum Science and Technology. 2016;**34**:429-436

[51] Mikkola V. Impact of concentration, particle size and thermal conductivity on effective convective heat transfer of nanofluids [MS thesis]. Aalto, Finland: Aalto University; 2015

[52] Zhu D, Wei L, Wang B, Feng Y. Aqueous hybrids of silica nanoparticles and hydrophobically associating hydrolyzed polyacrylamide used for EOR in high-temperature and high-salinity reservoirs. Energies. 2014;**7**: 3858-3871

[53] Choi SK, Son HA, Kim HT, Kim JW. Nanofluid enhanced oil recovery using hydrophobically associative zwitterionic polymer-coated silica nanoparticles. Energy & Fuels. 2017;**31**:7777-7782

[54] Anne-Archard D, d'Olce M, Tourbin M, Frances C. Aggregation of silica nanoparticles in concentrated suspensions under turbulent, shear and extensional flows. Chemical Engineering Science. 2013;**95**:184-193

[55] Berret J, Yokota K, Morvan M. Interactions between polymers and nanoparticles: Formation of

"supermicellar" hybrid aggregates. Soft Materials. 2004;**2**:71-84

[56] Brunelli A, Pojana G, Callegaro S, Marcomini A. Agglomeration and sedimentation of titanium dioxide nanoparticles (n-TiO_2) in synthetic and real waters. Journal of Nanoparticle Research. 2013;**15**:1684

[57] Pranami G. Understanding nanoparticle aggregation [Ph.D. thesis]. Ames, USA: Iowa State University; 2009

[58] Capco DG, Chen Y. Nanomaterial: Impacts on Cell Biology and Medicine. Dordrecht, The Netherlands: Springer; 2014

[59] Jiang W, Ding G, Peng H, Hu H. Modeling of nanoparticles' aggregation and sedimentation in nanofluid. Current Applied Physics. 2010;**10**:934-941

[60] Kang H, Zhang Y, Yang M, Li L. Molecular dynamics simulation on effect of nanoparticle aggregation on transport properties of a nanofluid. Journal of Nanotechnology in Engineering and Medicine. 2012;**3**: 021001

[61] Li LS. Effects of nanoparticle aggregation, particle size and temperature of nanofluids using molecular dynamics simulation [Ph.D. thesis]. Kuala Lumpur, Malaysia: University of Malaya; 2016

[62] Markus AA, Parsons JR, Roex EWM, de Voogt P, Laane RWPM. Modeling aggregation and sedimentation of nanoparticles in the aquatic environment. Science of the Total Environment. 2015;**506**:323-329

[63] Markutsya S. Modeling and simulation of nanoparticle aggregation in colloidal systems [Ph.D. thesis]. Ames, USA: Iowa State University; 2008

[64] Kohli I. Dynamics of gold nanoparticles in synthetic and biopolymer solutions [Ph.D. thesis]. Detroit, USA: Wayne State University; 2013

[65] Cardellini A, Fasano M, Bigdeli MB, Chiavazzo E, Asinari P. Thermal transport phenomena in nanoparticle suspensions. Journal of Physics-Condensed Matter. 2016;**28**:483003

[66] Sbai MA, Azaroual M. Numerical modeling of formation damage by two-phase particulate transport processes during Co_2 injection in deep heterogeneous porous media. Advances in Water Resources. 2011;**34**:62-82

[67] Omari RA, Aneese AM, Grabowski CA, Mukhopadhyay A. Diffusion of nanoparticles in semidilute and entangled polymer solutions. Journal of Physical Chemistry B. 2009; **113**:8449-8452

[68] Dong Y, Feng X, Zhao N, Hou Z. Diffusion of nanoparticles in semidilute polymer solutions: A mode-coupling theory study. Journal of Chemical Physics. 2015;**143**:024903

[69] Kohli I, Mukhopadhyay A. Diffusion of nanoparticles in semidilute polymer solutions: Effect of different length scales. Macromolecules. 2012;**45**: 6143-6149

[70] xian Li S, jun Jiang H, huai Hou Z. Diffusion of nanoparticles in semidilute polymer solutions: A multiparticle collision dynamics study. Chinese Journal of Chemical Physics. 2016;**29**: 549-556

[71] Ju B, Fan T. Experimental study on nanoparticles transport and its effects on two-phase flow behavior in porous networks. Particulate Science and Technology. 2013;**31**:114-118

[72] Ju B, Fan T. Experimental study and mathematical model of nanoparticle transport in porous media. Powder Technology. 2009;**192**:195-202

[73] Taborda EA, Franco CA, Ruiz MA, Alvarado V, Cortes FB. Experimental and theoretical study of viscosity reduction in heavy crude oils by addition of nanoparticles. Energy & Fuels. 2017;**31**:1329-1338

[74] Zhang T et al. Investigation of nanoparticle adsorption during transport in porous media. SPE Journal. 2015;**20**:667-677

[75] Metin C. Characterization of nanoparticle transport in flow through permeable media [Ph.D. thesis]. Austin, USA: University of Texas at Austin; 2012

[76] Ju B, Fan T, Ma M. Enhanced oil recovery by flooding with hydrophilic nanoparticles. China Particuology. 2006;**4**:41-46

[77] El-Amin MF, Kou J, Sun S, Salama A. An iterative implicit scheme for nanoparticles transport with two-phase flow in porous media. Procedia Computer Science. 2016;**80**:1344-1353

[78] El-Amin MF, Meftah R, Salama A, Sun S. Numerical treatment of two-phase flow in porous media including specific interfacial area. Procedia Computer Science. 2015;**51**:1249-1258

[79] El-Amin MF, Sun S, Salama A. Enhanced oil recovery by nanoparticles injection: Modeling and simulation. In: SPE Middle East Oil and Gas Show and Conference. Society of Petroleum Engineers; 2013

[80] Allen R, Sun S. Investigating the role of tortuosity in the Kozeny-Carman equation. In: International Conference on Numerical and Mathematical Modeling of Flow and Transport in Porous Media; Dubrovnik, Croatia. Vol. 29. 2014

[81] Ozgumus T, Mobedi M, Ozkol U. Determination of Kozeny constant based on porosity and pore to throat size ratio in porous medium with rectangular rods. Engineering Applications of Computational Fluid Mechanics. 2014;**8**:308-318

[82] Alpak FO, Lake LW, Embid SM. Validation of a modified Carman-Kozeny equation to model two-phase relative permeabilities. In: SPE Annual Technical Conference and Exhibition. Houston, USA: Society of Petroleum Engineers; 1999

[83] Jacquey AB, Cacace M, Bloecher G, Watanabe N, Scheck-Wenderoth M. Hydro-mechanical evolution of transport properties in porous media: Constraints for numerical simulations. Transport in Porous Media. 2015;**110**: 409-428

[84] Matyka M, Khalili A, Koza Z. Tortuosity-porosity relation in porous media flow. Physical Review E. 2008;**78**: 026306

[85] Safari M, Golsefatan A, Rezaei A, Jamialahmadi M. Simulation of silica nanoparticle flooding for enhancing oil recovery. Petroleum Science and Technology. 2015;**33**:152-158

Section 3

CFD Application in Mechanical Engineering

Chapter 8

Development of the Flight Dynamic Model (FDM) Using Computational Fluid Dynamic (CFD) Simulations for an Unknown Aircraft

Adil Loya, Shoaib Arif, Muhammad Arsalan, Siraj Anis, Arsalan Khan, Muhammad Saad Saeed and Abdul Hameed Siddiqui

Abstract

The usage of computational fluid dynamics (CFD) has enhanced 10-fold since the last decade, especially in the area of aerospace science. In this chapter, we will focus on determining the feasibility and validity of CFD results that are plugged in flight dynamic model (FDM) to that of actual flight of an aircraft. Flight data of an actual aircraft is used to determine the aerodynamic performance of the designed FDM. In addition to this, FDM consist of various systems integration of an aircraft; however, this study will focus on aerodynamic parameter optimization. Relative analysis is carried out to validate the FDM. This will enable readers to know how CFD can be a great tool for designing FDM of an unknown aircraft.

Keywords: flight dynamic modeling, computational fluid dynamic, aerodynamics, aircraft, FlightGear

1. Introduction

Stable flight dynamic modeling and designing of an aircraft is a crucial phase faced by the aviation industry. From these perspective, when it comes to on-ground training of pilots, simulative training is required before they can face actual dynamics of real flight. Smooth and stable flight is a necessary fact as the concerned pilot is not always alone, and also high risks are involved if the pilot is untrained with regard to aircraft dynamics. Therefore, ground training on flight simulators is given to pilots for a particular aircraft. However, for the particular aircraft to run on flight simulators, initially its flight dynamic model (FDM) is designed.

1.1 Why one should design FDM?

It's a question of interest, why one should design FDM?, because the FDM is a heart for flight simulator [1]. The current dire need of flight simulators is captured

IntechOpen

from the high rate of hiring of new pilots and indulgence of the airlines whether it is in Pakistan or international forum. As the hiring and training of new pilots is expensive, for this reason internationally, ground training with respect to particular aircraft is being catered by flight simulators. For this reason, high-fidelity FDM is a concerning aspect. In line with the above, it is significantly necessary to train new pilots from a realistic approach, keeping in mind the existing piloting reviews from old fellows of that aircraft to adopt for a dynamically changing environment whether that be in terms of standard operating procedures (SOPs) of flight, systems, or navigation [2]. Moreover, the need of designing a realistic FDM for flight simulators can also reduce the amount of actual flight time pilots put on aircrafts by which fuel and CO_2 emission can be saved. This effort will inline flight organizations to act according to the ICAO Programme of Action on International Aviation and Climate Change, which enforces the ways to save material cost, i.e., fuel economy, and protect the environment which are the key concerns [1] that can be a driving factor for designing flight simulators.

1.2 What actually is FDM?

The era of FDM has changed the pathway of flight simulations. FDM consist description of flight model of a certain aircraft, which are the propulsion, navigation, controls, avionics, and aerodynamic data. The major part of this FDM is the aerodynamic data, as it handles the attitude behavior of the aircraft during flight. This aerodynamic data is acquired from either wind tunnel testing or computational fluid dynamics (CFD). Acquiring from wind tunnel is a difficult task as it is time-consuming and requires scaling down of the design, as it is hard to test actual size model since they are bigger than the wind tunnel test sections. Therefore, CFD is preferred [2] (**Figure 1**).

1.3 CFD as a visual and graphical quantification tool

CFD is a tool for testing and quantifying fluid dynamics over an internal or external body where fluid flow is involved. CFD saves material cost and manufacturing time for analyzing aerodynamics of an initial design concept that has been created [3]. Aerodynamics is a study of airflow in either the internal or

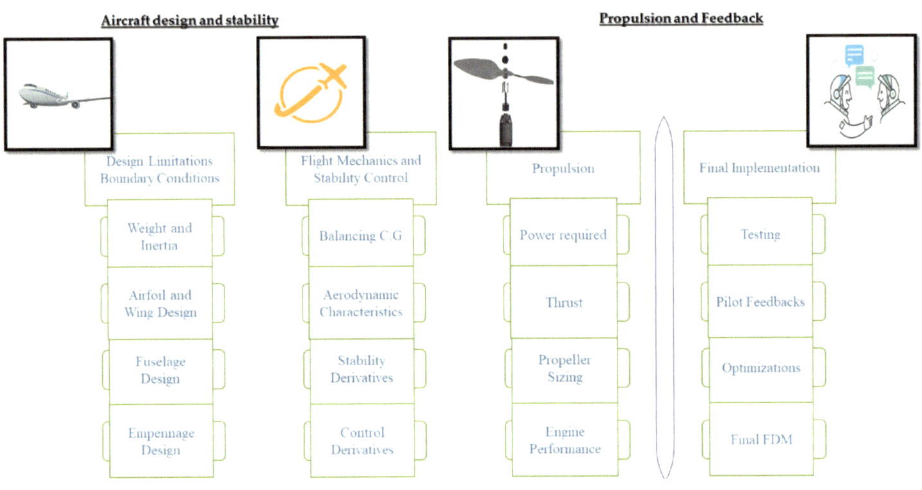

Figure 1.
Flowchart of FDM.

external side of the body. The aerodynamics concerned in this chapter is related to the external flow over an unknown aircraft body. CFD helps to simulate actual size of computer-aided design (CAD) model, in an enclosed control volume. Any property in the control volume is controlled using Reynolds transport theorem as shown in Eq. (1), and further this approach is applied on velocities over three-dimensional space and time using Eulerian technique as shown in Eq. (2):

$$\frac{DB_{sys}}{Dt} = \frac{\partial}{\partial t}\int_{cv} \rho b d\llap{-}V + \int_{cs} \rho b \mathbf{V} \cdot \hat{\mathbf{n}} dA \tag{1}$$

$$\frac{Du}{Dt} = u\frac{\partial u}{\partial x} + v\frac{\partial u}{\partial y} + w\frac{\partial u}{\partial z} + \frac{\partial u}{\partial t} \tag{2}$$

Partial differential equations are used for describing system of fluids (i.e., gasses and liquids) that are represented by the general laws of conservation of mass, momentum, and energy [4].

The principle of mass balance is used in light of law of conservation of mass for fluid element, and it is written in Eq. (3) [5].

$$\frac{\partial \rho}{dt} + \nabla \times (\rho \times v) = 0 \tag{3}$$

where $\frac{\partial \rho}{dt}$ with derivation of density with time change and $\nabla \times (\rho \times v)$ is time rate of change of volume of moving fluid.

The momentum equations in the x-, y-, and z-axes, respectively, are expressed in Eq. (4).

$$\begin{aligned}
\frac{\partial(\rho u)}{\partial t} + \nabla \times (\rho u V) &= -\frac{\partial p}{\partial x} + \frac{\partial \tau_{xx}}{\partial x} + \frac{\partial \tau_{yx}}{\partial y} + \frac{\partial \tau_{zx}}{\partial z} + \rho f_x \\
\frac{\partial(\rho v)}{\partial t} + \nabla \times (\rho v V) &= -\frac{\partial p}{\partial y} + \frac{\partial \tau_{xy}}{\partial x} + \frac{\partial \tau_{yy}}{\partial y} + \frac{\partial \tau_{zy}}{\partial z} + \rho f_y \\
\frac{\partial(\rho w)}{\partial t} + \nabla \times (\rho w V) &= -\frac{\partial p}{\partial z} + \frac{\partial \tau_{xz}}{\partial x} + \frac{\partial \tau_{yz}}{\partial y} + \frac{\partial \tau_{zz}}{\partial z} + \rho f_z
\end{aligned} \tag{4}$$

The energy equations used were derived from Navier-Stokes which depends upon the first law of thermodynamics [6]. The derivation for conservation of energy on a finite fluid element consists of a single equation which is expressed in Eq. (5).

$$\begin{aligned}
&\frac{\partial}{\partial t}\left[\rho\left(e + \frac{V^2}{2}\right)\right] + \nabla \times \left[\rho\left(e + \frac{V^2}{2}\right)V\right] = \rho\dot{q} \\
&+ \frac{\partial}{\partial x}\left(\mathrm{k}\frac{\partial T}{\partial x}\right) + \frac{\partial}{\partial y}\left(\mathrm{k}\frac{\partial T}{\partial y}\right) + \frac{\partial}{\partial z}\left(\mathrm{k}\frac{\partial T}{\partial z}\right) - \frac{\partial(up)}{\partial x} \\
&- \frac{\partial(vp)}{\partial y} - \frac{\partial(wp)}{\partial z} + \frac{\partial(u\tau_{xx})}{\partial x} + \frac{\partial(u\tau_{yx})}{\partial y} + \frac{\partial(u\tau_{zx})}{\partial z} \\
&+ \frac{\partial(v\tau_{xy})}{\partial x} + \frac{\partial(v\tau_{yy})}{\partial y} + \frac{\partial(v\tau_{zy})}{\partial z} + \frac{\partial(w\tau_{xz})}{\partial x} + \frac{\partial(w\tau_{yz})}{\partial y} \\
&+ \frac{\partial(w\tau_{zz})}{\partial z} + \rho f \times V
\end{aligned} \tag{5}$$

CFD is a cost-effective and easy to use method which empowers engineers to virtually simulate and visualize the experiments carried out using wind tunnels. As far as the visualization of flow is concerned, CFD helps in depicting pattern of the fluid flow, which is difficult with regular wind tunnel experiments. However, wind tunnel experiments are expensive to conduct, and their real flow characteristics are hard to analyze due to the limitation of size of the test section for which scaling down of the geometry is required. Moreover, to determine the forces and moments in a wind tunnel, several pressure orifices are required and mounted over the model of interest to determine the pressure distribution on the surface of the model [7], due to which it is hard to set up the experiment as compared to CFD. Now keeping in mind our application, i.e., related to aerospace industry, Menter's shear stress transport (SST) model initially developed by F.R. Menter in 1994 is suitable. Moreover, according to the F.R. Menter study [8], it is noted that the SST Kω model outperforms and predicts the reduction of kinematic eddy viscosity quantity due to the adverse pressure gradient profiles in very good agreement for all x-station of a flat plate with that of the experiments. Moreover, in his study, SST Kω model is capable of predicting the accurate velocity profile charts as acquired from experimental study. In addition to this, SST Kω solves two equations, viz., turbulent kinetic energy "k" and the eddy dissipation rate "ω" which are given in Eqs. (6) and (7), where the variables are in italic and constants are in non-italic format:

$$\frac{\partial(\rho \mathrm{k})}{\partial t}+\frac{\partial(\rho u_j \mathrm{k})}{\partial x_j}=\rho P-\upbeta^*\rho\omega \mathrm{k}+\frac{\partial}{\partial x_j}\left[(\mu+\upsigma_{\mathrm{k}}\mu_t)\frac{\partial \mathrm{k}}{\partial x_j}\right] \tag{6}$$

$$\frac{\partial(\rho\omega)}{\partial t}+\frac{\partial(\rho u_j\omega)}{\partial x_j}=\frac{\upgamma}{v_t}P-\upbeta^*\rho\omega^2+\frac{\partial}{\partial x_j}\left[(\mu+\upsigma_{\omega}\mu_t)\frac{\partial \mathrm{k}}{\partial x_j}\right]+2(1-\mathrm{F}_1)\frac{\partial \mathrm{k}}{\partial x_j}\frac{\partial\omega}{\partial x_j}\frac{\rho\omega^2}{\omega} \tag{7}$$

Variables in Eqs. (6) and (7) are as follows: i.e., ρ is the density; t is the time; u_j and x_j are the velocity and position vectors, respectively; μ_t is the eddy viscosity; ν_t is the kinematic eddy viscosity; γ is the intermittency factor; F_1 is a blending function; and β^*, σ_t, σ_ω are constants. It is basically a combination of the $K-\varepsilon$ model in the freestream and $K-\omega$ model near the walls of the geometry and is well suited for external aerodynamic flows around complex geometries and highly separated flows like airfoils at high angles of attack. In this study Siemens STAR-CCM+ software is used to carry out CFD analysis. This software includes numerous fluid dynamic models that are widely used in industry-level simulation requirements. Moreover, in STAR-CCM+ different wall treatment methods like all y+, low y+, and high y+ for treating boundary layer formation can be incorporated with the SST Kω model for true shear stress depiction.

For conducting CFD analysis, numerous test settings are permutated for gathering the aerodynamic data. However, out of different test scenarios, significance is given to cruise profile for which assumptions that are adopted for constraining our simulation are as follows:

Altitude: 1000 m
Air density: 0.9075 kg/m^3
Air viscosity: 1.581 m^2/s
Velocity: 115 m/s ~ 220 knots
Attitude cases: 372

Table 1 shows that the total number of cases conducted for CFD simulations were 372. In **Table 1**, "All" in first column first row means all control surfaces at zero deflection level. "Elevator" in first column second row shows the number of

Type	AoA	Beta	Flaps		
All	11	8	16		
Elevator	88				
Rudder	88	72			
Aileron	88				
Subtotal	275	80	16	**Total**	372

Table 1.
Total number of cases for CFD simulation.

elevator deflection cases conducted. "Rudder" in first column third row shows the number of rudder deflection cases conducted, and finally "Aileron" in first column fourth row shows the number of aileron cases conducted. These cases were carried out with properties already mentioned before in **Table 1**.

2. CFD setup

Generic settings required for setting up any CFD requires the three basic processes:

- Preprocessing
- Processing
- Post-processing

2.1 Preprocessing

The object of the study was C-130, and its CAD model was acquired from FlightGear database, which is an open-source platform. Preprocessing involves CAD import, generating and optimizing mesh using various techniques, physics, and environmental settings. The CAD model is imported using inbuilt feature which only supports listed file formats. In this study ".stl" format file is used from CAD software.

Note: It is worth noting here that STAR-CCM+ requires considerable computer hardware resources to work in a faster pace. Loading times, mesh generation, and simulation times are significantly reduced with improvement in hardware. It has been tested by running same simulations at different desktop configuration machines and noticed significant reduction with respect to elapsed time.

After successful importing, the CAD model is visible in current scene. The next step is to generate mesh. "STAR-CCM+ has all-around mesh generating feature that creates unstructured form fitted finite volume meshes of fluid and solid domains. Software is designed such that mesh generation is automatically informed by the surface tessellation and CAD elements defining the geometry, such as local curvature, surface proximity, and retained feature elements, and is further controlled by user-specified meshing parameters. The latter are organized into a hierarchy of global specifications and local refinements that enables precise control to achieve cell quality metrics, such as skewness, connectivity, conformity, near-wall cell properties, and growth rate with the smallest practical mesh even for exceptionally intricate geometries. STAR-CCM+ also employs a face-based solver technology uniquely designed to recognize arbitrary polyhedral cell topology [9]."

For meshing user-specified parameters that were assigned are shown in **Table 2**, these were values adopted for a four-engine turboprop transporter aircraft.

Tables 2 and **3** values were evaluated after repetitive attempts to achieve a well-defined and fine mesh at zero angle of attack steady flight. After running meshing method, software generated approximately 12 million cells in the mesh. The process of meshing was repeated with mild tweaks in values for every case of angle of attack and control surface deflections. For each case of flight profile, the number of cells in a mesh increases with changes in angle of attack, sideslip, and various configurations of control surfaces.

2.2 Processing

At this stage, the preprocessed settings are evaluated or computed using the solver which is tailored to our specific requirements with the options selected earlier. The processing depends on the stopping criteria for resolution of different number of iterations. Once the correct models and settings are chosen for physics

Continua meshing parameters	
Base size	1.0 m
Number of prism layers	12
Size type	Relative to base
Percentage of base	33.33%
Absolute size	0.3333 m
#Pts/circle	40.0
Curvature deviation distance	0.01 m
Thickness of near-wall prism layer	0.008 m

Table 2.
Continua meshing parameters for STAR-CCM+.

Aircraft body region parameters	
Number of prism layers	15
Size type	Relative to base
Percentage of base	25%
Absolute size	0.25 m
#Pts/circle	42.0
Curvature deviation distance	0.01 m
Relative minimum size: percentage of base	5%
Relative minimum size: absolute size	0.05 m
Relative target size: percentage of base	~6.0%
Relative target size: absolute size	0.08 m
Thickness of near-wall prism layer	0.006 m

Table 3.
Specific mesh optimization for aircraft body region parameters.

conditions, then the simulations are processed for examination of the applied conditions. Here, the air properties, other physical conditions, result extraction for aerodynamic coefficients, and graphical depiction of iterated data were selected. Physical condition and fluid dynamics models implemented are listed below:

- All y+ wall treatment
- Constant density
- Gas
- Gradients
- K-Omega turbulence
- Proximity interpolation
- Reynolds-averaged Navier-Stokes
- Segregated flow
- Segregated fluid isothermal
- SST (Menter) K-omega
- Steady
- Three-dimensional
- Turbulent
- Wall distance

The aerodynamic coefficients generated in the final report were as follows:

$$C_l = \text{roll moment coefficient}$$

$$C_m = \text{pitch moment coefficient}$$

$$C_n = \text{yaw moment coefficient}$$

$$C_L = \text{lift coeffeicient}$$

$$C_D = \text{drag coefficient}$$

$$C_Y = \text{side force coeffeceint}$$

These parameters were set up, and each case of simulation had certain control surface deflection and angle of attack. For a four-engine turboprop transporter aircraft, 372 cases were chosen. These cases proved to be sufficient for attaining high-fidelity flight dynamic model for aircraft simulator. The air speed was kept constant at 220 knots and at an altitude of 5000 feet. Air density and viscosity were set up accordingly. After all these steps, the CFD software is ready for analysis. By simply clicking on run button on the top pane, STAR-CCM+ starts running iteration and plots the results simultaneously.

The CFD software as mentioned earlier requires sufficient computer hardware to function properly. The elapsed time for analysis was almost 6 hours for a high-end desktop configuration machine in year 2017. Compared to this, same analysis was done within 3 h on a high-end machine in year 2018 with new specifications. For better and faster results, cluster computers and supercomputers are used to run CFD simulations for acquiring tremendous amount of data sets.

2.3 Post-processing

The results attained for aircraft aerodynamics from the CFD simulations are then used by the FDM's aerodynamic module. Results obtained are of forces and moment coefficients for six different axes, i.e., drag, lift, side force, roll, pitch, and yaw axes, with deflected surfaces at different angles.

3. Mesh independency study

To acquire accuracy in attaining the CFD results, and keeping in mind the computational power, it is necessary to analyze the geometry for mesh independency study. For this the model's physical properties were defined with similar inlet velocities, and physical boundary conditions were set similar for the different cases. The model chosen was SST Kω model. Grid convergence analysis was conducted on coarse, medium, and fine mesh specifications at which C_D, C_L, and C_M were analyzed. This is conducted to determine the effect of mesh quality on CFD results. The number of cells and simulation time for three different cases was simulated. First set was conducted with 0° angle of attack with all control surfaces non-deflected. The second set was with 0° angle of attack with elevator deflected by +5°. The results collected shown in **Table 4** depict that the number of cells has a huge impact on the mesh independence and time period required for conducting simulations. The six meshes shown in **Table 4** demonstrate that mesh independency was acquired as per deviation from coarse level mesh to fine settings has achieved a level of stagnation for estimated parameters of C_L, C_D, and C_M. For the fine mesh, base size reduction for different geometrical parts was set around 6% of actual geometry.

Mesh resolution	Coarse mesh	Medium mesh	Fine mesh
At +5° elevator on 0° AoA			
Number of cells	1,810,981	2,294,045	3,628,023
CFD simulation time	1 h 30 min	2 h 02 min	2 h 15 min
Estimated C_D	9.002894e−02	9.662265e−02	9.033574e−02
Estimated C_L	−6.521853e−02	−6.139491e−02	−7.067754e−02
Estimated C_M	−6.438546e−02	−6.970677e−02	−6.398511e−02
At 0° elevator on 0° AoA			
Number of cells	1,781,326	2,314,142	3,760,216
CFD simulation time	1 h 2 min	1 h 11 min	1 h 30 min
Estimated C_D	8.790877e−02	9.168335e−02	8.853459e−02
Estimated C_L	1.435455e−03	3.072429e−03	−8.159027e−03
Estimated C_M	7.883882e−02	5.123151e−02	5.492518e−02

Table 4.
Mesh independence test on different number of meshed cells, with CFD predicted results.

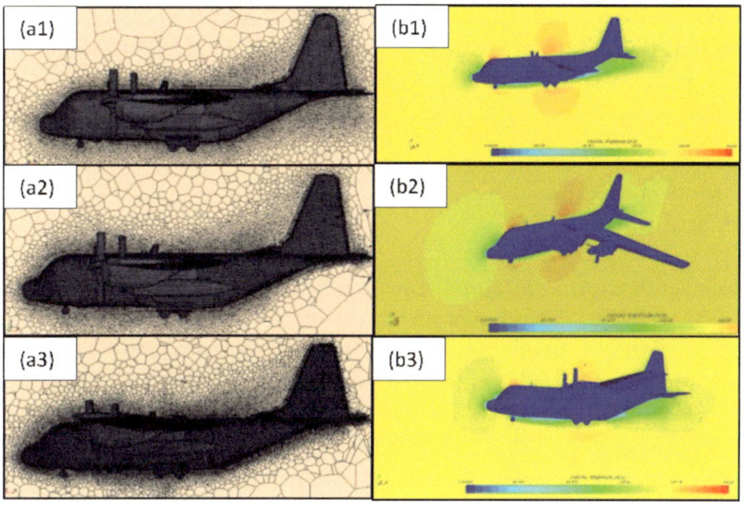

Figure 2.
Mesh independency study on a 0° AoA profile and 5° elevator deflection of the aircraft; (a1) coarse mesh, (a2) medium mesh, (a3) fine mesh, (b1) CFD output with coarse mesh, (b2) CFD output with medium mesh, (b3) CFD output with fine mesh.

Conducted CFD simulations for above mesh settings are demonstrated in **Figure 2**. It can be noted that the CFD results demonstrated significant velocity profiles and depiction of wake generation was considered but dominance was given to shearing stress, i.e., related to the near-wall stresses. Special focus was given to the surface shearing stress, as to capture the precise effect caused due to control surface deflections. Moreover, wake dominance can be optimized further by deploying more number of cells at the aft side of aircraft geometry with more computational power.

4. CFD to FDM integration

The FDM file is then processed using the FlightGear flight simulating software; this file is in .xml format. Moreover similar procedure is followed for checking under the JSBSim stand-alone module designed by Jon Berndt in 2004 [10]; however, it is embedded to the external image generation tool for visual effects. Input devices for aircraft control loading system used with JSBSim simulations were similar to the actual flight control loading system; however, with FlightGear simulation, Logitech extreme 3d edition flight joystick was used, this is further aggregated in **Table 5**.

Properties	Actual flight	FlightGear	JSBSim
Control loading system	Actual control loading system of aircraft	Logitech extreme 3d edition flight joystick	Similar to aircraft control loading system
Altitude	5000 fts	~5000 fts	4900–5300 fts
Image generation	Real	FlightGear	CIGI
Throttle condition	Max	Max	Max
Cruise velocity	210–220 knots	210–220 knots	210–220 knots

Table 5.
Control loading and basic settings set at different simulators and during actual flight mode.

After the hardware is set up and FDM .xml scripted file is tested, we are able to collect different flight data of interest for quantification of flight maneuvers. In the next section, results of responses generated for angular rates that are dependent on the aerodynamic coefficients plugged in FDM file are discussed.

5. Results and discussion

The pitch rate variation attained from flight data recorder (FDR) data, FlightGear, and JSBSim data shows similarities from the maxima and minima values. The pitch rate data of actual aircraft, i.e., PITR, from **Figure 3** depicts that during the cruise phase, aircraft pitching rate was within 1.5–1°/s. However, from the FDM perspective, i.e., for JSBSim and FG, it shows pitching motion rate maxima and minima between −2.25 to 3.5°/s and −2.25 to 1.5°/s, respectively. The depiction gives a clear understanding that the aircraft modeled using FDM is replicating the motion dynamics of the actual aircraft but with some deviation. Simulation results demonstrate greater rates than the actual aircraft because the CFD results plugged in the aircraft dynamic file are overpredicting the pitching rate of actual aircraft. Nevertheless, it does give insight to the reader that CFD can be helpful in the initial designing of FDM aerodynamic table [11]. The optimizations can be used to predict the actual behavior of the aircraft pitching rate. A note to remember for damping the motion and response dynamics of the aircraft from pitch axis, C_{mq} and $C_{m\dot{\alpha}}$ are the two variables that can be used for fine adjustments according to pilot's requirement. Before, it is necessary to have a correct initializing FDM model. The file that was being used with JSBSim-based simulator was introduced with the specific tables of C_{mq} and $C_{m\dot{\alpha}}$ for assisting for specific cases of landing and takeoff phase. This was implemented as per pilot's observations.

The roll rate variation attained from FDR, FlightGear, and JSBSim data shows high degree of resemblance from the maxima and minima values. The roll rate data of actual aircraft, i.e., ROLLR, from **Figure 4** depicts that during the cruise phase, aircraft rolling rate was within −4 to 4°/s. However, from the flight dynamic model perspective, i.e., for JSBSim and FG, it shows rolling motion rate minima and maxima between −1.25 to 6.25°/s and −2.25 to 2.15°/s, respectively. The depiction

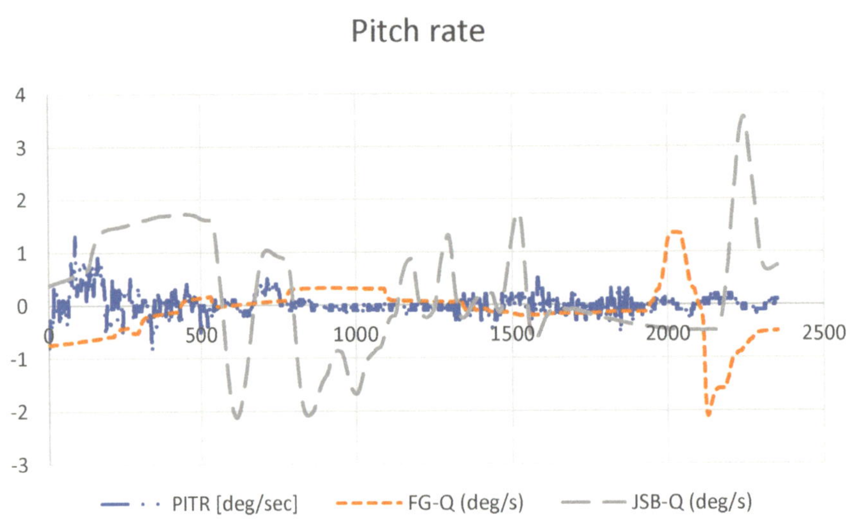

Figure 3.
Variation of pitch rate obtained from different flight data.

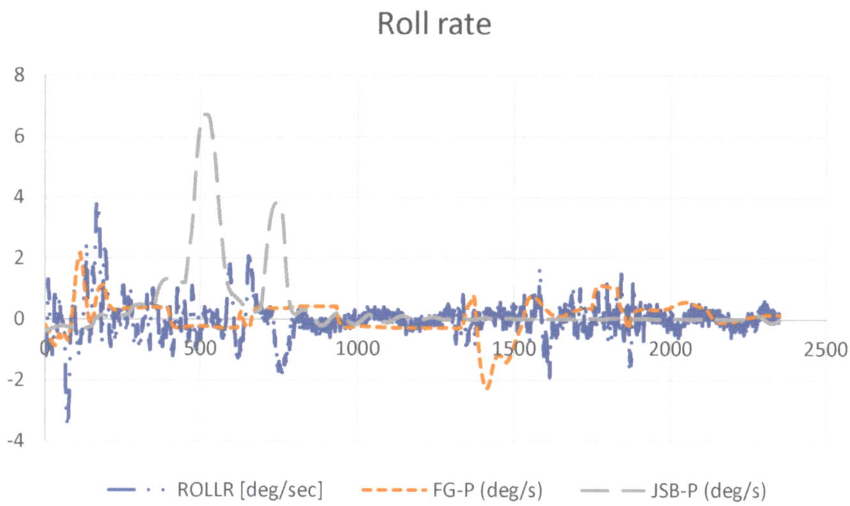

Figure 4.
Variation of roll rate obtained from different flight data.

gives a clear understanding that the aircraft modeled using FDM is replicating the motion dynamics of the actual aircraft. In addition, the simulative results of FG are in high degree of agreement to actual aircraft roll rate performance. However, JSBSim demonstrates greater roll rates, specifically in positive direction, than the actual aircraft, and this can be because the control dynamics on the JSBSim were being operated using a feedback-based control loading system; however, in an actual aircraft, relative wind component affects the feedback felt by pilot on the stick, hence adding an extra variability to aircraft control. Nevertheless, the roll rate performance was better than the pitching rate. For controlling the oscillation and damping effects in roll axis, C_{lp} (i.e., coefficient of rolling moment due to damping) and C_{lr} (i.e., effect of yaw rate on coefficient of rolling moment) are the major variables. Mostly, C_{lp} is used for altering the damping effect. The effectiveness in rolling moment of the control surface is catered by C_{lda} where da denotes the change in aileron deflection angle. In addition to C_{lda}, the FDM file was constrained in such a way that the code for right and left ailerons was separately designed for giving correct effect of aerodynamic deflections. Nevertheless, in pitch similar steps were also followed for defining aerodynamics of positive elevator separately to that of the negative elevator deflection using C_{mde} variable; however this was causing oscillatory modes at maximum deflection of positive elevators, for this two major variables were given full consideration during the pilot phase optimization which were C_{mq} and $C_{m\dot{\alpha}}$. These two variables assisted in changing the damping and oscillations caused due to pitching motion.

The yaw rate variation attained from FDR, FlightGear, and JSBSim data shows high degree of resemblance from the maxima and minima values. The yaw rate data of actual aircraft, i.e., YAWR, from **Figure 5** depicts that during the cruise phase, aircraft yaw rate was within −2 to 2°/s. However, from the flight dynamic model perspective, i.e., for JSBSim and FG, it shows yaw rate minima and maxima between −2.5 to 1.05°/s and −6.45 to 2.45°/s, respectively. The depiction gives a clear understanding that the aircraft modeled using FDM is replicating the motion dynamics of the actual aircraft. In addition, the simulative results of JSBSim are in high degree of agreement to the actual aircraft yaw rate performance. However, FG demonstrates greater yaw rates, specifically in negative direction, than the actual aircraft, and this can be because the control loading used on the FG was being

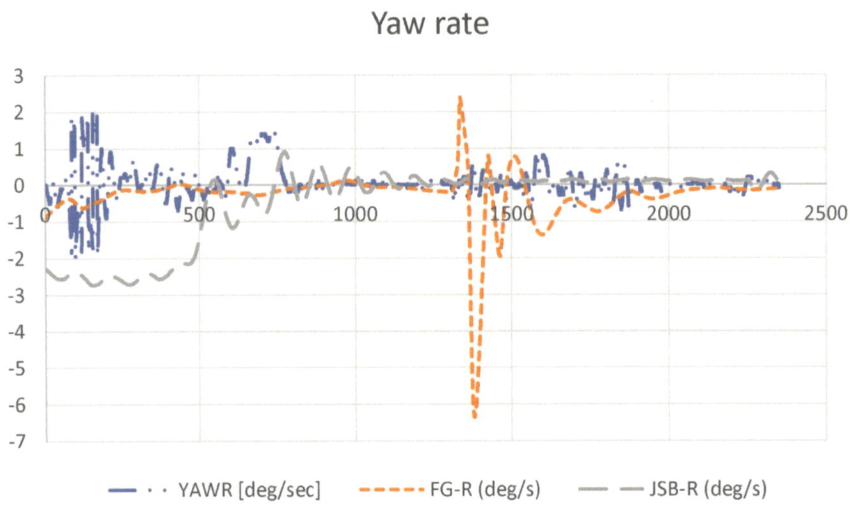

Figure 5.
Variation of yaw rate obtained from different flight data.

operated using a joystick control. Logitech extreme 3d edition flight joystick has a small moment in yawing direction, i.e., the maxima and minima is reached in just a slight deflection, causing it hard to deflect gradually, hence adding an extra difficulty during yawing moment while controlling through FlightGear. Therefore, it was found that the yaw rate demonstrated better performance using JSBSim than the FG. The stability derivatives that are used for optimizing the yaw performance of the aircraft were C_{nr}, C_{ndr}, and $C_{n\beta}$, where C_{nr} is coefficient of yaw moment due to yaw rate, C_{ndr} is the yaw moment caused due to rudder deflection, and $C_{n\beta}$ is yaw moment caused due to sideslip angle. $C_{n\beta}$ and C_{nr} were adjusted using specific values; however, for C_{ndr} table was defined as per to assist for landing and ground run effectiveness of the yaw moment due to rudder deflection.

5.1 Data quantification of the results

For the steady-state cases, the linear model with required changes according to pilot's input, satisfactory estimation of the aerodynamic response is achievable. However, if the pilot induces large-amplitude maneuvers or rapid divergences from the steady-state conditions, then nonlinear parameters need to be considered with the basic aerodynamic model. Klein and Morelli in their research state two ways of doing this: (a) using Taylor series expansion for defining nonlinear stability derivatives and (b) combining static terms and treating stability and control derivatives of aircraft as a function of explanatory variables, i.e., angle of attack, angle of sideslip, and Mach number [12]. Moreover, system identification technique is also good for validating the required results of the aerodynamic coefficients from FDR data by reverse engineering the states using observation matrix with specific input and outputs [13–16]. However, in this study as we are considering the steady-state flight dynamics, some mathematical techniques like standard deviation can be used for evaluating performance of the CFD attained variables. This quantification was carried out using Excel.

Quantification of **Figures 5–13** was conducted using Excel through which the local maxima and minima were depicted on different charts for observing the exact value from specific plots. By the help of the local maxima and minima, standard deviation (SD) was also calculated using Eq. (8):

$$\sigma = \sqrt{\frac{\sum (x_i - \mu)^2}{N}} \tag{8}$$

where σ is the population standard deviation, N is the size of the population, and x_i and μ are the population mean.

It was found that the SD lied on a discrepancy of about 1 °/s from mean values, giving us a generic idea that the results deviation was nominal as seen from **Table 6**. It is found that the max deviation is mostly found in the JSBSim-based results. However, FG results deviation is similar to the deviation of the actual aircraft. From JSBSim perspective, it can be depicted that the atmospheric model was not completely integrated to give the effects as it was found in FlightGear.

Figure 6 shows the local maxima and minima of the actual aircraft roll rate during a steady-state flight condition. The figure gives an idea of the local maxima and minima in red and green color, respectively. Using this SD was calculated to understand the difference between different operating platforms.

Figure 7 shows the local maxima and minima of the JSBSim aircraft roll rate during a steady-state flight condition. The figure gives an idea of the local maxima and minima in red and green color, respectively. Using this SD was calculated to understand the difference between different operating platforms.

Figure 8 shows the local maxima and minima of the FlightGear aircraft roll rate during a steady-state flight condition. The figure gives an idea of the local maxima and minima in red and green color, respectively. Using this SD was calculated to understand the difference between different operating platforms.

Rates	Actual (°/s)	JSB (°/s)	FG (°/s)
P	0.64380827	1.24255506	0.56483048
Q	0.19825922	1.08843922	0.49144324
R	0.45549470	1.05413173	0.73910523

Table 6.
Standard deviation of the angular rates.

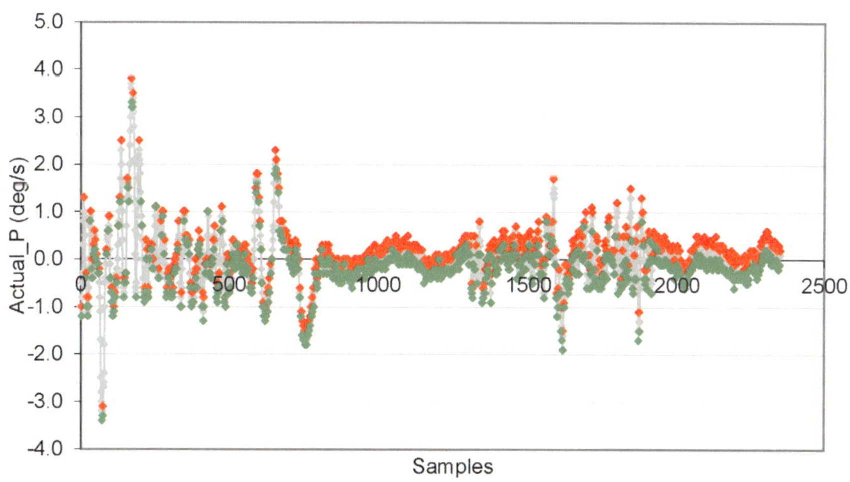

Figure 6.
Local maxima and minima on the actual (P) roll rate in °/s chart, where green are the minima and red are the maxima.

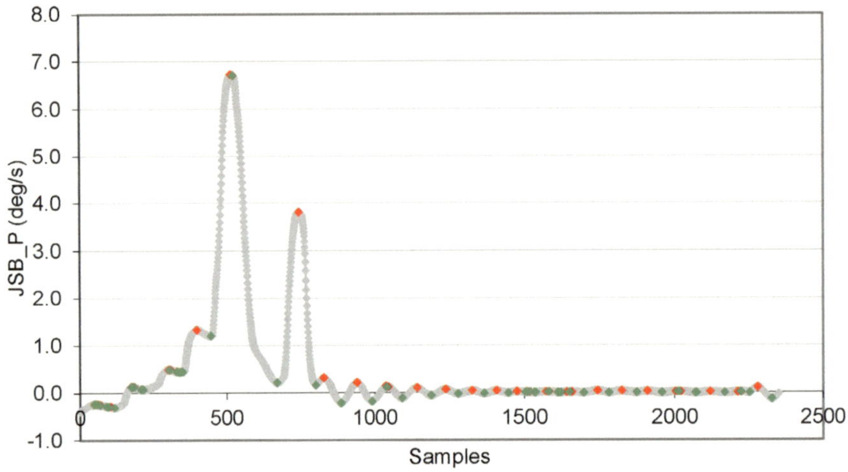

Figure 7.
Local maxima and minima on the JSB (P) roll rate in °/s chart, where green are the minima and red are the maxima.

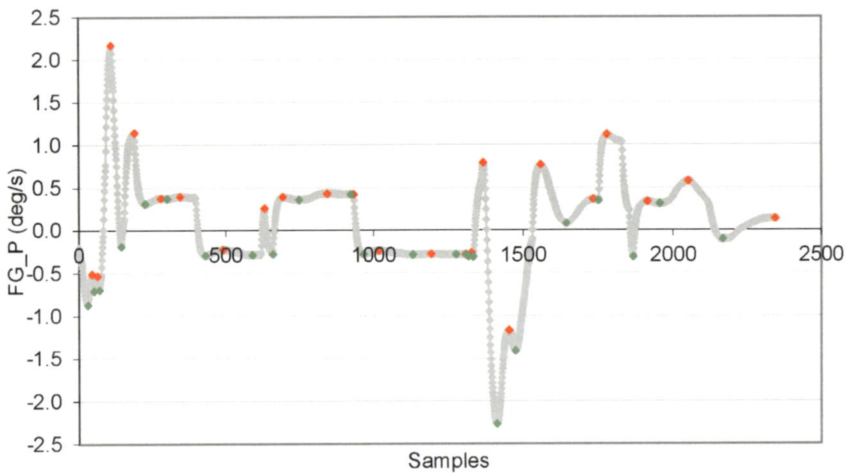

Figure 8.
Local maxima and minima on the FG (P) roll rate in °/s chart, where green are the minima and red are the maxima.

Figure 9 shows the local maxima and minima of the actual aircraft pitch rate during a steady-state flight condition. The figure gives an idea of the local maxima and minima in red and green color, respectively. Using this SD was calculated to understand the difference between different operating platforms.

Figure 10 shows the local maxima and minima of the JSBSim aircraft pitch rate during a steady-state flight condition. The figure gives an idea of the local maxima and minima in red and green color, respectively. Using this SD was calculated to understand the difference between different operating platforms.

Figure 11 shows the local maxima and minima of the FlightGear aircraft pitch rate during a steady-state flight condition. The figure gives an idea of the local maxima and minima in red and green color, respectively. Using this SD was calculated to understand the difference between different operating platforms.

Figure 12 shows the local maxima and minima of the actual aircraft yaw rate during a steady-state flight condition. The figure gives an idea of the local maxima

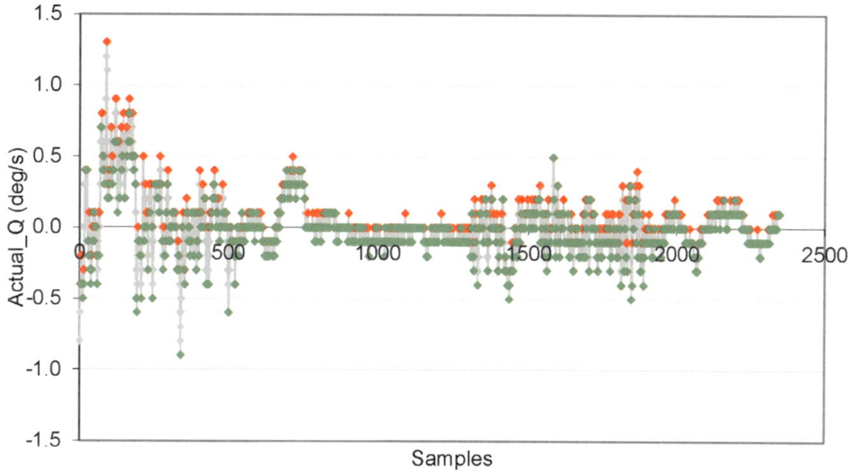

Figure 9.
Local maxima and minima on the actual (Q) pitch rate in °/s chart, where green are the minima and red are the maxima.

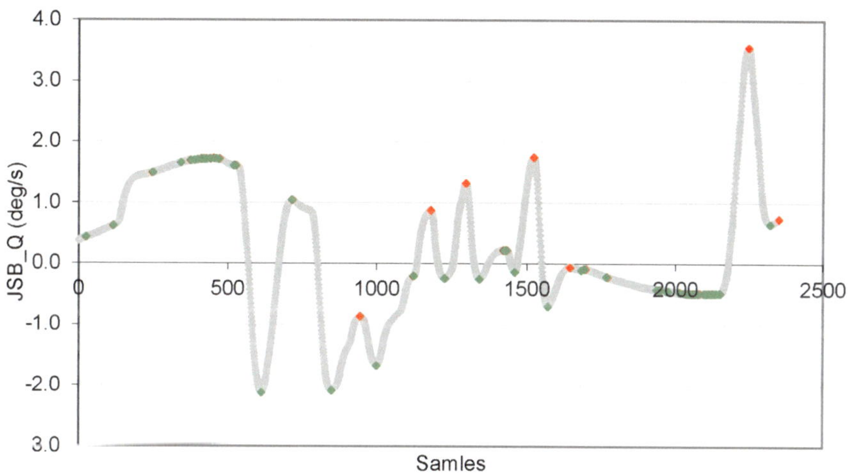

Figure 10.
Local maxima and minima on the JSB (Q) pitch rate in °/s chart, where green are the minima and red are the maxima.

and minima in red and green color, respectively. Using this SD was calculated to understand the difference between different operating platforms.

Figure 13 shows the local maxima and minima of the JSBSim aircraft yaw rate during a steady-state flight condition. The figure gives an idea of the local maxima and minima in red and green color, respectively. Using this SD was calculated to understand the difference between different operating platforms.

Figure 14 shows the local maxima and minima of the FlightGear aircraft yaw rate during a steady-state flight condition. The figure gives an idea of the local maxima and minima in red and green color, respectively. Using this SD was calculated to understand the difference between different operating platforms.

After observing the SD, standard error measure was also calculated. **Table 7** demonstrates values of the standard error measure of angular rates of actual, JSBSim, and FlightGear responses. Standard error measure shows how varied data is acquired from the actual responses. It can be seen that the response variation from

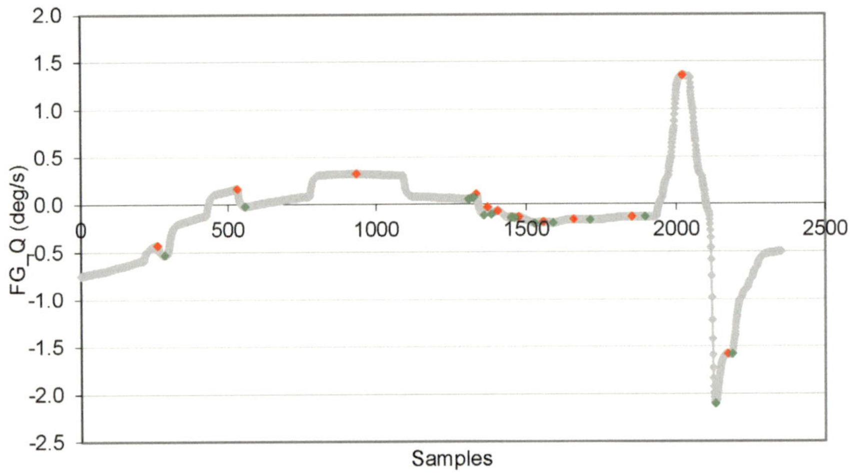

Figure 11.
Local maxima and minima on the FG (Q) pitch rate in °/s chart, where green are the minima and red are the maxima.

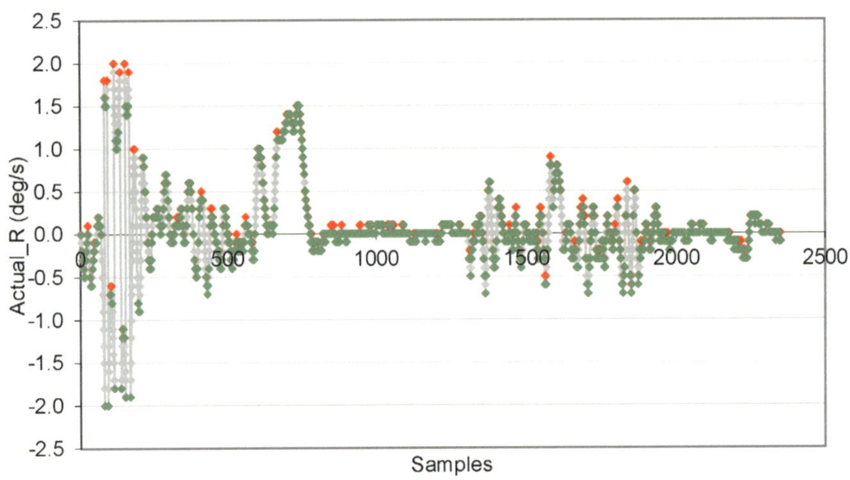

Figure 12.
Local maxima and minima on the actual (R) yaw rate in °/s chart, where green are the minima and red are the maxima.

mean is in similar constraints to each other. The standard error deviation for pitch rate (i.e., Q) is higher for JSB and FG to that of the actual model because the values chosen for the aerodynamic coefficients are generating greater pitching moment; nevertheless, here the control column perspective should not be neglected as this has also a greater impact on variability of the results.

6. Conclusions

The study has compiled a way for designing FDM tables using CFD obtained results and pilots response from FDR data. The performance of CFD-designed FDM was tested and compared with the FDR data with similar steady-state conditions as prevailed during actual flight scenario. It was found that the response

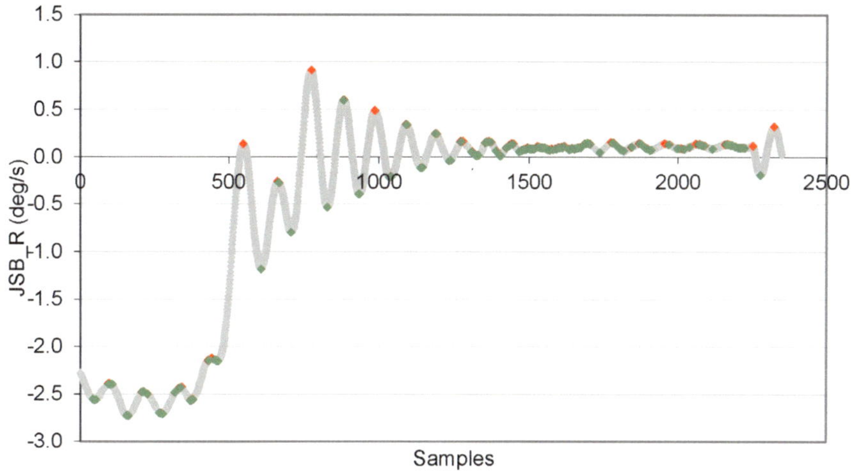

Figure 13.
Local maxima and minima on the JSB (R) yaw rate in °/s chart, where green are the minima and red are the maxima.

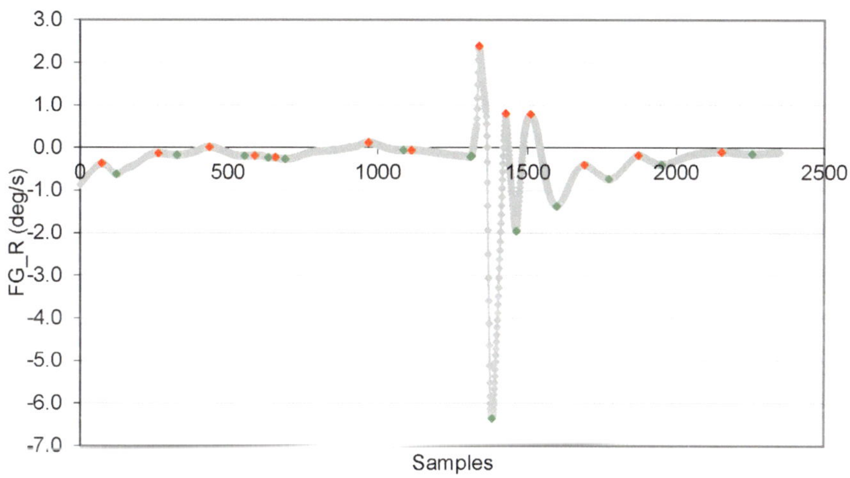

Figure 14.
Local maxima and minima on the FG (R) yaw rate in °/s chart, where green are the minima and red are the maxima.

Rates	Actual	JSB	FG
P	0.0132808	0.025632	0.0116516
Q	0.0040898	0.0224528	0.0101355
R	0.0093961	0.0217451	0.0152466

Table 7.
Standard error measure of the angular rates.

characteristics of the simulated angular rates were in correspondence with the actual rates. In addition to this, a standard deviation error measure was below 0.1 for all the rates from the mean position, giving us a confidence on the values of the aerodynamic coefficients plugged in the FDM. Moreover, 1°/s SD of the angular rates explained that the CFD data is useful for initially designing the FDM;

however, further modifications using system identification with least square method (LSM), Taylor series expansion, and filtering methods can be employed for increasing the accuracies of the results.

Acknowledgements

Thanks to PAF Karachi Institute of Economics and Technology for providing extra time for research.

Conflict of interests

There is no conflict of interest between authors, and authors are well aware of the submission.

Nomenclature

CFD	computational fluid dynamics
FDM	flight dynamic model
FDR	flight data recorder
AoA	angle of attack
FDR	flight data recorder
V	velocity vector
$\frac{DB_{sys}}{dt}$	change in system material derivative with time
$\frac{DU}{dt}$	change in forward velocity with time
$\frac{\partial\tau_{xx}}{dx}$	change in shear stress in x-direction with time
$\frac{DT}{dx}$	temperature gradient
e	symmetric stress tensor
f_x	body forces in x-direction due to gravity
f_y	body forces in x-direction due to gravity
f_z	body forces in x-direction due to gravity
$\frac{\partial u}{dx}$	velocity gradient in x-direction
$\frac{\partial p}{dx}$	pressure gradient
$\frac{\partial k}{dx}$	dissipation rate gradient in x-direction
$\frac{\partial p}{dt}$	pressure gradient with respect to time
$\frac{\partial \rho}{dt}$	derivation of density with time change
$\nabla \times (\rho \times v)$	time rate of change of volume of moving fluid
ρ	density
P	pressure
t	time
u_j and x_j	velocity and position vectors, respectively
μ_t	Eddy viscosity
ν_t	kinematic eddy viscosity
γ	intermittency factor
F_1	blending function
β^*, σ_t, σ_ω	shearing constants
$K-\varepsilon$	K-epsilon model
$K-\omega$	K-omega model
K	turbulent kinetic energy

ω	Eddy dissipation rate
u	forward velocity in x-direction
v	lateral velocity in y-direction
w	normal velocity in z-direction
p	roll rate
q	pitch rate
r	yaw rate
C_L	lift coeffeicient
C_D	drag coefficient
C_Y	side force coeffeceint
C_l	roll moment coefficient
C_{lp}	roll moment coefficient due to roll rate
C_{lr}	roll moment coefficient due to yaw rate
C_{lda}	roll moment coefficient due to aileron input
C_{ldr}	roll moment coefficient due to rudder input
$C_{l\beta}$	roll moment coefficient due to rudder input
C_{mq}	pitch moment coefficient
$C_{m\acute{\alpha}}$	pitch moment coefficient
C_{mde}	pitch moment coefficient
C_n	yaw moment coefficient
C_{np}	roll moment coefficient due to roll rate
C_{nr}	roll moment coefficient due to yaw rate
C_{nda}	roll moment coefficient due to aileron input
C_{ndr}	roll moment coefficient due to rudder input
$C_{n\beta}$	roll moment coefficient due to rudder input
$C_{y\beta}$	roll moment coefficient due to rudder input
SD	standard deviation
σ	the population standard deviation
N	the size of the population
x_i and μ	the population mean
SDE	standard deviation error

Author details

Adil Loya[1*], Shoaib Arif[2], Muhammad Arsalan[3], Siraj Anis[1], Arsalan Khan[1], Muhammad Saad Saeed[1] and Abdul Hameed Siddiqui[2]

1 PAF Karachi Institute of Economics and Technology, Karachi, Pakistan

2 National University of Science and Technology, Risalpur, Pakistan

3 Middle East Technical University, Ankara, Turkey

*Address all correspondence to: adil.loya@pafkiet.edu.pk

IntechOpen

© 2020 The Author(s). Licensee IntechOpen. This chapter is distributed under the terms of the Creative Commons Attribution License (http://creativecommons.org/licenses/by/3.0), which permits unrestricted use, distribution, and reproduction in any medium, provided the original work is properly cited.

References

[1] Gu H, Wu D, Liu H. Development of a novel low-cost flight simulator for pilot training. World Academy of Science, Engineering and Technology. 2009;**60**:685-689

[2] Longridge T, Burki-Cohen JS, Go TH, Kendra A. Simulator fidelity considerations for training and evaluation of today's airline pilots (in English). Proceedings. 2001

[3] Mason WH, Knill DL, Giunta AA, Grossman B, Watson LT, Haftka RT. Getting the full benefits of CFD in conceptual design. In: 16th AIAA Applied Aerodynamics Conference, Albuquerque, NM, Paper No. AIAA. 1998 pp. 98-2513

[4] Barnes DI. Understanding pulverised coal, biomass and waste combustion—A brief overview. Applied Thermal Engineering. 2015;**74**:89-95

[5] Anderson JD. Fundamentals of Aerodynamics. McGraw; 2009

[6] Houghton EL, Carpenter PW. Aerodynamics for Engineering Students. 6th edition. Butterworth-Heinemann; 2012. ISBN-10: 0080966322, ISBN-13: 978-0080966328

[7] Cosentino GB. CFD to flight: Some recent success stories of x-plane design to flight test at the NASA Dryden flight research center. 2007

[8] Menter F. Zonal two equation kw turbulence models for aerodynamic flows. In: 23rd Fluid Dynamics, Plasmadynamics, and Lasers Conference. 1993. p. 2906

[9] Brewster R, Carpenter C, Volpenhein E, Baglietto E, Smith J. Application of CD-adapco Best Practices to NESTOR OMEGA MVG Benchmark Exercises using STAR-CCM+. In: Proc. of NURETH-16, Chicago, IL. 2015

[10] Berndt J. JSBSim: An open source flight dynamics model in C++. In: AIAA Modeling and Simulation Technologies Conference and Exhibit. 2004. p. 4923

[11] Ghoreyshi M, Da Ronch A, Badcock KJ, Dees J, Bérard A, Rizzi A. Aerodynamic Modelling for Flight Dynamics Analysis of Conceptual Aircraft Designs. 2009

[12] Klein V, Morelli EA. Aircraft System Identification: Theory and Practice. Reston, VA: American Institute of Aeronautics and Astronautics; 2006

[13] Larsson Cahlin S. Real-Time Estimation of Aerodynamic Parameters. Doctoral Dissertation. 2016

[14] Luo Z. Lateral Post-stall Aerodynamic Modeling using Flight Test Data. Diss. Doctoral Dissertation. 2016

[15] Vitale A, Corraro F. Multi-step estimation approach for aerospace vehicle system identification from flight data [Ph.D. thesis]. University of Naples Federico II; 2013

[16] Trankle TL, Bachner SD. Identification of a nonlinear aerodynamic model of the F-14 aircraft. Journal of Guidance, Control, and Dynamics. 1995;**18**(6):1292-1297

Chapter 9

Computational Fluid Dynamic Simulation of Vertical Axis Hydrokinetic Turbines

Edwin Lenin Chica Arrieta and Ainhoa Rubio Clemente

Abstract

Hydrokinetic turbines are one of the technological alternatives to generate and supply electricity for rural communities isolated from the national electrical grid with almost zero emission. These technologies may appear suitable to convert kinetic energy of canal, river, tidal, or ocean water currents into electricity. Nevertheless, they are in an early stage of development; therefore, studying the hydrokinetic system is an active topic of academic research. In order to improve their efficiencies and understand their performance, several works focusing on both experimental and numerical studies have been reported. For the particular case of flow behavior simulation of hydrokinetic turbines with complex geometries, the use of computational fluids dynamics (CFD) nowadays is still suffering from a high computational cost and time; thus, in the first instance, the analysis of the problem is required for defining the computational domain, the mesh characteristics, and the model of turbulence to be used. In this chapter, CFD analysis of a H-Darrieus vertical axis hydrokinetic turbines is carried out for a rated power output of 0.5 kW at a designed water speed of 1.5 m/s, a tip speed ratio of 1.75, a chord length of 0.33 m, a swept area of 0.636 m^2, 3 blades, and NACA 0025 hydrofoil profile.

Keywords: renewable energy, hydrokinetic turbine, numerical simulation, power coefficient, tip-speed ratio

1. Introduction

Nowadays, both the developed and developing countries have been actively promoting renewable energy for their environmental and economic benefits such as the reduction of greenhouse gas emissions, decarbonization, and the diversification of the energy supply in order to reduce the dependence on fossil fuels, especially for the production of electricity, creating economic development and jobs in manufacturing and installation of the systems [1–4]. Several countries have developed policies and programs to promote the use of renewable energy sources. In general, the policies and programs of developing countries are limited in scope, focusing mainly on regulatory measures, with little attention to other relevant aspects such as research and development, market liberalization, information campaigns, and training [3–6].

The most common renewable power technologies include solar (photovoltaic and solar thermal systems), wind, biogas, geothermal, biomass, low-impact

IntechOpen

hydroelectricity, and emerging technologies such as wave and hydrokinetic systems. These technologies have unquestionably started to transform the global energy landscape in a promising way [7, 8].

The hydrokinetic systems convert kinetic energy of canal, river, and tidal or ocean water currents into mechanical energy and, consequently, in electrical energy without the use of large civil structures, being the energy provided by the hydrokinetic system more valuable and predictable than that supplied by wind and solar devices [9].

Even though the hydrokinetic systems are still in early stages of development, the referred technology is attractive as renewable energy sources, and it could be a response to the decentralization of energy markets due to the low operating costs, good predict ability, and minimal environmental impact associated [10]. Although the hydrokinetic turbines have relatively small-scale power production, they can be installed as multiunit arrays like wind farms to increase energy extraction. Therefore, these turbines can be used for rural electrification of isolated and river-side communities [10].

The two most common types of hydrokinetic turbines extract energy utilizing horizontal or vertical axis rotor with blade moving through the water [11]. The horizontal and vertical axis turbines have axes parallel and perpendicular to the fluid flow, respectively. Horizontal axis turbines are widely used in tidal energy converters [11, 12]. In contrast, the hydrokinetic turbines with vertical axis are generally used for small-scale power generation because they are less expensive and require lower maintenance than horizontal axis turbines [12, 13]. On the other hand, the rotor of vertical axis hydrokinetic turbines can rotate regardless of the flow direction, which constitutes an advantage. In turn, horizontal axis hydrokinetic turbines typically reach higher tip speeds, making them more prone to cavitation, which reduce the efficiency and create surface damage [11–13]. In spite of vertical axis turbine being not as efficient as the horizontal axis turbines because they exhibit a very low starting torque as well as dynamic stability problems, the interest in implementing vertical axis turbines is increased which drives further research into the development of improved turbine designs [11–13]. Gney and Kaygusu reported a detailed comparison of various types of hydrokinetic turbines and concluded that vertical axis turbines are more suitable for the cases where water flow rate is relatively limited [14].

There are two different types of vertical axis turbines: (i) those ones based on the drag force, which are included in the former group and (ii) those ones mainly based on the lift force, corresponding to the second group. Savonius turbines have a drag-type rotor and helical turbine (Gorlov turbine). In turn, Darrieus and H-shaped Darrieus turbines have a lift-type rotor since lift force is the main driving force for these kinds of machines. Generally, drag-based turbines are considered to be less efficient than their lift-based counterparts. The power coefficient of Darrieus hydrokinetic turbines is comparatively higher than that of the Savonius hydrokinetic turbines. However, Darrieus hydrokinetic turbines have poor starting characteristics [15].

Due to the high cost of the harvesting energy from water current associated with the use of these turbines, choosing a turbine with an optimum performance at the selected site is utmost importance. The hydrodynamics and performance of these turbines are governed by several parameters such as (i) the tip-speed ratio (TSR or $\lambda = R\psi/V$, which is defined as the ratio between the blade tip speed and the fluid speed (V), being R the turbine radius and ψ the rpm); (ii) the aspect ratio (AR), defined as the ratio between the hydrokinetic turbine height (H) and its diameter (D); and (iii) the solidity ($\sigma = Bc/2R$), which refers to the ratio of the product of the blade chord length (c) and the number of blades (B), the blade profiles, and the

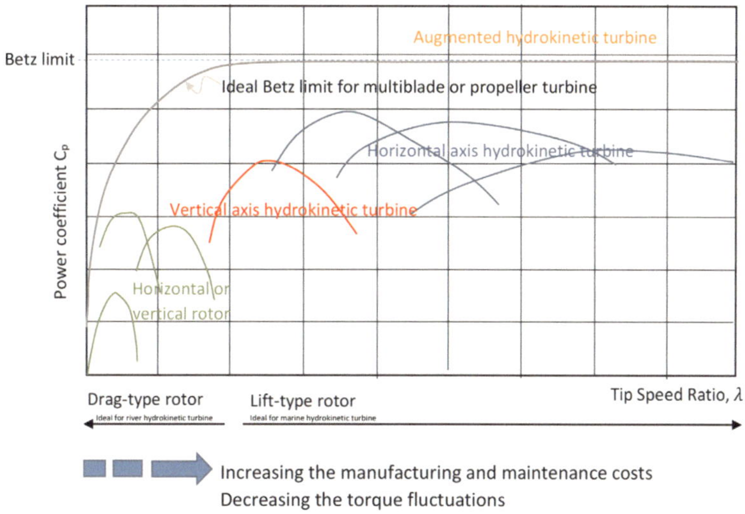

Figure 1.
Power coefficients for different rotor designs.

chord Reynolds number ($Re = \rho Vc/\mu$), where ρ and μ are the water density and viscosity, respectively. In the literature, there are several studies including experimental, numerical (computational fluid dynamics) and theoretical studies, focused on optimizing the hydrokinetic turbine geometric parameters, mainly the blade profiles, using different techniques [8, 16–23]. **Figure 1** shows the typical *power* curves for the most common types of turbine. In **Figure 1**, it can be observed that horizontal axis hydrokinetic turbines are typically more hydrodynamically efficient and operate at much higher TSR values than vertical axis hydrokinetic turbines. **Figure 1** is adapted from the behavior of wind turbines; from the best of the authors' knowledge, there is no overall graph regarding the behavior of such types of turbines considered as hydrokinetic turbines.

The objective of this chapter is to present a methodology for the efficient design and numerical simulation of a H-Darrieus vertical axis hydrokinetic turbine using CFD simulation and the commercial software ANSYS Fluent. The designed hydrokinetic turbine is expected to generate 500 W output power at 1.5 m/s water velocity. The hydrodynamic performance (power coefficient) was analyzed for several TSR values.

2. Vertical axis hydrokinetic turbine hydrodynamic models

For the design of vertical axis hydrokinetic turbine, several numerical models have been used, each of them with their own strengths and weaknesses, to accurately predict the performance of a hydrokinetic turbine depending on their configuration. The most common models are (i) blade element momentum (BEM) models, (ii) vortex models, and (iii) computational fluid dynamics models [24–32]. BEM are analytical models that combine the blade element and the momentum theory in order to study the behavior of the water flow on the blades and related forces. It is a technique that was pioneered by Glauert [24], Strickland [25], and Templin [26]. BEM models can be classified into three submodels: (a) single stream tube model, (b) multiple stream tube model, and (c) double multiple stream tube (DMS) model [28, 30], respectively, according to the increasing order of complexity.

The hydrokinetic turbine is placed inside a single streamtube, and the blade revolution is translated in an actuator disk when the single stream tube model is used. The water speed in the upstream and downstream sides of the turbine is considered constant. Additionally, the effects outside the streamtube are assumed negligible. It is highlighted that this model does not deliver good accuracy due to the assumptions required to be made. The obtained results in a number of cases provide higher estimate values. In contrast, it has fast processing time compared to other models.

A variation of the single streamtube modeling is the multiple streamtube modeling, which divides the single streamtube in several parallel adjacent streamtubes that are independent from each other and have their own undisrupted, wake, and induced velocities. Although this model is not accurate, the predicted performance is close to field test values, tending to give values a little higher for high TSR with a fast processing time [25, 28].

On the other hand, a variation of the multiple streamtube modeling is the double multiple streamtube modeling. This model divides the actuator disk into two half cycles in tandem, representing the upstream and downstream sides of the rotor, respectively. This model has suffered several improvements throughout time and offers a good performance prediction and, however, presents convergence problems for high solidity turbines, giving high power prediction for high TSR and having a high processing time [31–35].

A number of studies have been carried out to simplify numerical models [36–40]. Nevertheless, the use of numerical methods associated with different algorithms of CFD to analyze complex fluid flow applications has grown in recent years due to the rapid development of computer technology. Therefore, the analysis of vertical axis hydrokinetic turbines is also being conducted using CFD simulation [18–23].

CFD is found in some commercial software tools such as ANSYS Fluent, FLACS, Phoenix, and COMSOL, among others. The CFD tools transform the governing equations of the fundamental physical principles of the fluid flow in discretized algebraic forms, which are solved to find the flow field values in time and space. The simulation results can be very sensitive to the wide range of computational parameters that must be set by the user; for a typical simulation, the user needs to select the variables of interest, turbulence models, computational domain, computational mesh, boundary conditions, methods of discretization, and the convergence criteria, among other simulation setup parameters [20–23]. In general, the CFD tools compute the flow field values by the equations of the fundamental physical principles of a fluid flow. The physical aspects of any fluid flow are governed by three principles: mass is conserved, Newton's second law (momentum equation) is fulfilled, and energy is conserved. These principles are expressed in integral equations or partial differential equation (continuity, momentum and energy equation), being the most common form of momentum equation, the Navier-Stokes equations for viscous flows and the Euler equations for inviscid flows [20–23].

The use of CFD simulation of the turbulent flow surrounding the hydrokinetic turbine is either done by statistical modeling of turbulence based on the Reynolds-averaged Navier-Stokes (RANS) equations, large eddy simulation (LES), or direct numerical simulation (DNS) [41]. RANS are time-averaged equations of motion for fluid flow; therefore, when the flow is not a stationary one, RANS equation is not the same, due to the simulation must be a time-dependent one. This significantly increases the computational time over conventional RANS. The use of LES and DNS methods for this case requires a careful application and significant computational resources [42, 43]. Additionally, since LES and DNS methods require very fine grids for wall-bounded flows, they are still in development stages for high Reynolds number flow; therefore, they are not practical for modeling the flow around the hydrokinetic turbine. An alternative for CFD simulation is the use of unsteady Reynolds-averaged Navier-Stokes (URANS) turbulence models, which are computationally economical and of great use among both industry and academia in comparison with LES models [41, 42, 44–50].

Recently, hybrid unsteady RANS/LES method has been increasingly used for certain classes of simulations, including separated flows; nevertheless, the techniques that combine the near-wall RANS region with the outer, large-eddy simulation region need further development [51].

The specific case of the URANS equations that govern incompressible and isothermal turbulent flow around the turbine blade is given by Eqs. (1) and (2) [52]:

$$\frac{\partial u_i}{\partial x_j} = 0 \tag{1}$$

$$\frac{\partial \rho u_i}{\partial t} + \frac{\partial \rho u_i u_j}{\partial x_j} = -\frac{\partial p}{\partial x_i} + \frac{\partial}{\partial x_i}\left(\mu \frac{\partial u_i}{\partial x_j}\right) + \frac{\partial}{\partial x_j}(\tau_{ij}) + \rho f_i \tag{2}$$

where $i = 1, 2, 3;\ j = 1, 2, 3;\ u$ and p represent the time-averaged velocity and pressure, respectively; μ is the dynamic viscosity of water; ρ represents the density of water; f_i expresses the body force per unit of volume, which may represent the Coriolis and centrifugal contributions; x_j denotes the spatial coordinate component; t is the time; and τ_{ij} refers to the Reynolds stress, which is required to be modeled for approximating the above equations. Nowadays, turbulent flows may be divided into four groups (**Figure 2**). The first one relies on the Boussinesq assumption according to which Reynolds stress is proportional to the mean strain rate. These models are often called eddy viscosity models. The second group includes the so-called nonlinear eddy viscosity models, which allow for the modeling of the turbulent stress as a nonlinear function of mean velocity gradients. Within this group, turbulent scales are determined by solving transport equations (usually, $k+$ one other physical quantity). The model is set to mimic response of turbulence to certain important types of strain. The third group involves modeling Reynolds stress transport equations known as a second-moment closure method. The last group concerns the DNS, LES, and the detached eddy simulation (DES). These models rely on a direct resolution of the Navier-Stokes equation for some class of the flows (DNS) or direct resolution of the Navier-Stokes equation for large scales and modeling sub-grid scale (LES, DES). DES is a hybrid model of URANS and LES

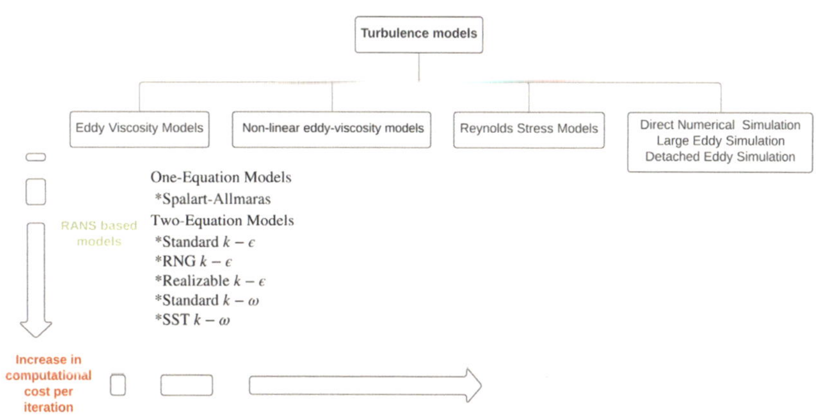

Figure 2.
Classification of turbulence models.

models, being the former one for modeling the wall region and the latter one for the other regions. The motivation of researchers for utilizing DES is the high cost of LES in the boundary layer region; the strategy of using DES aims at accurately modeling the large vortices in the separated shed wake while keeping the simulation of the near-wall region at a limited cost [51].

Eddy viscosity models use the Boussinesq approximation, which is a constitutive relation to compute the turbulence stress tensor, τ_{ij}, defined in Eqs. (3) and (4):

$$\overline{\rho}\tau_{ij} = 2\mu_t\left(\tilde{S_{ij}} - \frac{1}{3}\tilde{S_{kk}}\delta_{ij}\right) - \frac{2}{3}\rho k\delta_{ij} \quad (3)$$

$$\tau_{ij} \overset{def}{=} \overline{-u_i'u_j'} \quad (4)$$

where $\tilde{S_{ij}}$ is given by Eq. (5)

$$\tilde{S_i}j = \frac{1}{2}\left(\frac{\partial\tilde{u_i}}{\partial x_j} + \frac{\partial\tilde{u_j}}{\partial x_i}\right) \quad (5)$$

In Eq. (5), k is the turbulent kinetic energy, μ_t refers to the dynamic eddy viscosity, ρ is the water density, τ_{ij} denotes the shear stress, δ_{ij} is the Kronecker delta, and u expresses the velocity. Based on dimensional analysis, μ_t can be determined from a turbulence time scale (or velocity scale) and a length scale: (i) turbulent kinetic energy (k), represented by Eq. (6); (ii) turbulence dissipation rate (ϵ), described by Eq. (7); and (iii) specific dissipation rate (ω), expressed in Eq. (8):

$$k = \overline{-u_i'u_i'}/2 \quad (6)$$

$$\epsilon = \overline{\upsilon\partial u_i'/\partial x_j\left(\partial u_i'/\partial x_j + \partial u_j'/\partial x_j\right)} \quad (7)$$

$$\omega = \epsilon/k \quad (8)$$

Each turbulence model calculates μ_t differently. Within the group of eddy viscosity models, it is possible to identify the following one- and two-equation turbulence models: Spalart-Allmaras, standard $k-\epsilon$ model, renormalization group (RNG) $k-\epsilon$ model, realizable $k-\epsilon$ model, standard $k-\omega$ model, and shear stress transport (SST) $k-\omega$ model. The choice of turbulence model will depend on several considerations such as the physics encompassed in the flow, the established practice for a specific class of problem, the level of accuracy required, the available computational resources, and the amount of time available for the simulation. To select the most appropriate model for each application, the capabilities and limitations of such as options are needed to be understood [20, 43, 52–60].

The turbulence models most usually utilized for the CFD analysis of a vertical axis hydrokinetic turbine are RANS approach because LES models require a highly refined mesh and are, therefore, extremely expensive from a computational point of view [53]. For vertical axis hydrokinetic design, the most commonly used turbulence models include the referred models mentioned previously, e.g., $k-\epsilon$ model, realizable $k-\epsilon$ model, standard $k-\omega$ model, and shear stress transport (SST) $k-\omega$ model. The SST model was developed to provide an answer to the necessity of models which can handle aeronautical flows with strong adverse pressure gradients and separation of the boundary layers. For this purpose, SST model combines the $k-\omega$ model in regions near to solid walls, with $k-\epsilon$ model for the free stream flow out of the boundary layers. SST model has been found to give very good results for CFD analysis of horizontal and vertical axis hydrokinetic turbines [20, 54–60].

For this reason, an SST turbulence model is adopted in the design of this vertical axis hydrokinetic turbine proposed here in the same way used by other authors [22, 61–65].

In general, BEM and vortex models previously described must adopt several assumptions and corrections to account for the full three-dimensional, turbulent flow dynamics around the turbine blades. The use of CFD simulation eliminates the need for many, but not all, of these assumptions. It also has the advantage of resolving the full time-dependent flow field allowing for a better understanding of the flow at the blade wall as well as in the wake of the turbine. Although a much greater computation cost is associated with the use of CFD simulation, it has demonstrated an ability to generate results that can be favorably compared with experimental data [19, 58, 66–69].

3. Geometrical modeling of a Darrieus hydrokinetic turbine

In the vertical axis category, H-Darrieus turbines are the most common options because their installation is simple, they are less expensive, and they require less maintenance than other types of vertical axis hydrokinetic turbines [17, 20]. During the design of Darrieus turbines, a high rotor efficiency is desirable for increased water energy extraction and should be maximized within the limits of affordable production. H-Darrieus turbine consists of two or more hydrofoil blades around a shaft perpendicular to the direction of the water flow. The available hydrokinetic power (P) of this turbine can be computed from the water flow and the turbine dimension, as represented by Eq. (9) [11, 12]:

$$P = \frac{1}{2}\rho C_p A V^3 \tag{9}$$

where A is the swept area and C_p refers to the power coefficient of the turbine. It is widely known that a physical limit exists to quantify the amount of energy that can be extracted, which is independent on the turbine design. The energy extraction is maintained in the flow process through the reduction of kinetic energy and subsequent velocity of the water. The magnitude of energy harnessed is a function of the reduction water speed over the turbine. Total extraction, i.e., 100% extraction, would imply zero final velocity and, therefore, zero flow. The zero-flow scenario cannot be achieved; hence all the water kinetic energy may not be utilized. This principle is widely accepted and indicates that hydrokinetic turbine efficiency cannot exceed 59.3%. This parameter is commonly known as C_p, being the maximum C_p equal to 0.593 is referred to as the Betz limit. It is important to note that the Betz theory assumes constant linear velocity, inviscid, and without swirl flow. Therefore, any rotational forces such as wake rotation and turbulence caused by drag or vortex shedding (tip losses) will further reduce the maximum efficiency. Efficiency losses are generally reduced by avoiding low tip-speed ratios which increase wake rotation, selecting hydrofoils which have a high lift-to-drag ratio and specialized tip geometries. Making an analysis of data from literature, the value of maximum C_p has been found to be usually ranging between 0.15 and 0.586 [22]. This coefficient only considers the mechanical energy converted directly from water energy, i.e., it does not consider the mechanical conversion into electrical energy [70, 71].

The power output of the hydrokinetic turbine given by Eq. (9) is also limited by mechanical losses in transmissions and electrical losses [72, 73]. Because of these losses and inefficiencies, one more variable is added to Eq. (9). The variable is η, which is a measure of the efficiency of the gearbox, the electrical inverter, and the generator. It takes into account all the friction, slippage, and heat losses associated with the interior mechanical and electrical components. Values for η can greatly differ among the turbine models. Some smaller turbines do not have transmission; therefore, the electric generator is directly moved by the turbine shaft, while larger turbines have transmission [74]. The range of values of η is presented in the literature, highlighting the study conducted by Hagerman et al., which states a range of efficiencies between 95 and 98%. However, for the design of the blade, a reasonable and conservative value of η around 70% was used in this work [75].

Adding η into Eq. (9), Eq. (10) is obtained:

$$P = \frac{1}{2}\rho C_p A V^3 \eta \tag{10}$$

Eq. (10) is considered the power equation for the hydrokinetic turbine. P is the net power derived from the water after accounting for losses and inefficiencies. The hydrokinetic turbine parameters considered in the design process mentioned previously are (i) (A), (ii) (P) and C_p, (iii) TSR or λ, (iv) blade profile and c, (v) B, and (vi) σ and AR.

The swept area is the section of water that encloses the turbine in its movement. Its shape depends on the rotor configuration; thus, the swept area of horizontal axis hydrokinetic turbine is a circular shaped, while for a straight-bladed vertical axis hydrokinetic turbine, the swept area has a rectangular shape. Therefore, for a H-Darrieus vertical axis hydrokinetic turbine, the swept area depends on both (D) and (H) [76, 77]. Therefore, A is given by Eq. (11):

$$A = DH = 2RH \tag{11}$$

A limits the volume of water passing throughout the turbine. The rotor converts the energy contained in the water in a rotational movement, so as the area is larger, the power output under the same water operating conditions is also larger. Furthermore, C_p is strongly dependent on λ, defined in Eq. (12). This parameter is the ratio between the tangential speed at the blade tip and the water speed:

$$\lambda = \frac{R\psi}{V} \tag{12}$$

Each rotor design has an optimal TSR at which the maximum power extraction is achieved. On the other hand, the number of blades (B) has a direct effect in the smoothness of the rotor operation as they can compensate cycled hydrodynamic loads [76–78]. For easiness of the turbine manufacturing, three blades have been considered in the rotor design. In general, three blades are used for the turbine system to keep the dynamic balance and minimize the fatigue effect [75–77].

In turn, σ is defined as the ratio between the total blade area and the projected turbine area [76, 77]. It is a crucial non-dimensional parameter affecting self-starting capabilities. For a straight-bladed hydrokinetic turbine, σ is calculated from Eq. (13):

$$\sigma = \frac{Bc}{2R} \tag{13}$$

The factor AR is defined as the ratio between H and R [76, 77]. It can be defined as represented by Eq. (14):

$$AR = \frac{H}{R} \tag{14}$$

The optimum range for the rotor AR for a vertical axis wind turbine was obtained using the double multiple streamtube analytical approach by Ahmadi-Baloutaki et al. [77], and it was found to be $1 < H/R < 4$. The hydrokinetic turbine of the mentioned work was designed for a rotor with an AR of 1.5.

In relation to the blade profile, a great number of hydrofoil families and thicknesses have been informed to be suitable for Darrieus turbines. Since it is impossible to analyze all of them, the choices must be narrowed in some way. Symmetrical hydrofoils have been traditionally selected because the energy capturing is approximately symmetrical throughout the turbine axis.

In this sense, the majority of the previously conducted research works on vertical axis hydrokinetic turbines has been focused on straight-bladed vertical hydrokinetic turbines equipped with symmetric hydrofoils (such as NACA four-digit series of 0012, 0015, 0018, and 0025). In the current work, NACA 0025 hydrofoil profile was selected [66, 76, 79]. The maximum C_p of a turbine with NACA 0025 hydrofoil profile was numerical and experimentally analyzed by Dai and Lam [101]. The results showed that a maximum average C_p equal to 25.1% was obtained for a TSR of 1745. Consequently, this C_p value was used for the design of the blade.

Nevertheless, before using Eq. (10), V must be determined or assumed. In this research, a V value of 1.5 m/s was assumed since it constitutes an average speed of the great rivers in Colombia. In turn, P, η, and ρ equal to 500 W, 70%, and 997 kg/m^3(at 25°C) were calculated. R and H were equal to 1.13 m and 0.75 m, respectively. It is important to note that long blades can cause many natural frequency of vibration, which must be avoided during the operation [78, 80, 81].

Subsequently, given the rotor design parameters (e.g., R, H, hydrofoil profile and water current velocity, among others), the main task during the blade design procedure is to determine c, which can be obtained by equaling the thrust on the rotor, determined from the actuator disk theory, with the thrust obtained from the BEM theory, considering the drag different from zero [72, 81]. From the actuator disk theory, the axial thrust (I) on the disk can be written as represented by Eq. (15):

$$I = 2\rho A a(1-a)V^2 \tag{15}$$

where a is the axial induction factor, which is defined as the fractional water velocity decrease between the free stream and the rotor plane. On the other hand, because the Darrieus turbine is primarily a lift force drive turbine, the resultant tangential components of the lift and the drag forces on the hydrofoil provide the driving force to the turbine. The value of a can be determined from Eq. (16) that relates C_p with a [82]:

$$C_p = \frac{4a(1-a)^2}{\eta} \tag{16}$$

Therefore, for C_p equal to 0.25 and η of 70%, three values for axial induction can be obtained (1.1916, 0.76, 0.0483). The thrust coefficient for an ideal turbine is equal to zero when a is equal to 1. On the other hand, the momentum theory is no longer valid for axial induction factors greater than 0.5 because V in the far wake would be negative. In practice, as the axial induction factor increases above 0.5, the flow patterns through the hydrokinetic turbine become much more complex than those predicted by the momentum theory. However, when axial induction is equal

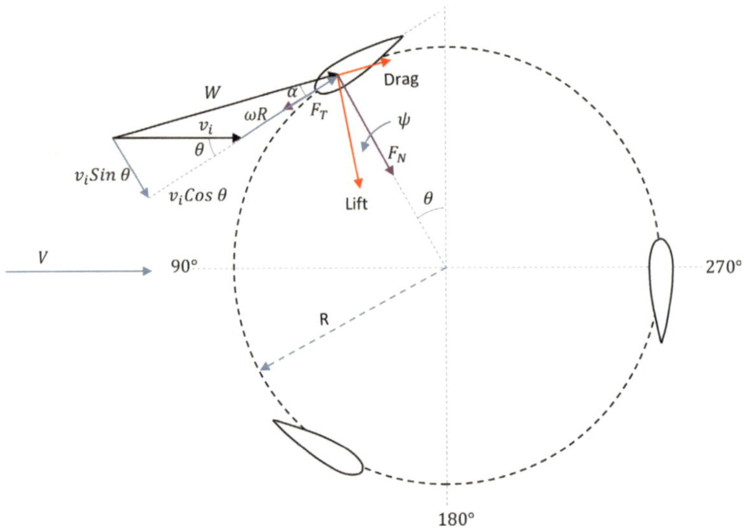

Figure 3.
Vertical axis hydrokinetic turbine flow velocities and blades.

to a small value, the chord length values can be also smaller, generating a decrease in the surface where the water acts and subsequently complicating the structural integrity of the blade; therefore, a value of axial induction equal to 0.76 was used for the design of the vertical axis hydrokinetic turbine proposed here.

Figure 3 shows a basic schematic representation of the hydrodynamic design of a H-Darrieus turbine. A feature of H-Darrieus turbine is that the angle of attack (α) on a blade is a function of the blade azimuth angle (θ). In **Figure 3**, the tip velocity vector and the lift and drag vectors generated by the rotation of the turbine blade are illustrated. As it can be seen from **Figure 3**, the relative velocity (w) can be obtained from the tangential and normal velocity components given by Eq. (17):

$$w = \sqrt{(v_i Sin(\theta))^2 + (v_i Cos(\theta) + \psi R)^2} \tag{17}$$

where v_i is the induced velocity through the rotor. w can be written in a non-dimensional form using free stream velocity. Therefore, Eq. (17) can be expressed as Eq. (18):

$$\frac{w}{V} = \sqrt{\left(\frac{v_i}{V} Sin(\theta)^2\right) + \left(\frac{v_i}{V} Cos(\theta) + \frac{\psi R}{V}\right)^2} \tag{18}$$

The v_i can be written in terms of a, as described by Eq. (19):

$$v_i = V(1 - a) \tag{19}$$

Using Eqs. (12) and (19), Eq. (18) can be rewritten as Eq. (20):

$$\frac{w}{V} = \sqrt{\left((1 - a)Sin(\theta)^2\right) + ((1 - a)Cos(\theta) + \lambda)^2} \tag{20}$$

From the geometry of **Figure 3**, W can also be represented by Eq. (21):

$$w = \frac{v_i Sin(\theta)}{Sin(\alpha)} \tag{21}$$

Using Eq. (19), Eq. (21) can be rewritten as Eq. (22):

$$w = \frac{V(1-a)Sin(\theta)}{Sin(\alpha)} \tag{22}$$

From the geometry of **Figure 3**, the local angle of attack can be expressed as Eq. (23):

$$tan\ (\alpha) = \frac{v_i Sin(\theta)}{v_i Cos(\theta) + \psi R} \tag{23}$$

Eq. (23) can be put in a non-dimensional form using Eq. (19); therefore, Eq. (24) is obtained:

$$\alpha = tan^{-1}\left(\frac{(1-a)Sin(\theta)}{(1-a)Cos(\theta) + \lambda}\right) \tag{24}$$

As shown in Eq. (24), α, is affected by λ. In **Figure 4**, the variation of α as the blade rotates according to λ different values is illustrated. It is noteworthy that α is the angle between the vector of the relative velocity and the direction of c. As the vector of the relative velocity changes, a positive and a negative values in the upstream and downstream region of the rotor, respectively, are observed. The larger λ, the smaller α throughout the blade rotation. The larger the blade tip velocity vector is, the larger the tip speed ratio is.

The normal (F_N) and tangential (F_T) force for a single blade at a single azimuthal location can be expressed as Eqs. (25) and (26), respectively:

$$F_N = \frac{1}{2}\rho w^2 Hc(C_L Cos(\alpha) + C_D Sin(\alpha)) \tag{25}$$

$$F_T = \frac{1}{2}\rho w^2 Hc(C_L Sin(\alpha) - C_D Cos(\alpha)) \tag{26}$$

where C_L is the lift coefficient and C_D is the drag coefficient for α. The instantaneous thrust force, which is the force of the water on the turbine

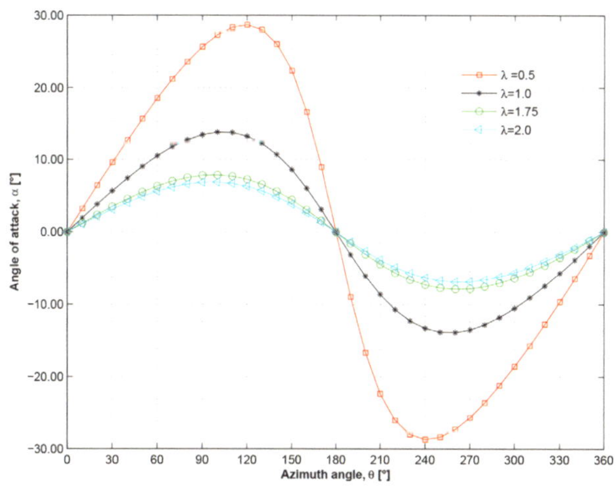

Figure 4.
Angle of attack variation in a blade revolution for different tip-speed ratios.

experienced by one-blade element in the direction of the water flow can be calculate using Eq. (27):

$$I_i = F_N Sin(\theta) - F_T Cos(\theta) \tag{27}$$

Substituting Eqs. (25) and (26) in Eq. (27), Eq. (28) is obtained, which represents the instantaneous thrust force:

$$I_i = \frac{1}{2}\rho w^2 Hc[(C_L Cos(\alpha) + C_D Sin(\alpha))Sin(\theta) - (C_L Sin(\alpha) - C_D Cos(\alpha))Cos(\theta)] \tag{28}$$

By equating the thrust value from Eq. (15) and Eq. (28) and substituting Eq. (22), c can be found using Eq. (29):

$$c = \frac{8aRSin^2\alpha}{(1-a)Sin^2\theta(C_D Cos(\alpha-\theta) - C_L Sin(\alpha-\theta))} \tag{29}$$

C_L and C_D depend on the shape of the blade, α, and Reynolds number under a given operating condition. C_L and C_D values of 0.5097 and 0.0092 were used. These values were found in the literature for the hydrofoil NACA 0025 for α equal to 5° [79].

Although α is a variable in Eq. (29) and depends on θ and α, in order to calculate c, α was considered as a constant and was equaled to angle that maximizes C_L and C_D. In this case, α was equal to 5° in order to find the blade c for any θ. Therefore, with the function defined in Eq. (29) and once C_L, C_D, a, and R were found, the c response was possible to be evaluated at different θ values (**Figure 5**). It is observed that c changes with azimuthal position, being a cyclic distribution every 180° with asymptotic values at 0° and 180°. For θ values between 0° and 60° and between 120° and 180°, c is very large, even much larger than R. Therefore, by averaging θ values between 60° and 120°, an optimal c value equal to 0.33 m is achieved.

Using Eq. (13), σ can be found, obtaining a value of 0.66. A lower σ requires less material and subsequently lower effective manufacturing cost.

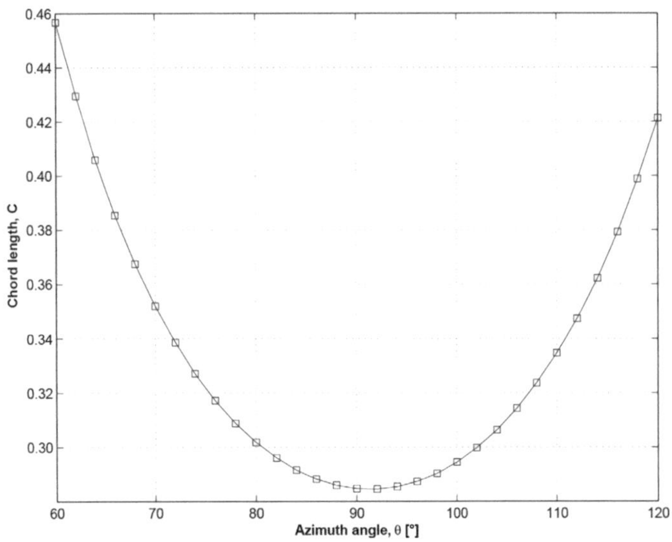

Figure 5.
Chord length for different azimuth angles.

On the other hand, the blade aspect ratio (AB) is defined as the ratio between the blade length and c [76], as expressed by Eq. (30):

$$AB = \frac{H}{c} \tag{30}$$

For blades with smaller AB, the lift-to-drag ratio can be reduced. Any decrease in the lift-to-drag ratio will reduce tangential forces. This would decrease the turbine overall torque and, consequently, the turbine output power. Some researchers have demonstrated an increase in the maximum C_p with the blade AB increasing a value of about 15. In spite of this fact, utilizing blades with relatively high AB has some drawbacks. Relatively long blades add a great amount of weight to the turbine, which increases the manufacturing and maintenance costs and creates a need for a more complex bearing. Moreover, such blades are exposed to larger bending moments. In this work, the blade AB parameter was fixed at 3.42.

After finding out the value of c for every section, the next step was to multiply this value by the non-dimensional coordinates of NACA 0025 hydrofoil profile. The values of x and y profile coordinates for each section were exported to parametric 3D design software. From the cross sections of the blade and using the Loft command, a 3D model of the whole blade was produced. The resulting image is shown in **Figure 6**. In general, the geometric specifications of the turbine are given in

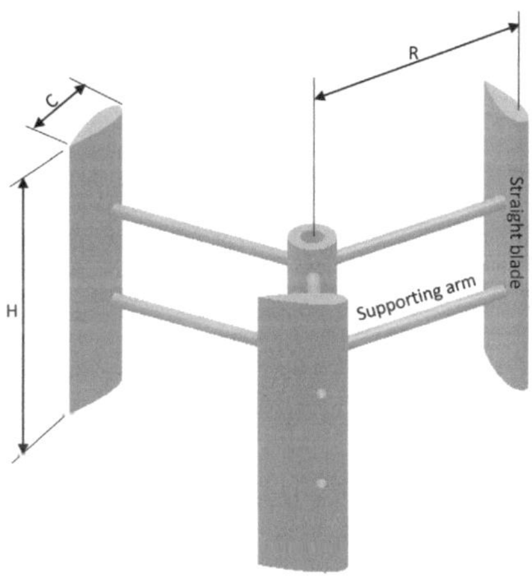

Figure 6.
H-Darrieus vertical axis hydrokinetic rotor with three blades.

Parameter	Value (units)
Power output (P)	500 W
Blade's profile NACA	0025
Chord length (c)	0.33 m
Number of blades (N)	3
Solidity (σ)	0.66
Turbine height (H)	1.13 m
Turbine radius (R)	0.75 m

Table 1.
Geometric specifications of the considered turbine.

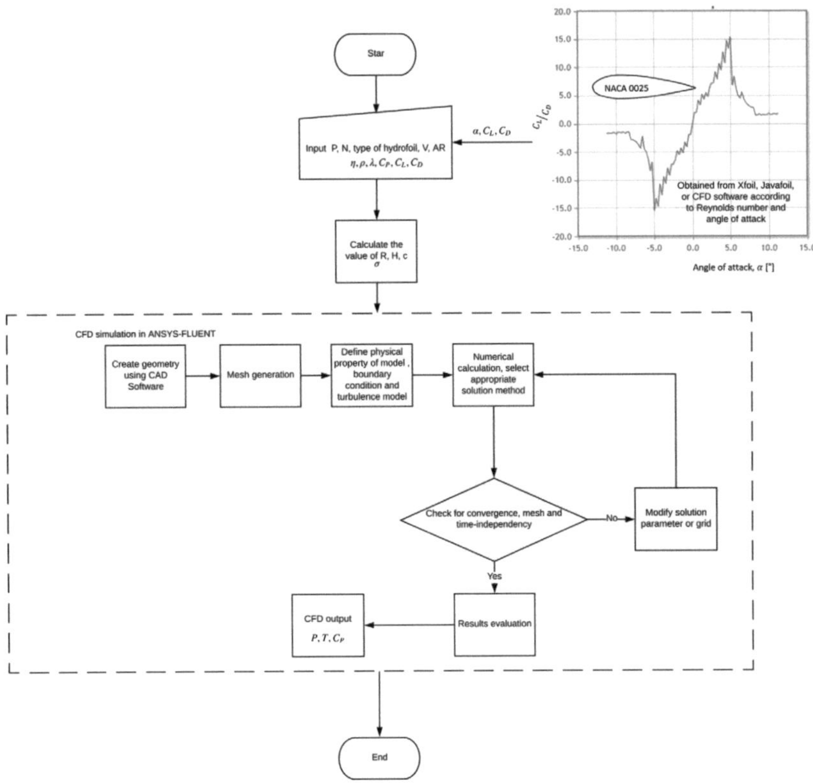

Figure 7.
Design procedure of a vertical axis hydrokinetic turbine.

Table 1. In order to calculate those parameters, the flowchart shown in **Figure 7** can be followed.

4. Computational fluid dynamics of a Darrieus hydrokinetic turbine

A computational domain is a region in the space in which the numerical equations of the fluid flow are solved by CFD. In order to solve the physics of the flow field around the blade, dividing the flow domain in a set of small subdomains (rotating and stationary domains) is required, which implies the generation of a mesh of cells, also defined as control volumes. Three fundamental aspects to evaluate the accuracy and the resolution time of a simulation are the geometry and size of the cells coupled to the numerical method used to solve the governing equations. The mesh size must be sufficiently small to provide an accurate numerical approximation, but it cannot be so small that the solution is impractical to be obtained with the available computational resource. Thus, the mesh is usually refined in the regions of interest around the main obstacles affecting the flow [83]. In the particular case of the vertical axis hydrokinetic turbines, these obstacles are the blades.

The mesh can be structured or unstructured. In the first case, the mesh consists of quadrilateral cells in 2D or hexahedral cells in 3D. In turn, the unstructured mesh usually consists of triangles in 2D and tetrahedral in 3D, but cells of the mesh can adopt any form. Structured mesh usually implies shorter time resolution; nevertheless, the unstructured meshes may better represent the geometry. Yasushi [84] presents a discussion about the development of efficient computational analysis using unstructured mesh. On the other hand, Luo and Spiegel [85] propose a method to generate a hybrid mesh (coupling structured and unstructured mesh). In some occasions, the hybrid meshes consist of quadrilateral or hexahedral elements generated in layers near the wall surfaces to capture the boundary layer and of triangular or tetrahedral elements that fill the remainder of the domain. Several concepts related to mesh generation are found in [86], and a detailed discussion about the influence of the mesh in CFD applications is presented by Thompson et al. [83].

The CFD package, ANSYS Fluent version 18.0, was used for all the simulations performed in this study. 2DCFD model with less computational cost than 3D model was utilized to represent the vertical hydrokinetic turbine and the water domain. Based on the review of relevant works [78, 87–89], the use of a 2D model has been reported to be enough for revealing the factors that influence the performance of the turbine and the majority of factors affecting the flow physics surrounding the vertical axis hydrokinetic turbine, such as the hydrofoil profile, B, σ, and D. In the 2D numerical study of this work, the effects from supporting arms were not taken into consideration. URANS equations were solved using the SIMPLE algorithm for pressure velocity coupling [34, 37, 90, 91].

The vertical hydrokinetic turbine studied was a three-bladed Darrieus rotor with a NACA 0025 blade profile. c was set at 0.33 m with R equal to 0.75 m. And central axis with a D of 0.025 m was placed in the rotation axis. As previously discussed, the mesh is a critical part of a CFD simulation for engineering purposes. It has to be coarse enough so that the calculation is affordable but also fine enough so that each important physical phenomenon is captured and simulated. In this sense, the domain mesh was created around each hydrofoil, and the surrounding water channel geometry was defined based on studies of the boundary extents. There is an inner circular rotating domain connected to a stationary rectangular domain via a sliding interface boundary condition that conserves both mass and momentum. The domain extents were also selected from a series of sensitivity tests to determine the appropriate distance of the walls, inlet and outlet boundaries from the R rotor. The domain extends three upstream and 8R downstream of the center of the turbine and two 5R laterally to either side of the turbine. The circular rotating domain has an overall diameter of two 8R. Because the computational domain is 2D, the turbine blades were implicitly assumed to be infinitely long.

Unstructured meshes were applied to both the rotor away from the near surface region and the outer grids. Finer meshes were used around the blades and regions in the wake of the blades. Particularly, regions at the leading and trailing edge and in the middle of blade were finely meshed in order to capture the flow field more accurately. The outer mesh was coarsened in regions expanding away from the rotor in order to minimize the central processing unit (CPU) time. The different mesh zones used for the present simulations are illustrated in **Figure 8**, while various mesh details are shown in **Figure 9a** and **b**. A number of simulations were

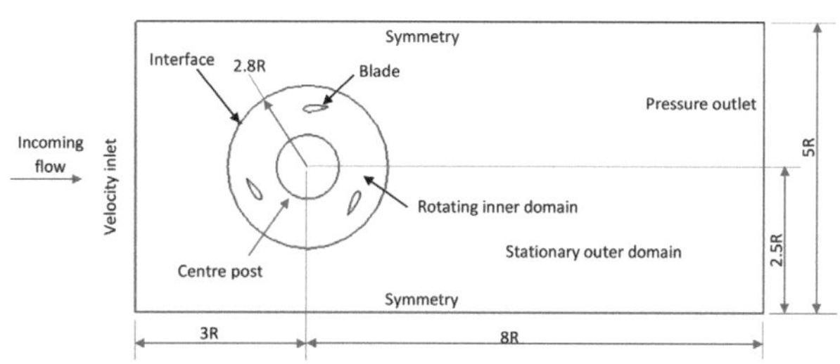

Figure 8.
Illustration of the 2D *numerical domain.*

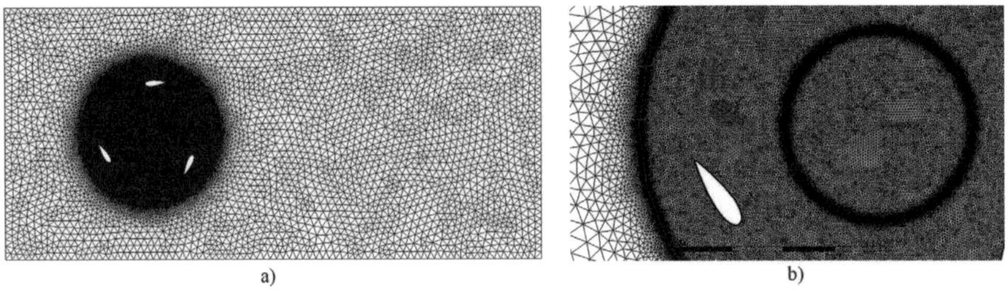

Figure 9.
(a) Detail of the area mesh around the blades and (b) detail of the rotor mesh around the interface area.

λ	0.5	1.00	1.75	2.00	2.50
ψ (rad/s)	0.997	1.994	3.489	3.987	4.984

Table 2.
Flow conditions in CFD *analysis.*

carried out in order to determine how the mesh quality affected the CFD results. In this sense, the torque was calculated for each grid using the Fluent solver. The objective was to select the most appropriate mesh that can guarantee low computational costs and good result accuracy. It is widely known that the region near the blade plays an important role on the hydrokinetic operation, since it has the highest gradient of static pressure and velocity. Additionally, the near wake flow, which can extent up to downstream of the blade, has a great effect on the power [56, 92, 93]. It is well known that a refinement in the boundary layer and a sensitivity study of y^+ are very important, since both of them have an effect on the turbine hydrodynamics. Therefore, a common parameter to identify the subparts of the boundary layer is the dimensionless distance from the wall y^+ [94], defined by $y^+ = (\Delta y u^+/v)$, where Δy is the distance of the first node from the wall, u^+ is the wall shear velocity, and v is the kinematic viscosity. In this regard, the near-wall region can be roughly subdivided into three layers: the viscous layer ($y^+ < 5$), the buffer layer ($5 < y+ < 30$), and the fully turbulent layer ($y^+ > 30$) [95]. Therefore, the quality of the mesh was also checked, as well as the y^+ values around the blades, which is important for the turbulence modeling.

Symmetrical boundary conditions were defined at the top plane and in the bottom plane. Additionally, a uniform pressure on the outlet boundary was set, and a uniform velocity on the inlet boundary with a magnitude of 1.5 m/s was used for the TSR and the corresponding ψ, as shown in **Table 2**. $k - \omega$ shear stress transport (SST) turbulence model was employed for turbulence modeling since it showed better performance for complex flows including adverse pressure gradients and flow separations like occurs in vertical axis hydrokinetic turbines. The no-slip boundary condition was applied on the turbine wall blades. To simulate the rotation of the rotor, the circular turbine mesh with embedded blades allowed the relative movement to the outer inertial fixed domain. An interface wall was introduced between the fixed and rotating domains. The origin of the reference frame is the center of the rotor. The simulation methods used in this study are similar to the methods used in other numerical studies. Here, the rotational speed of the turbine axis is specified by user input.

To solve the viscous sublayer of the turbulence model used, the values of y^+ generally must be less than 1. Subsequently, several meshes were made by increasing the refinement of the computational domain near the blade, achieving a good resolution of the boundary layer. The mesh 4 reached a convergence of the results

Mesh	Nodes	y^+	Maximum power coefficient C_{pmax}	Error (%)
Mesh 1	143,672	0.246	0.56	10.71
Mesh 2	167,896	0.221	0.64	14.28
Mesh 3	187,630	0.283	0.63	1.56
Mesh 4	298,878	0.261	0.62	1.58

Table 3.
Mesh convergence study.

with 298,878 nodes and 1.58% error on C_p (**Table 3**). In this regard, further refinement might not improve the numerical results. The numerical results obtained from mesh 4 were compared with several results available in the literature.

The modeling of vertical axis hydrokinetic turbines can be done in steady state or transient modes depending on the objectives of the research and the available computational resources. If computational resources are scarce, relatively simple, steady flow models can be used to model the turbine blades in different azimuthal positions [96]. A more common approach is the transient modeling of the moving blades through the use of URANS approach [19, 97–99].

In this work, transient analyses were carried out to characterize the performance of the investigated NACA 0025 hydrofoil profile. Performances are described in terms of P, T, and C_p, which are calculated according to Eq. (31). Because the flow over Darrieus turbine is a periodic one, sufficient temporal resolution is necessary to ensure proper unsteady simulation of the vertical axis hydrokinetic turbine and to obtain an independent solution on the time step. Different time step sizes Δt that are equivalent to specific rotational displacements along the azimuth were tested. The chosen time step size was $\Delta t = 0.1°$, which properly captures the vortex shedding. Time step convergence was monitored for all conserved variables, and it was observed that acceptable levels of residuals (less than 1×10^{-6}) were attained after six rotations of the hydrokinetic turbine. This means that periodic convergence was also achieved:

$$C_p = \frac{P}{0.5\rho A V^3} \tag{31}$$

C_p variation over a range of TSR values was studied for the selection of the best TSR, leading to the optimal performance of NACA 0025 hydrofoil profile, which was used in the blade geometrical design. The power versus azimuthal position of different TSRs during rotation is shown in **Figure 10**. The instantaneous power generated by the turbine is equal to the product of the turbine ψ and T acting on it. A nearly sinusoidal curve was obtained with three positive maxima at each turn and three positive minima (when TSR is equal to 1.75) or negative minima (when TSR is equal to 0.50, 1.00, 2.00); meaning that during a revolution, there are periods of time where the turbine produces torque on the fluid. It is observed that C_p increases and decreases approximately until 50 and 120° azimuthal position, respectively, i.e., the main power production occurs between 0 and 120° azimuthal positions for the first blade when TSRs are 1.75 and 2.00. In the figure, it can be seen that the maximum torque for the first blade is achieved at the azimuth angle around 50°. After the peak, the drag begins to increase as the blade enters into a dynamic stall, and the drag starts to be dominant up to a θ of 120°. Then, the second blade repeats the motion of the first blade, and the power production is completed with the same motion of the third blade for one rotation of the turbine. Plot of C_p versus TSR

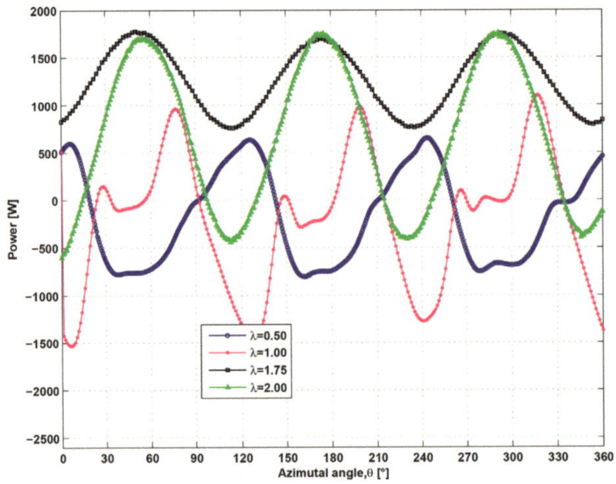

Figure 10.
Power output for different TSRs.

shows positive values of TSR close to 1.75, meaning that fluid is providing torque to the turbine. Beyond 1.75 TSR, C_p is negative, indicating that the turbine, rotating at a constant ψ, exerts torque on the fluid. This can be explained because a high TSR implies a high turbine angular speed and, in such as case, kinetic energy contained in flow is not enough to deliver torque to the turbine and make it to rotate with the same ψ.

In the literature, optimum values of TSR for a Darrieus turbine were reported. For example, Kiho et al. [100], using a Darrieus turbine diameter of 1.6 m, found that the highest efficiency was achieved at 0.56 at 1.1 m/s V and two TSRs. In turn, Torri et al. [101] also found that the peak C_p of a three-blade straight vertical axis hydrokinetic turbine was about 0.35 at a TSR of around 2. However, Dai et al. [22] performed numerical and experimental analyses with four sets of a three-bladed rotor where blades were designed with NACA 0025 with four different c values. These four sets of rotors were tested in a tank at different V and R. The highest C_p was achieved by the rotor with a c of 162.88 mm, R of 450 mm, and TSR of 1.745. Furthermore, authors investigated the same rotor at different V and found 1.2 m/s as the most effective velocity. They concluded that larger rotors are more efficient, while flow velocity has little effect on the turbine efficiency. **Figure 10** shows that there is a high fluctuation with each revolution. The fluctuation decreases when λ is equal to 1.75.

The same tendencies of C_p are found in the case of TSR of 1.75 and 2.00, as it can be seen in **Figure 10**. The variation amplitude of C_p is lower when TSR is equal to 1.75. This result confirms that the blade with TSR of 1.75 has better performance than any other TSRs. When TSR is equal to 1.75, the maximum C_p is 0.62, and the variation amplitude of C_p is near 0.37. On the other hand, when TSR is 2.00, the turbine achieved a maximum C_p near 0.61 and an amplitude of only 0.83. Therefore, a turbine with a TSR of 1.75 has advantages over a turbine with a TSR of 2.00 in terms of the reduced fluctuation of its torque curve. C_p decreases for lower values of TSR because at low values of TSR, the flow surrounding the blade is separated, implying low lift and high drag. As a result, transferred torque from fluid to turbine decreases. In general, the fluctuations in C_p and, therefore, in T can produce a high amount of vibration in the turbine. The effect of these vibrations directly influences the fatigue life of the turbine blades and that of the generated power.

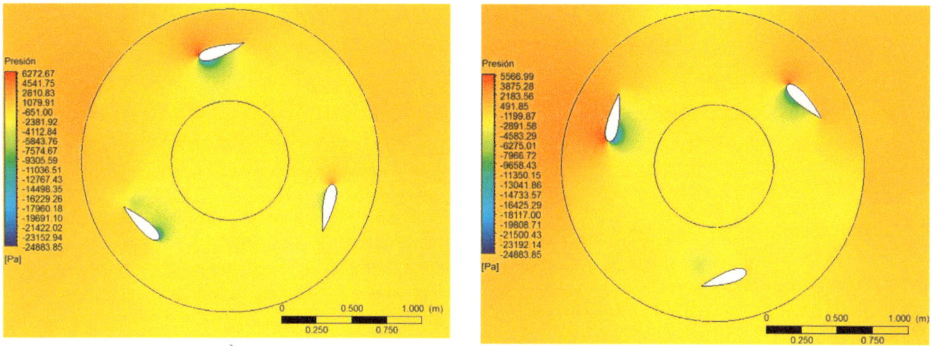

a) b)

Figure 11.
The contours of pressure magnitude for the rotor (a) $\theta = 0°$ and (b) $\theta = 30°$.

Figure 11 shows pressure contours at θ are equal to 0° and 30°. The contour plot shows that there is an increase of pressure from the upstream side to the downstream side across the blade. Additionally, the pressure in the outer area of the profile is observed to be larger (intrados) than that of the inner zone (extrados). The difference of pressure and velocity causes the overall lift for the turbine. The pressures are negative near the ends of the blade but positive at the exit of the computational domain.

In order to perceive the effect of TSR on the turbine performance, the average C_p achieved for different TSRs can be found. C_p is not constant because the torque and velocity are not constant in a Darrieus turbine. Hence, the average C_p per cycle is calculated as the product of the average values of these terms per cycle. TSR and C_p are in a direct relationship when the TSR is between 0.5 and 1.75. However, TSR and C_p have an opposite tendency for TSR greater than 1.75. The maximum value of C_p was achieved when TSR was 1.75; the value of the average C_p was 44.33% per cycle. Similar results were obtained by Dai et al. [22] for NACA 0025 blade profile. Additionally, Lain and Osorio [102] used experimental data of Dai and Lams work [22] and developed numerical models. They performed analysis on CFX solver, DMS model, and Fluent solver and achieved efficiencies of 58.6, 46.3, and 52.8%, respectively, when TSR was 1.745. Results observed by using Fluent solver were quite accurate.

5. Conclusions

The design and simulation of a vertical axis hydrokinetic turbine were presented in this work. The most common modeling method for vertical axis hydrokinetic turbine is CFD. The numerical simulations allow analyzing many different turbine design parameters, providing an optimal configuration for a given set of design parameters. Before the simulations can be performed, the determination of the optimal grid and time step sizes required must be conducted. Finer grids and smaller time steps might give a more accurate solution, but they increase the computational cost. Thus, finding optimal values is also required. For this purpose, the turbulence model generally used is k-ω shear stress transport (SST) turbulence model.

The TSR is a significant parameter that affects the performance of hydrokinetic turbines. Consequently, the performance of the turbine was investigated with a simplified 2D numerical model. From the 2D model, C_p was computed for various TSRs. During a turbine revolution, the blade of the turbine may experience large, as well as rapid variation, in C_p. A C_p maximum of 62% was achieved when TSR was equal to 1.75. The value of the average C_p was 44.33% per cycle. Though Darrieus turbine is very simple to be constructed, it has some disadvantages when compared to axial turbines since they exhibit a lower power coefficient and a variation in the torque produced within the cycle, leading to periodic loading on the components of the turbine.

Acknowledgements

The authors gratefully acknowledge the financial support provided by the Colombia Scientific Program within the framework of the call Ecosistema Cientifico (Contract No. FP44842-218-2018).

Abbreviations

CFD	computational fluid dynamics
TSR	tip-speed ratio
BEM	blade element momentum theory
DMS	double multiple streamtube
RANS	Reynolds-averaged Navier-Stokes
LES	large eddy simulation
DNS	direct numerical simulation
URANS	unsteady Reynolds-averaged Navier-Stokes
DES	detached eddy simulation

Nomenclature

P	power required
ρ	water density
V	water speed
θ	Azimuth angle
ψ	turbine rotational speed
λ	blade tip-speed ratio (TSR)
α	blade angle of attack
AR	blade aspect ratio
H	turbine height
σ	solidity
R	turbine radius
D	turbine diameter
B	number of blades
c	blade chord length
a	axial induction factor
A	swept area of the turbine
C_D	drag coefficient
C_L	lift coefficient
C_P	power coefficient
T	turbine torque
I	AXIAL thrust

Author details

Edwin Lenin Chica Arrieta[1*†] and Ainhoa Rubio Clemente[1,2†]

1 Departamento de Ingeniería Mecánica, Facultad de Ingeniería, Universidad de Antioquia UdeA, Medellín, Colombia

2 Facultad de Ingeniería, Tecnólogico de Antioquia-Institución Universitaria TdeA, Medellín, Colombia

*Address all correspondence to: edwin.chica@udea.edu.co

† These authors contributed equally.

IntechOpen

© 2019 The Author(s). Licensee IntechOpen. This chapter is distributed under the terms of the Creative Commons Attribution License (http://creativecommons.org/licenses/by/3.0), which permits unrestricted use, distribution, and reproduction in any medium, provided the original work is properly cited.

References

[1] Hansen K, Breyer C, Lund H. Status and perspectives on 100% renewable energy systems. Energy. 2019;**175**: 471-480

[2] Mamat R, Sani MSM, Sudhakar K. Renewable energy in Southeast Asia: Policies and recommendations. Science of the Total Environment. 2019;**670**: 1095-1102

[3] Nicolli F, Vona F. Energy market liberalization and renewable energy policies in OECD countries. Energy Policy. 2019;**128**:853-867

[4] Djφrup S, Thellufsen JZ, Sorkns P. The electricity market in a renewable energy system. Energy. 2018;**162**: 148-157

[5] Sinha A, Shahbaz M, Sengupta T. Renewable energy policies and contradictions in causality: A case of next 11 countries. Journal of Cleaner Production. 2018;**197**:73-84

[6] Jefferson M. Renewable and low carbon technologies policy. Energy Policy. 2018;**123**:367-372

[7] Yang X, Liu N, Zhang P, Guo Z, Ma C, Hu P, et al. The current state of marine renewable energy policy in China. Marine Policy. 2019;**100**:334-341

[8] Dos Santos IFS, Camacho RGR, Tiago Filho GL, Botan ACB, Vinent BA. Energy potential and economic analysis of hydrokinetic turbines implementation in rivers: An approach using numerical predictions (CFD) and experimental data. Renewable Energy. 2019;**143**:648-662

[9] Fraenkel PL. Marine current turbines: Pioneering the development of marine kinetic energy converters. Proceedings of the Institution of Mechanical Engineers, Part A: Journal of Power and Energy. 2007;**221**(2):159-169

[10] Wang WQ, Yin R, Yan Y. Design and prediction hydrodynamic performance of horizontal axis micro-hydrokinetic river turbine. Renewable Energy. 2019;**133**:91-102

[11] Rourke FO, Boyle F, Reynolds A. Marine current energy devices: Current status and possible future applications in Ireland. Renewable and Sustainable Energy Reviews. 2010;**14**(3):1026-1036

[12] Vermaak HJ, Kusakana K, Koko SP. Status of micro-hydrokinetic river technology in rural applications: A review of literature. Renewable and Sustainable Energy Reviews. 2014;**29**: 625-633

[13] Khan MJ, Iqbal MT, Quaicoe JE. River current energy conversion systems: Progress, prospects and challenges. Renewable and Sustainable Energy Reviews. 2008;**12**(8):2177-2193

[14] Gney MS, Kaygusu K. Hydrokinetic energy conversion systems: A technology status review. Renewable and Sustainable Energy Reviews. 2010; **14**(9):2996-3004

[15] Patel V, Eldho TI, Prabhu SV. Performance enhancement of a Darrieus hydrokinetic turbine with the blocking of a specific flow region for optimum use of hydropower. Renewable Energy. 2019;**135**:1144-1156

[16] Rostami AB, Fernandes AC. The effect of inertia and flap on autorotation applied for hydrokinetic energy harvesting. Applied Energy. 2015;**143**: 312-323

[17] Khan M, Islam N, Iqbal T, Hinchey M, Masek V. Performance of Savonius rotor as a water current

turbine. The Journal of Ocean Technology. 2009;**4**(2):71-83

[18] Kumar A, Saini RP. Performance parameters of Savonius type hydrokinetic turbine a review. Renewable and Sustainable Energy Reviews. 2016;**64**:289-310

[19] Lanzafame R, Mauro S, Messina M. 2D CFD modeling of H-Darrieus wind turbines using a transition turbulence model. Energy Procedia. 2014;**45**:131140

[20] López O, Meneses D, Quintero B, Laín S. Computational study of transient flow around Darrieus type cross flow water turbines. Journal of Renewable and Sustainable Energy. 2016;**8**(1): 014501

[21] Chen B, Nagata S, Murakami T, Ning D. Improvement of sinusoidal pitch for vertical-axis hydrokinetic turbines and influence of rotational inertia. Ocean Engineering. 2019;**179**: 273-284

[22] Dai YM, Lam W. Numerical study of straight-bladed Darrieus-type tidal turbine. Proceedings of the Institution of Civil Engineers-Energy. 2009;**162**: 6776

[23] Ma Y, Hu C, Li Y, Li L, Deng R, Jiang D. Hydrodynamic performance analysis of the vertical axis twin-rotor tidal current turbine. Water. 2018; **10**(11):1694

[24] Glauert H. The Elements of Airfoil and Airscrew Theory. 2nd ed. England: Cambridge University Press; 1926. 243 p

[25] Strickland JH. The Darrieus turbine: A performance prediction model using multiple streamtubes. In: Technical Report SAND75-0431. Albuquerque, USA: Sandia National Laboratories; 1975

[26] Templi RJ. Aerodynamic theory for the NRC vertical-axis wind turbine. NRC of Canada, Report No. TR LTR-LA-160. 1974

[27] Sharpe DJ. A general momentum theory applied to an energy-extracting actuator disc. Wind Energy. 2004;7: 177188

[28] Read S, Sharpe DJ. An extended multiple streamtube theory for vertical axis wind turbines. In: Proceedings of the 2nd BWEA Workshop; Cranfield, UK. 1980. p. 6572

[29] Strickland JH, Webster BT, Nguyen T. A vortex model of the Darrieus turbine: An analytical and experimental study. Journal of Fluids Engineering. 1979;**101**:500505

[30] Paraschivoiu I. Double-multiple streamtube model for Darrieus wind turbines. In: Second DOE/NASA Wind Turbines Dynamics Workshop, NASA CP-2186; Cleveland, OH, USA. 1981. p. 1925

[31] Paraschivoiu I. Wind Turbine Design: With Emphasis on Darrieus Concept. Presses Inter Polytechnique; Montreal. 2002

[32] Larsen HC. Summary of a vortex theory for the cyclogiro. In: Proceedings of the Second US National Conferences on Wind Engineering Research; Colorado State University. 1975

[33] Batista NC, Melcio R, Mendes VM, Figueiredo J, Reis AH. Darrieus wind turbine performance prediction: Computational modeling. In: Doctoral Conference on Computing, Electrical and Industrial Systems. Vol. 2013. Berlin, Heidelberg: Springer; 2013, April. pp. 382-391

[34] Islam M, Ting DSK, Fartaj A. Aerodynamic models for Darrieus-type straight-bladed vertical axis wind turbines. Renewable and Sustainable Energy Reviews. 2008;**12**(4):1087-1109

[35] Hirsch IH, Mandal AC. A cascade theory for the aerodynamic performance of Darrieus wind turbines. Wind Engineering. 1987;**11**(3):164-175

[36] Coiro DP, De Marco A, Nicolosi F, Melone S, Montella F. Dynamic behaviour of the patented kobold tidal current turbine: Numerical and experimental aspects. Acta Polytechnica. 2005;**45**:7784

[37] Dai YM, Gardiner N, Sutton R, Dyson PK. Hydrodynamic analysis models for the design of Darrieus-type vertical-axis marine current turbines. Proceedings of the Institution of Mechanical Engineers, Part M: Journal of Engineering for the Maritime Environment. 2011. 2011;**225**(3): 295-307

[38] Li Y, Calisal SM. Three-dimensional effects and arm effects on modeling a vertical axis tidal current turbine. Renewable Energy. 2010;**35**:23252334

[39] Li Y, Calisal SM. Modeling of twin-rotors systems with vertical axis tidal current turbines: Part I. Power output. Ocean Engineering. 2010;**37**: 627637

[40] Li Y, Calisal SM. Modeling of twin-rotors systems with vertical axis tidal current turbine: Part II. Torque fluctuation. Ocean Engineering. 2011; **38**:550558

[41] Kang S, Borazjani I, Colby JA, Sotiropoulos F. Numerical simulation of 3D flow past a real-life marine hydrokinetic turbine. Advances in Water Resources. 2012;**39**:33-43

[42] Khan J, Bhuyan G, Moshref A, Morison K, Pease JH, Gurney J. Ocean wave and tidal current conversion technologies and their interaction with electrical networks. In: Power and Energy Society General Meeting Conversion and Delivery of Electrical Energy in the 21st Century. Pittsburgh, PA, USA: IEEE; 2008. pp. 1-8

[43] Simo Ferreira CJ. The near wake of the VAWT: 2D and 3D views of the VAWT aerodynamics [Ph.D. Thesis]. Delft University of Technology. 2009

[44] Calaf M, Meneveau C, Meyers J. Large eddy simulation study of fully developed wind-turbine array boundary layers. Physics of Fluids. 2010;**015110**:22

[45] Calaf M, Parlange MB, Meneveau C. Large eddy simulation study of scalar transport in fully developed wind-turbine array boundary layers. Physics of Fluids. 2011;**126603**:23

[46] Miller VB, Schaefer LA. Dynamic modeling of hydrokinetic energy extraction. Journal of Fluids Engineering. 2010;**132**:091102

[47] Gaden D, Bibea E. A numerical investigation into upstream boundary-layer interruption and its potential benefits for river and ocean kinetic hydropower. Renewable Energy. 2010; **35**:2270-2278

[48] Turnock S, Phillips AB, Banks J, Nichollos-Lee R. Modelling tidal current turbine wakes using a coupled RANS-BEMT approach as a tool for analysing power capture of arrays of turbines. Ocean Engineering. 2011;**38**:1300-1307

[49] Zahle F, Sϕrensen NN. Characterization of the unsteady flow in the nacelle region of a modern wind turbine. Wind Energy. 2011;**14**:271-283

[50] Sezer-Uzol N, Long LN. 3-D time-accurate CFD simulations of wind turbine rotor flow fields. AIAA Paper. 2006;**394**:1-23

[51] Chaouat B. The state of the art of hybrid RANS/LES modeling for the simulation of turbulent flows. Flow, Turbulence and Combustion. 2017; **99**(2):279-327

[52] Anderson JD, Wendt J. Computational Fluid Dynamics. New York: McGraw-Hill; 1995

[53] Iida A, Kato K, Mizuno A. Numerical simulation of unsteady flow and aerodynamic performance of vertical axis wind turbines with LES. 2007. pp. 1295-1298

[54] Menter FR, Kuntz M, Langtry R. Ten years of industrial experience with the SST turbulence model. Turbulence, Heat and Mass Transfer. 2003;**4**(1): 625-632

[55] Maître T, Achard J.L, Guittet L, Ploesteanu C. Marine turbine development: Numerical and experimental investigations. In: Workshop on Vortex Dominated Flows. Achievement and Open Problems. 2005. pp. pas-prcis

[56] Moshfeghi M, Song YJ, Xie YH. Effects of near-wall grid spacing on SST-K-ω model using NREL Phase VI horizontal axis wind turbine. Journal of Wind Engineering and Industrial Aerodynamics. 2012;**107**:94-105

[57] Mo JO, Lee YH. CFD investigation on the aerodynamic characteristics of a small-sized wind turbine of NREL PHASE VI operating with a stall-regulated method. Journal of Mechanical Science and Technology. 2012;**26**(1):81-92

[58] Amet E, Maître T, Pellone C, Achard JL. 2D numerical simulations of blade-vortex interaction in a Darrieus turbine. Journal of Fluids Engineering. 2009;**131**(11):111103

[59] Nabavi Y. Numerical study of the duct shape effect on the performance of a ducted vertical axis tidal turbine [M. Sc. thesis]. British Columbia University; 2008

[60] Untaroiu A, Wood HG, Allaire PE, Ribando RJ. Investigation of self-starting capability of vertical axis wind turbines using a computational fluid dynamics approach. Journal of Solar Energy Engineering. 2011;**133**(4): 041010

[61] Silva PA, Oliveira TF, Brasil Junior AC, Vaz JR. Numerical study of wake characteristics in a horizontal-axis hydrokinetic turbine. Anais da Academia Brasileira de Ciências. 2016;**88**(4):2441-2456

[62] Gorle JMR, Chatellier L, Pons F, Ba M. Flow and performance analysis of H-Darrieus hydroturbine in a confined flow: A computational and experimental study. Journal of Fluids and Structures. 2016;**66**:382-402

[63] McLaren K, Tullis S, Ziada S. Computational fluid dynamics simulation of the aerodynamics of a high solidity, small scale vertical axis wind turbine. Wind Energy. 2012;**15**(3): 349-361

[64] Lee T, Gerontakos P. Investigation of flow over an oscillating airfoil. Journal of Fluid Mechanics. 2004;**512**: 313-341

[65] Danao LA, Qin N, Howell R. A numerical study of blade thickness and camber effects on vertical axis wind turbines. Proceedings of the Institution of Mechanical Engineers, Part A: Journal of Power and Energy. 2012;**226**(7): 867-881

[66] Laneville A, Vittecoq P. Dynamic stall: The case of the vertical axis wind turbine. Journal of Solar Energy Engineering. 1986;**108**(2):140-145

[67] Fraunie P, Beguier C, Paraschivoiu I, Brochier G. Water channel experiments of dynamic stall on Darrieus wind turbine blades. Journal of Propulsion and Power. 1986;**2**(5): 445449

[68] Fujisawa N, Shibuya S. Observations of dynamic stall on Darrieus wind turbine blades. Journal of Wind Engineering and Industrial Aerodynamics. 2001;**89**:201214

[69] Georgescu AM, Georgescu SC, Cosoiu CI, Alboiu N, Hamzu A. Velocity

field in the wake of a hydropower farm equipped with Achard turbines. In: IOP Conference Series: Earth and Environmental Science. Vol. 12(1). 'Politehnica' University of Timişoara, Timişoara, Romania: IOP Publishing; 2010. p. 012108

[70] Vennell R. Exceeding the Betz limit with tidal turbines. Renewable Energy. 2013;**55**:277-285

[71] Ragheb M, Ragheb AM. Wind turbines theory-the Betz equation and optimal rotor tip speed ratio. Fundamental and Advanced Topics in Wind Power. 2011;**1**(1):19-38

[72] Biadgo MA, Simonovic A, Komarov D, Stupar S. Numerical and analytical investigation of vertical axis wind turbine. FME Transactions. 2013; **41**:49-58

[73] Chica E, Prez F, Rubio-Clemente A. Rotor structural design of a hydrokinetic turbine. International Journal of Applied Engineering Research. 2016;**11**(4): 2890-2897

[74] Anyi M, Kirke B. Hydrokinetic turbine blades: Design and local construction techniques for remote communities. Energy for Sustainable Development. 2011;**15**(3):223-230

[75] Hagerman G, Polagye B, Bedard R, Previsic M. Methodology for estimating tidal current energy resources and power production by tidal in-stream energy conversion (TISEC) devices. In: Rep. EPRI-TP-001 NA Rev 2, Electr. Power Res. Inst., Palo Alto, CA. 2006. pp. 1-57

[76] Mohamed MH. Performance investigation of H-rotor Darrieus turbine with new airfoil shapes. Energy. 2012;**47**(1):522-530

[77] Ahmadi-Baloutaki M, Carriveau R, Ting DS-K. Straight-bladed vertical axis wind turbine rotor design guide based on aerodynamic performance and loading analysis. Proceedings of the Institution of Mechanical Engineers, Part A: Journal of Power and Energy. 2014;**228**(7):742759

[78] Khan M, Bhuyan G, Iqbal M, Quaicoe JE. Hydrokinetic energy conversion systems and assessment of horizontal and vertical axis turbines for river and tidal applications: A technology status review. Applied Energy. 2009;**86**(10):1823-1835

[79] Robert E, Sheldahl P, Klimas C. Aerodynamic characteristics of seven symmetrical airfoil sections through 180-degree angle of attack for use in aerodynamic analysis of vertical axis wind turbines. In: SAND80-2114. Unlimited Release. Albuquerque, NM (USA): Sandia National Laboratories; 1981

[80] van Els RH, Junior ACPB. The Brazilian experience with hydrokinetic turbines. Energy Procedia. 2015;**75**: 259-264

[81] Manwell JF, McGowan JG, Rogers AL. Aerodynamics of wind turbines. In: Manwell JF, McGowan JG, Rogers AL, editors. Wind Energy Explained: Theory, Design and Application. 2nd ed. UK: John Wiley & Sons; 2009. pp. 91-155

[82] Chica E, Rubio-Clemente A. Design of zero head turbines for power generation. In: Renewable Hydropower Technologies. Rijeka - Croatia: IntechOpen; 2017. pp. 25-52

[83] Thompson JF, Soni BK, Weatherill NP. Handbook of Grid Generation. Boca Raton, Florida (USA): CRC Press; 1998

[84] Yasushi I. Challenges in unstructured mesh generation for practical and efficient computational fluid dynamics simulations. Computer & Fluids. 2012, 2012:1-6

[85] Luo H, Spiegel S. Hybrid grid generation method for complex geometries. American Institute of Aeronautics and Astronautics Journal. 2010;**48**(11):1-9

[86] Anderson JDJ. Computational Fluid Dynamics—The Basics with Applications. Maryland: McGraw-Hill; 1995. p. 547

[87] Carrigan TJ, Dennis BH, Han ZX, Wang BoP. Aerodynamic shape optimization of a vertical-axis wind turbine using differential evolution. ISRN Renewable Energy. 2012; Article ID 528418:1-16

[88] Ghatage SV, Joshi JB. Optimisation of vertical axis wind turbine: CFD simulations and experimental measurements. The Canadian Journal of Chemical Engineering. 2011;**90**(5): 1186-1201

[89] Sabaeifard P, Razzaghi H, Forouzandeh A. Determination of vertical axis wind turbines optimal configuration through CFD simulations. In: International Conference on Future Environment and Energy, Singapore 22–25 February 2012. pp. 109-113

[90] Jung HJ, Lee JH, Rhee SH, Song M, Hyun B-S. Unsteady flow around a two-dimensional section of a vertical axis turbine for tidal stream energy conversion. International Journal of Naval Architecture and Ocean Engineering. 2009;**1**(2):64-69

[91] Raciti CM, Englaro A, Benini E. The Darrieus wind turbine: Proposal for a new performance prediction model based on CFD. Energy. 2011;**36**(8): 4919-4934

[92] Bassi F, Botti L, Colombo A, Ghidoni A, Rebay S. Investigation of near-wall grid spacing effect in high-order discontinuous Galerkin RANS computations of turbomachinery flows. In: Spectral and High Order Methods for Partial Differential Equations-ICOSAHOM 2012. Cham: Springer; 2014. pp. 125-134

[93] Bouhelal A, Smaili A, Guerri O, Masson C. Numerical investigation of turbulent flow around a recent horizontal axis wind turbine using low and high Reynolds models. Journal of Applied Fluid Mechanics. 2018;**11**(1): 151-164

[94] Wilcox DC. Reassessment of the scale-determining equation for advanced turbulence models. AIAA Journal. 1988;**26**(11):1299-1310

[95] Schlichting H, Gersten K. Boundary-Layer Theory. Berlin Heidelberg, Germany: Springer; 2016

[96] Gupta R, Biswas A. Computational fluid dynamics analysis of a twisted three-bladed H-Darrieus rotor. Journal of Renewable and Sustainable Energy. 2010;**2**(4):043111

[97] Howell R, Qin N, Edwards J, Durrani N. Wind tunnel and numerical study of a small vertical axis wind turbine. Renewable Energy. 2010;**35**(2): 412-422

[98] Almohammadi KM, Ingham DB, Ma L, Pourkashanian M. CFD sensitivity analysis of a straight-blade vertical axis wind turbine. Wind Engineering. 2012; **36**(5):571-588

[99] Balduzzi F, Bianchini A, Ferrara G, Ferrari L. Dimensionless numbers for the assessment of mesh and timestep requirements in CFD simulations of Darrieus wind turbines. Energy. 2016; **97**:246-261

[100] Kiho S, Shiono M, Suzuki K. The power generation from tidal currents by Darrieus turbine. Renewable Energy. 1996;**9**(1–4):1242-1245

[101] Torii T, Ookubo H, Yamane M, Sagara K, Seki K, Sekita K. A study on

effectiveness of straight-wing vertical-axis hydro turbine generation system in the tidal current. In: The Seventeenth International Offshore and Polar Engineering Conference. Lisbon, Portugal: International Society of Offshore and Polar Engineers; 2007

[102] Laín S, Osorio C. Simulation and evaluation of a straight-bladed Darrieus-type cross flow marine turbine. Journal of Scientific & Industrial Research. 2010;**69**:906-912

Chapter 10

The Multiphase Flow CFD Analysis in Journal Bearings Considering Surface Tension and Oil-Filler Port Flow

Masayuki Ochiai, Fuma Sakai and Hiromu Hashimoto

Abstract

This chapter presents the multiphase computational fluid dynamics (CFD) analysis on oil-lubricated high-speed journal bearings considering the oil-filler port. Journal bearings are widely used for high-speed rotating machinery such as turbines, compressors, pumps, automobiles, and so on. They can support the rotating shaft utilizing the oil lubrication film wedge effects used in the bearing clearance. Previously, in the analysis of journal bearings, which clearance is very narrow compared with shaft diameter, the Reynolds equation has been used on journal bearing analysis because of its applicability on the narrow space calculation and its low calculation cost. However, the gaseous-phase area generated in the journal bearing and the effect of oil-filler port cannot be reproduced accurately using the method. Under these backgrounds, some researchers use the CFD analysis to calculate the journal bearing characteristics in recent years. In this chapter, the authors describe the multiphase flow CFD analysis on journal bearing based on our previous studies . At first, the multiphase CFD calculation model on journal bearing and the experimental method are explained. Then, four types of calculation results under flooded and starved lubrication conditions are compared to the experimental ones. Additionally, the effect of surface tension on journal bearing characteristics is discussed. Finally, the CFD thermal analysis results under two types of supply oil conditions are shown.

Keywords: journal bearing, multiphase flow analysis, oil-filler port, cavitation, temperature, stability

1. Introduction

High-speed rotating machinery such as compressors, pumps, gas turbines, and automobiles is used all over the world [1, 2]. Oil-lubricated journal bearings are used widely as a support element of high-speed rotating shafts for reducing friction, enhancing the rotating accuracy. On the journal bearing, many researchers and engineers have been interested in the gaseous phase generated in the bearing clearance, and they tried to predict the existence of the gaseous phase because the bearing characteristics are strongly affected by the gaseous-phase areas. For example, Hashimoto and Ochiai clarified the stability characteristics under starved

IntechOpen

lubrication both theoretically and experimentally, and they propose a stabilization method utilizing the starved lubrication [3]. Furthermore, Naruse and Ochiai have experimentally studied the relation between gaseous-phase area and temperature distribution [4]. However, these observations have not been theoretically investigated [5], because the calculation method for the detailed gaseous-phase area has not been proposed. In addition, in actual bearing systems, if the lubricant is not able to supply the bearing sufficiently, it means there are starved lubrication condition and also a high possibility of serious erosion damage or serious seizure on the bearing surfaces. Therefore, it is believed that predicting the gaseous area in journal bearings is very important.

Generally, to analyze the oil-lubricated journal bearing, the Reynolds equation is used, and the half-Sommerfeld condition or the Swift-Stieber condition has been applied [6, 7] in determining the gaseous areas for the Reynolds equation. However, in these models, it is assumed that the negative-pressure areas exist only in the gas phase and there is no oil in the bearing clearance area. Therefore, the flow-rate conservation does not hold in the calculations. As a more advanced method, Coyne and Elrod's condition [8, 9] is used. This model assumes oil-film rupture calculates the surface tension between the oil film and gaseous phase which is ignored in the half-Sommerfeld or the Swift-Stieber conditions. However, it is impossible to estimate the cavitation area of the entire journal bearing including the oil-filler port. Therefore, boundary condition models that consider the cavitation have been studied. For example, Ikeuchi and Mori have analyzed the oil-film cavitation areas while using the modified Reynolds equation [10, 11]. In this method, the two-phase flow is considered as an averaged single-phase flow of oil and gas. However, in the case of high eccentricity ratio and starved lubrication condition, they did not conduct the experimental verification with the cavitation area. On the other hand, considering the finger-type cavitation, analytical methods have been proposed by Boncompine et al. [12] and Hatakenaka et al. [13]. However, the estimations of the variation of the gaseous-phase area against the changing of amount of oil lubricant have not been able to for bearing engineers or researchers. Furthermore, it is reported by Hashimoto, Ochiai, and Sakai that the oil-filler port and supply oil quantity affect the journal bearing characteristics [14, 15]. However, the internal flow of the oil-filler port has not been mentioned specifically. Therefore, a different approach is required to analyze the journal bearing while considering the internal flow of an oil-filler port.

The computational fluid dynamics (CFD) treating a two-phase flow has been proposed [16] recent years. The volume of fluid (VOF) method has the advantage of convergency and calculation times comparing with other methods [17]. Moreover, the VOF method also has the merit of reproducibility of slag flow in journal bearings. Therefore, there are some studies treating a two-phase flow CFD analysis utilizing the VOF method. For example, Zhai et al. and Dhande et al. [18, 19] studied the effect of vapor pressure and rotational speed on the gaseous-phase area of a journal bearing.

However, the experimental verifications have not been done in this study. On the other hand, a combination of the Reynolds equation and CFD analysis considering the two-phase flows was reported by Egbers et al. [20]. Furthermore, the gaseous-phase area in the oil-filler port and the opposite load side that was obtained by the analytical method in this study have been compared with that obtained in the experimental results. However, the analytical results cannot precisely produce the gaseous-phase scale and shape, because the influence of the surface tension has not been calculated.

In this situation, the authors have tried to reproduce the gaseous-phase area on both the bearing surface and the oil-filler port in a small-bore journal bearing under the flooded and starved lubrication conditions while using a CFD model. Further, effects of VOF, surface tension, and vapor pressure of the setting condition were studied, and these analytical results were compared with the experiment, and the applicability was verified. Then, the authors considered the influence of surface tension on journal bearing from the Weber number, W_e. Furthermore, thermal analysis results under two types of supply oil conditions are shown. The effect of supply oil on bearing temperature characteristics was discussed from the results of calculation and experimental results.

2. Theory and calculation model

As mentioned above, the VOF method has some advantages of calculation cost, convergency, and easy to handle compared with other methods [21–24]. Therefore, the authors selected the VOF method, and actually ANSYS FLUENT 15.0 is used in this study. The used methodology is explained below.

2.1 Governing equations

Instead of the Reynolds equation, we applied the Navier–Stokes equation considering the surface tension to the journal bearing analysis in this study. The mass conservation equation and the momentum equation are shown as follows:

$$\nabla \cdot \vec{u} = 0 \tag{1}$$

$$\rho\left[\frac{\partial \vec{u}}{\partial t} + \nabla\left(\vec{u} \cdot \vec{u}\right)\right] = -\nabla p + \mu \nabla^2 \vec{u} + \rho \vec{g} + \left(\nabla \sigma + \sigma \gamma \vec{n}\right)\delta_{int} \tag{2}$$

where ρ is the fluid density, $\vec{u}$ velocity vector, p fluid pressure, μ fluid viscosity, $\rho\vec{g}$ gravitational force, σ surface tension, $\vec{n}$ normal vector, γ curvature of the boundary surface, and δ_{int} Dirac's delta function.

Moreover, $\nabla$is a differential operator, defined by

$$\nabla = \left(\frac{\partial}{\partial x}, \frac{\partial}{\partial y}, \frac{\partial}{\partial z}\right) \tag{3}$$

In this VOF calculation model, the fluid density and fluid viscosity are expressed as follows:

$$\rho = F\rho_1 + (1 - F)\rho_2 \tag{4}$$

$$\mu = F\mu_1 + (1 - F)\mu_2 \tag{5}$$

where F is the volume fraction and subscription 1 means oil and 2 means gaseous phase.

This CFD analysis can analyze the internal flow of an oil-filler port with the bearing clearance simultaneously because the inertia term is considered in the basic equation.

On the other hand, the energy equation is shown as follows:

$$\rho c_p\left[\vec{u}\nabla(T)\right] = \lambda \nabla T + S_h + \mu\left(\vec{u}\cdot\nabla^2\vec{u}\right) \tag{6}$$

where c_p is the specific heat and λ thermal conductivity. Moreover, the second term in the right side means the volume heat term and the third term in the right side means the viscous dissipation term.

2.2 Surface tension and cavitation model

To consider the effect of surface tension, the continuum surface force (CSF) model proposed by Brackbill et al. [25] was used as the surface tension model implemented in FLUENT out-of-the-box. The last term of $\left(\nabla\sigma + \sigma\gamma\vec{n}\right)\delta_{int}$ in Eq. (2) represents the surface tension. In Brackbill et al.'s CSF model, the effect of surface tension is included as the surface tension term in the Navier–Stokes equation.

In addition, the cavitation model proposed by Schnerr and Sauer [20] was also used. The equation for the volume fraction of fluid is as follows:

$$\frac{\partial}{\partial t}(F\rho_2) + \nabla(F\rho_2) = \frac{\rho_1\rho_2}{\rho}\frac{DF}{Dt} \tag{7}$$

The vapor volume fraction ? is related to the number of bubbles n_b per unit volume of liquid and bubble radius R_b as shown in the following equation:

$$F = \frac{n_b \frac{4}{3}\pi {R_b}^3}{1 + n_b \frac{4}{3}\pi {R_b}^3} \tag{8}$$

where n_b is the number of bubbles which was set as 10^{13} in this study.

While considering the vapor pressure, the volume of air that is dissolved in oil expanded caused by a negative pressure which was observed in the journal bearing. Therefore, the vapor pressure was set to zero in this study. Moreover, the flow is laminar, and the analysis was conducted in a steady-state condition.

2.3 Calculation model

Figure 1 depicts the outline of a bearing treating in this study. We chose a full circular-type journal bearing in this study. The upper position of the bearing is provided with an oil-filler port, allowing the lubricating oil to flow into the bearing clearance by using gravity. **Table 1** lists its major dimensions. This model is one of the typical bearings for a small-size rotating machinery.

The model of bearing CFD calculation in this study is depicted in **Figure 2**. The bearing clearance with oil-filler port and oil-supply groove were set as flow calculation regions. The symmetrical configuration against the bearing center is used in reducing the calculation cost. The clearance around the minimum part contains

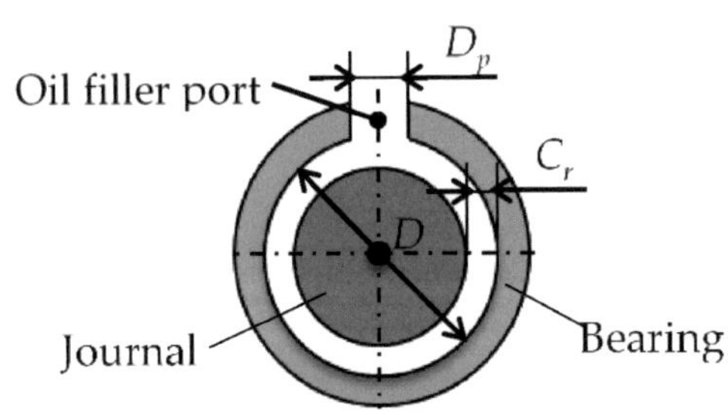

Figure 1.
Geometry of the test journal bearing [1].

Diameter (D) [mm]	25.0
Length (L) [mm]	14.5
Clearance (C_r) [mm]	0.125
Width-diameter ratio (L/D)	0.58
Diameter of oil-filler hole (D_p) [mm]	8.2

Table 1.
Specifications of bearing.

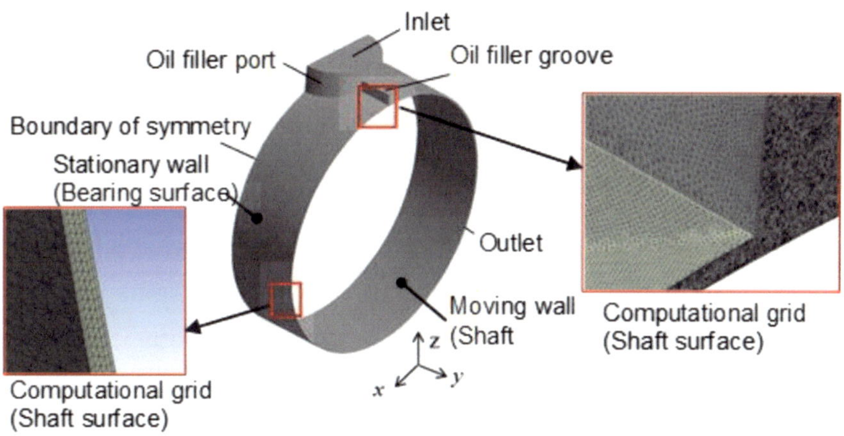

Figure 2.
Calculation model of the journal bearing [1].

Multiphase model			**Volume of fluid**
Calculation procedure			**Implicit method**
Calculation conditions	Vaporization pressure (P_v) [Pa]		0
	Surface tension (S) [N/m]		0.04
Fluid property	Density (ρ) [kg/m^3]	Oil	860
		Air	1.23
	Viscosity (μ) [Pa·s]	Oil	0.019
		Air	1.75×10^{-5}
	Thermal conductivity (f), [W/m·K]	Oil	0.13
		Air	0.024
	Specific heat (Cp), J/kg·K	Oil	19.5×10^{2}
		Air	10.1×10^{2}

Table 2.
Calculation conditions.

six-layer grids in the direction of the bearing clearance. The total mesh number under flooded and starved lubrication conditions were 64×10^5 and 18×10^6, respectively. The confirmation of mesh size was conducted, and enough calculation results were obtained as a pre-test study.

Table 2 shows the calculation conditions. The tension between oil and air was also considered while performing the calculations. The vapor pressure is set zero the same as the ambient pressure. Because, in this study, the side of the bearing and side of the oil-supply groove are open to the outside, the outside gas is easy to flow into the bearing and easy to generate the gaseous-phase cavitation at the position of negative-pressure generation. The surface tension was set to 0.04 N/m, which was measured while using the du Noüy method (ASTM 971–50).

3. Experiment setup and experimental method

The schematic of the experimental test rig is depicted in **Figure 3**. This rig consists of a rotor installed at its center and a rotating shaft supported by two bearings on its left- and right-hand sides. The right one is the test bearing for visualizations. It is manufactured while using transparent acryl, which allows us to observe the formation of the oil film and the generation of the gaseous phase. The shaft is driven with a DC motor that can control the rotational speed continuously by an inverter. The displacements in the horizontal and vertical directions of the journal are measured with eddy current-type proximity probes. The lubricating oil is supplied from the oil tank positioned on the top of the bearing through a control valve. The leaking oil from the side end of the bearing is returned using the pump. The VG22 oil is used and the oil temperature in the oil tank is fixed to 40°C with a heater.

In this study, we also conducted the measurement of temperature distribution in the journal bearing clearance using sheathed thermocouples. **Figure 4** shows the positions of the thermocouples in the bearing. Two lines of bearing centerline and halfway between the center and side end of the bearing are installed, and they are positioned 45° apart on the bearing's circumference. Moreover, the thermocouples were only installed on one side of the bearing in order to be able to visualize the gaseous phase in the bearing clearance at the same time. It was found in the previous experiments that the temperature measurement error was almost negligible in the case of the obliquely installed thermocouple. The thermocouples were secured by feedthroughs, and oil leakage through the insertion hole was prevented by applying a sealant. As the experimental method, the temperature of the supplied oil was fixed at 40°C, while the rotating speed of the shaft was increased to 7500 rpm. Moreover, the ambient temperature was fixed at 25°C. In this study, the temperature in the bearing clearance was measured under two kinds of supply oil conditions.

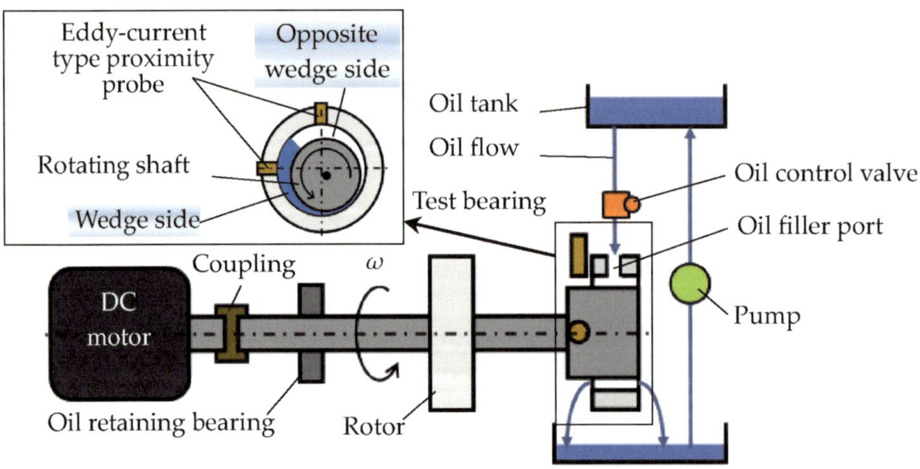

Figure 3.
Geometry of an experimental test rig ([1] partially modified).

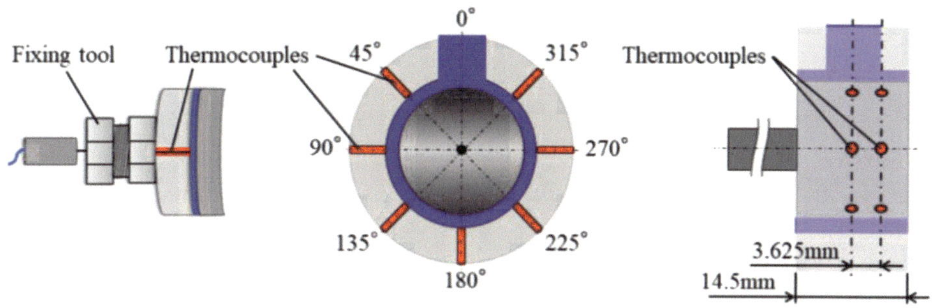

Figure 4.
Positions of thermocouples [2].

4. Results and discussions

4.1 Gaseous-phase area reproducibility

4.1.1 Under flooded lubrication conditions in the journal bearing

In this study, four types of calculations were calculated to clarify the effects of vapor pressure and surface tension. Results of (i) analysis of the volume fraction F distribution of oil and (ii) experimental visualization under flooded lubrication conditions are depicted in **Figure 5**. The red color in **Figure 5(i)** indicates the phase of complete oil, whereas the blue color indicates that of complete gas. Further, the solid and dotted black lines perpendicular to the circumferential direction of the bearing indicate the maximum and the minimum clearance, respectively. The 0° means the position of the most upper part of the bearing and the oil-filler port of the bearing exists at this position. The black arrow means the rotational direction and, consequently, the main flow direction of lubrication oil is the same.

In this study, we have presented the results for the surface of the rotating shaft. Moreover, in **Figure 5(ii)** of the experimental result, the yellow areas represent the gaseous phase, and the remaining areas indicate the oil film. The conditions of these calculations and experiment are as shown below. The rotational speed was n = 3500 rpm. The volume of oil supply was q = 2.6 cm^3/s, the eccentricity ratio ε = 0.54, and the attitude angle ϕ = 72.9°. These values are based on the experimental result.

In the case of VOF and VOF with surface tension as shown in **Figure 5(i-a, i-b)**, the volume fraction around the side end of the bearing decreases between 270° and 135°, and the volume fraction of the remaining area is approximately 1, which means complete oil. Around the oil-filler port, the volume fraction is most decreased. In the case of VOF with vapor pressure as depicted in **Figure 5(i-c)**, the volume fraction around the side end of the bearing slightly decreases between 300 and 100°. The range of the volume-fraction decrease in this case is smaller than that observed in the case of VOF alone or in the case of VOF with surface tension. On the other hand, in the case of VOF with vapor pressure and surface tension as shown in **Figure 5(i-d)**, the volume fraction around the bearing side end is approximately 1 between 300 and 0°. This tendency is quite different from other cases. Moreover, between 0 and 135°, the value of volume fraction decreases, and the decreased area which means the gaseous-phase area is wider than that of VOF with vapor pressure shown in **Figure 5(c-1)**. The tendency found in **Figure 5(i-d)** of the gaseous-phase area is also found in the experiment. Therefore, it is concluded that **Figure 5(i-d)** is in good agreement with the experiment compared with the other cases.

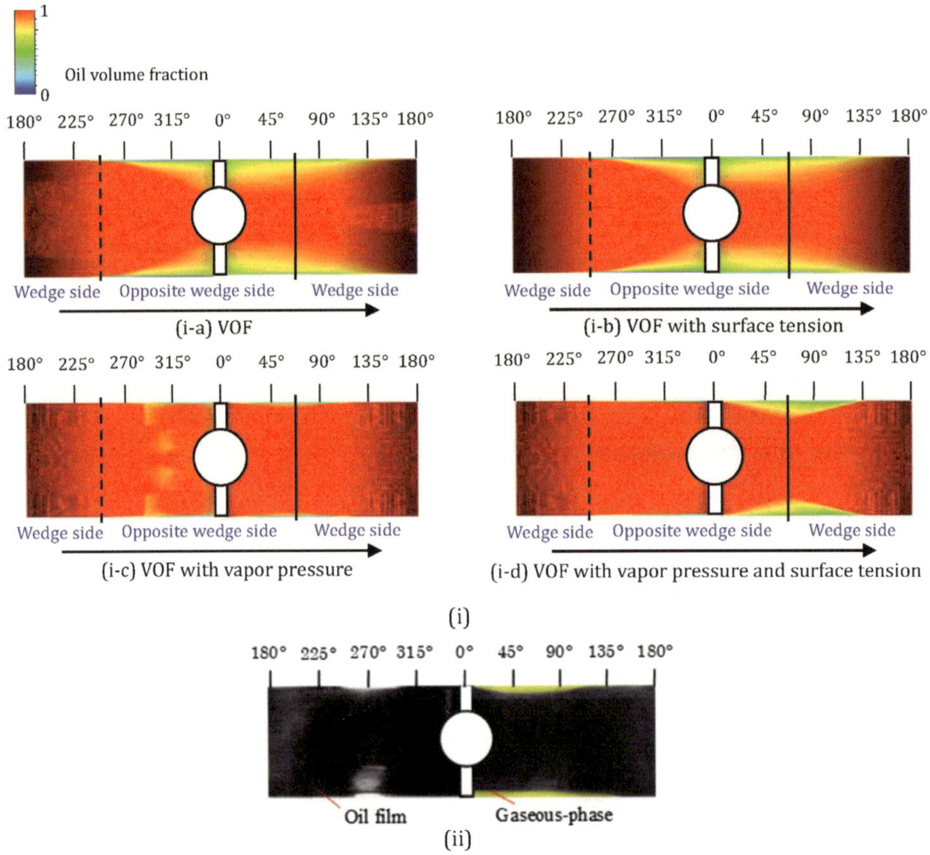

Figure 5.
Comparison of calculation and experimental results under flooded lubrication conditions ([1] partially modified). (i) Calculation results and (ii) Experimental result.

4.1.2 Under starved lubrication conditions in the journal bearing

The calculation results of oil volume fraction and experimental visualization result under starved lubrication conditions are depicted in **Figure 6**. Further, the volume of oil supply under this condition was q = 0.5 cm^3/s, the eccentricity ratio was ε = 0.76, and the attitude angle was ϕ = 71.5°.

From these results, it is found that the volume-fraction distribution and gaseous area under starved lubrication differ from the results under flooded lubrication. In the case of VOF, the volume fraction as shown in **Figure 6(i-a)** is one which means a complete oil phase at the minimum clearance of 180°, and the range of the decrease of volume fraction increases for the remaining ranges. In a wide range of around the side end of the bearing, the volume fraction decreases compared to that of the flooded lubrication conditions.

In **Figure 6(i-b)**, in the case of VOF with surface tension, a similar tendency is observed in the case of VOF. In contrast, from **Figure 6(i-c)**, the volume fraction increases at the side end of the opposite wedge side, and the volume fraction between 0 and 130° decreases at a greater rate than that observed in the cases of VOF and VOF with surface tension. Further, the volume fraction of the opposite wedge is observed to moderately decrease between 180 and 295°.

On the other hand, in the case of VOF with vapor pressure and surface tension shown in **Figure 6(i-d)**, the different tendency is observed. The volume fraction in the vicinity of bearing centerline and side end decreases from 270 to 0°, and

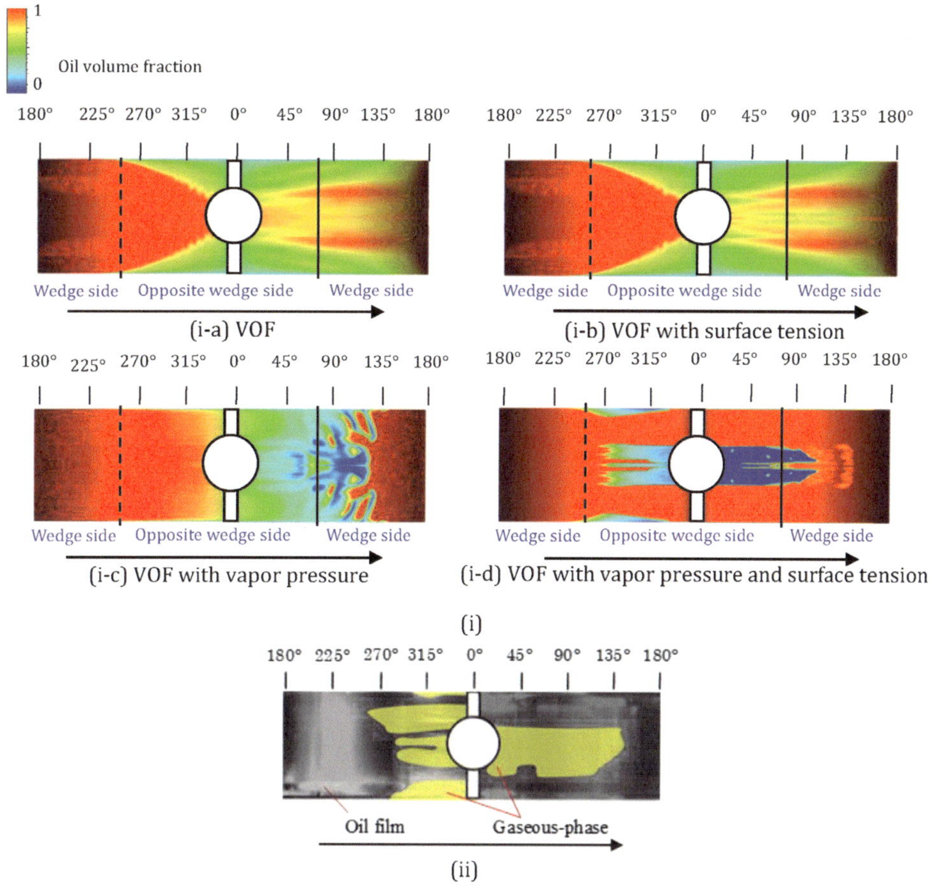

Figure 6.
Comparison of calculation and experimental results under starved lubrication conditions ([1] partially modified). (i) Calculation results and (ii) Experimental result.

between them, striped bands of oil are observed. Additionally, the volume of fraction in the vicinity of the bearing center between 0 and 120° is zero, and this range is observed to be the full air phase. From the above results, it is found that the gaseous phase changes before and behind of oil-filler port.

In the case of VOF with vapor pressure and surface tension as shown in **Figure 6(i-d)**, the tendency of the gaseous-phase area is in good agreement with that of the experiment shown in **Figure 6(ii)**. On the other hand, other calculation results which are shown in **Figure 6(i-a–c)** are very different from the experiment. The differences are more clear than the case of the flooded lubrication condition. Therefore, it is important to consider both vapor pressure and surface tension in the case of starved lubrication conditions especially. From these results, we were interested in the oil-film rupture shape at the end of the wedge side in the case of starved lubrication conditions as shown in **Figure 7**.

Figure 8 depicts the calculation and experimental visualization results of the oil-film rupture under starved lubrication conditions. It is found that the volume fraction near the moving shaft surface is one, and the value is decreasing gradually as approaching the bearing surface in the case of VOF with vapor pressure, as shown in **Figure 8(i-a)**. This tendency is the same as the model of the Coyne and Elrod which define the gas–liquid boundary between the bearing gap. However, in this case, the strong fluctuation is found between the boundaries of the volume fraction.

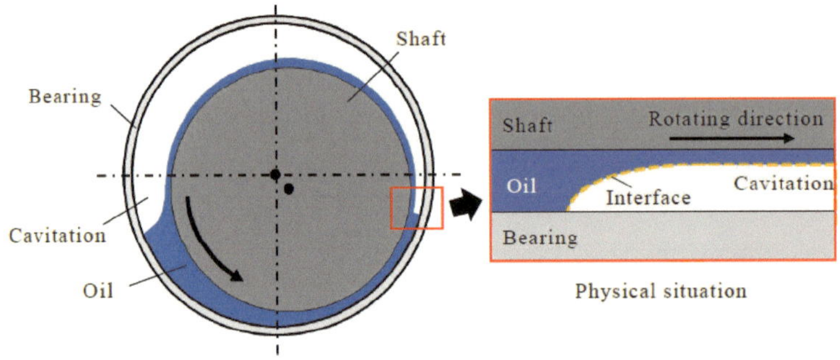

Figure 7.
Oil-film rupture position.

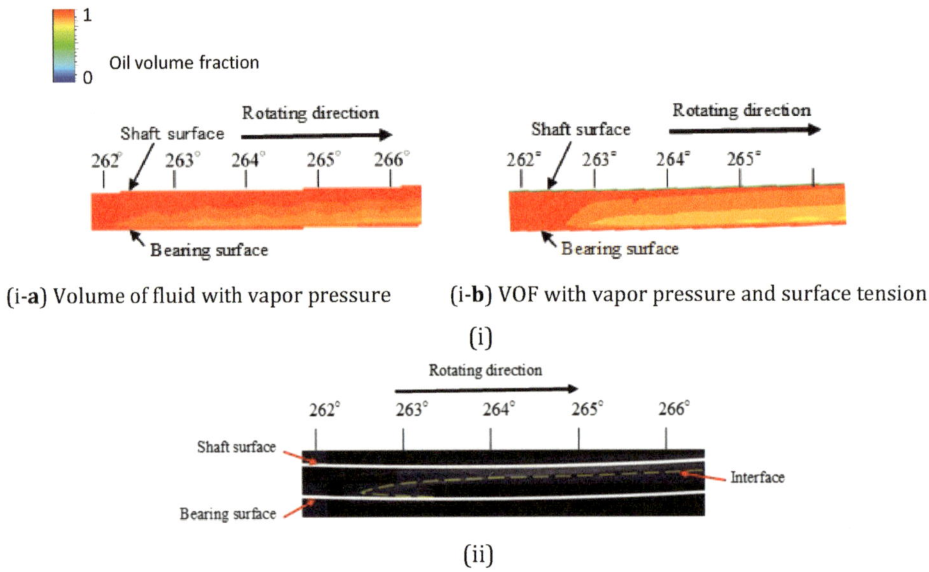

Figure 8.
Calculation and experimental results of the oil-film rupture under starved lubrication [1]. (i) Comparison of calculations and (ii) Experimental visualization result.

In the case of VOF with vapor pressure and surface tension as shown in **Figure 8(i-b)**, a two-phase flow exhibits a similar tendency as that exhibited by VOF with vapor pressure shown in **Figure 8(i-a)**. However, the interface between the oil film and gaseous phase is smoothly curved with an increase in clearance. From these results, it is confirmed that our proposed 3D CFD calculation model considering the vapor and the surface tension can reproduce the oil-gas boundary and it is in good agreement with the experiment shown in **Figure 8(ii)**.

4.2 Inner flow difference on the oil-filler port

From the abovementioned results, we focused on the inner state of the oil-filler port, because the gaseous area around the oil-filler port including the supply groove was large especially under the starved lubrication conditions.

Figure 9 depicts the results at a cross section of the oil-filler port in the case of under flooded lubrication conditions of **Figure 9(a)** and under starved lubrication

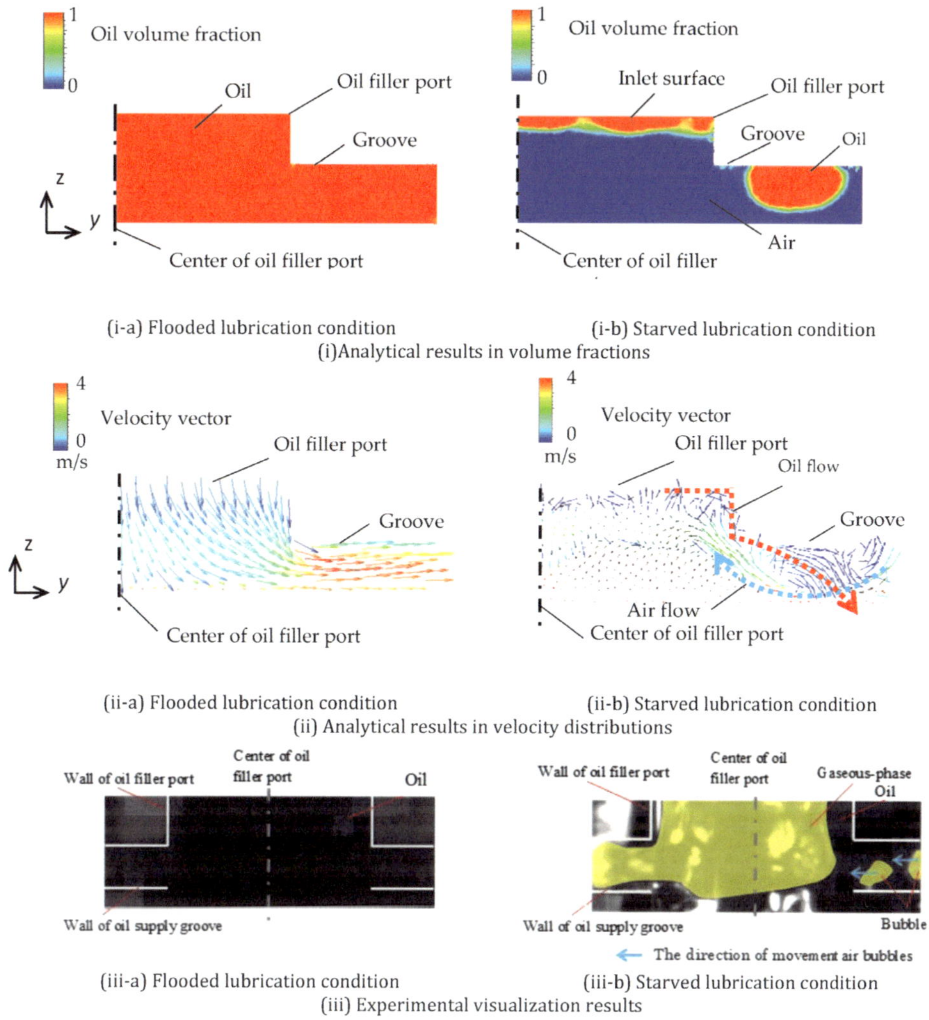

Figure 9.
Results of inner flow on the oil-filler port [1].

conditions **Figure 9(b)**. **Figure 9(i)** indicates the calculation results in volume fraction, **Figure 9(ii)** indicates the calculation results in velocity distribution, and **Figure 9(iii)** indicates the experimental visualization results. These figures depict the view from the front. The dash-dotted lines indicate the surface of symmetry.

From **Figure 9(i-a)**, under flooded condition, the oil phase is observed throughout the structure of the oil-filler port and the oil-supply groove. Furthermore, from **Figure 9(ii-a)**, it is observed that the velocity vectors are directed from the entrance surface of the oil-filler port to the oil-supply groove and bearing clearance in strong momentum. Hence, it is considered to be a less gaseous phase in the bearing clearance under flooded conditions. Here, the comparison between the analytical and experimental visualization results shows that the obtained results are close to the experimental results under the flooded lubrication conditions.

On the other hand, under starved lubrication conditions as shown in **Figure 9(i-b)**, it is observed that the gaseous phase exists in the major area of the oil-filler port whereas the oil exists around the inlet of oil-filler port and at the upper center area of oil-supply groove. Further, from the velocity vector result of **Figure 9(i-b)**, it is found that the oil flows from the top surface of the oil-filler port to the oil-supply groove along the walls whereas outer air flows into the oil-filler port through the oil-supply groove. One of the main causes of these flows is considered due to the surface tension of oil and air, and the effects appear to be significant especially in the case of starved lubrication.

Furthermore, oil is supplied from the oil-supply groove to the bearing, thereby causing the occurrence of the gaseous phase at the center of the bearing of the wedge side of the journal bearing under starved lubrication conditions.

The same tendencies are confirmed from the experimental visualization results of **Figure 9(iii)**.

4.3 Influence of surface tension in the journal bearing

In the previous section, it is confirmed that considering the surface tension to calculate the gaseous-phase areas accurately in journal bearings especially under starved lubrication is important. In the next step, we considered the influence of surface tension on journal bearing from the viewpoint of Weber number W_e. The Weber number is expressed in the following equation:

$$W_e = \frac{\rho U^2 H}{\sigma} \tag{9}$$

where U represents the speed and H represents representative length.

In the Weber number W_e, the meaning of the numerator is the fluid inertia force and that of the denominator is the surface tension. Generally, it is known that if W_e is less than one, the influence of surface tension is strong.

First, we examined the surface tension influence on bearing clearance. In this case, it was assumed that the representative speed U is the peripheral speed of moving the journal surface. Since the rotational speed is high, the Weber number extremely exceeds, and the influence of surface tension is negligible. This result is reasonable. Because many previous studies have been neglected, the surface tension effect and the reliability have been verified. However, as mentioned above, it is found that the influence of surface tension on gaseous-phase areas under starved lubrication is significant. Therefore, we focused on the internal flow of the oil-filler port. As the representative speed, the internal flow speed which is determined from supply oil flow rate was used.

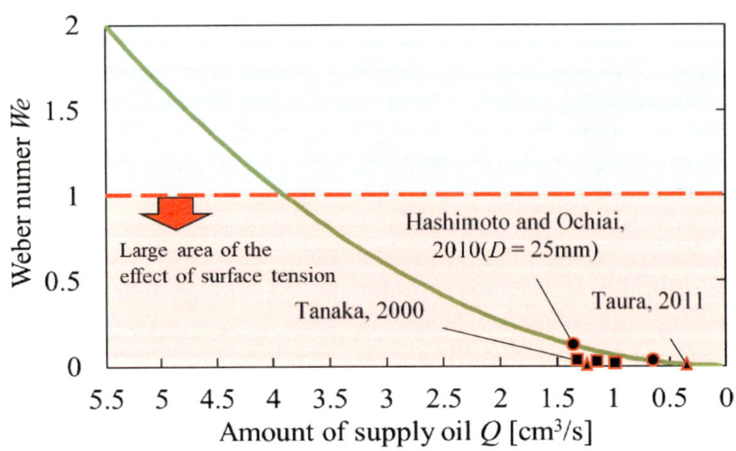

Figure 10.
Relation between the Weber number and amount of supply oil [1].

Figure 10 depicts the calculation result of the Weber number W_e against the oil flow rate. The continuous line indicates the value, whereas several plots indicate the value in other studies of journal bearings under the starved lubrication conditions [3, 26, 27]. Focusing on the continuous line, the Weber number W_e drop below 1 approximately under the amount of supply oil of 4 cm^3/s, and the influence of surface tension in the internal flow of the oil-filler port becomes too large to ignore.

Moreover, it is found that the Weber number W_e of another research of journal bearing under starved lubrication is remarkably smaller than one. Therefore, it is concluded that the cause of surface tension influence is related to supply velocity from the oil-filler port.

4.4 Temperature analysis

In this section, the thermal CFD analysis in journal bearing under two kinds of lubrication conditions is discussed. In our previous study [3, 4], it was found that the supply oil quantity affects the journal bearing stability, and the critical oil-supply quantity of transition state is determined. Therefore, two types of oil-supply quantity, one is transition state and the other one is starved lubrication condition, were selected in this chapter. The conditions of supply quantity and journal center positions are shown in **Table 3**. These conditions were decided by the experiment.

4.4.1 Oil-film temperature and volume fraction

Figure 11 shows the thermal CFD analysis results under two conditions of supply oil quantity. **Figure 11(i)** indicates the volume-fraction distribution of oil under both conditions. The red color means full oil and the blue color means full air. **Figure 11(ii)** indicates the analytical results of the temperature distribution of bearing. From **Figure 11(i-a)**, under transition condition, the volume fraction of the wedge side becomes zero at the side end. Thus, these areas are the gaseous phase. Moreover, the volume fraction increases at the centerline of bearing. On the other hand, the volume fraction of the inverse wedge side decreases with increasing of clearance. From **Figure 11(ii-a)**, the temperature slightly rises caused by the shear friction, but it is found that the temperature is almost 40°C at the full area of the bearing. Thus, it is found that the heat quantity by the shear friction is small in the case of the transition region.

On the other hand, from **Figure 11(i-b)**, the volume fraction of the wedge side increases at the center of bearing, while it is zero around the side end. The volume fraction of the inverse wedge side decreases with increasing of clearance, while it becomes zero around the oil-filler port. From **Figure 11(ii-b)**, it is found that the temperature around the oil-filler port is about 38°C while it decreases compared to the temperature of supply oil. Moreover, the temperature of the around centerline on the bearing is smaller compared around the side end.

Figure 12 shows the theoretical results of temperature distribution on the transition region and starved conditions at the point of centerline under (a) transition

	Oil supply amount Q, cm^3/s	Eccentricity ε	Attitude angle φ, deg.
Transition region	3.7	0.47	66.9
Starved lubrication	0.5	0.83	43.3

Table 3.
Specifications of calculation conditions [2].

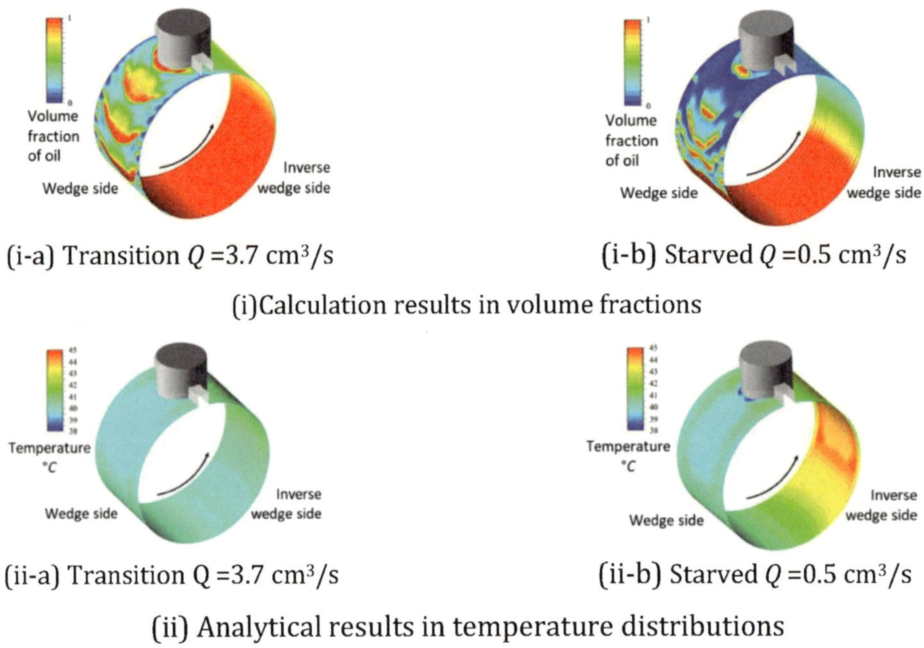

Figure 11.
Results of thermal CFD analysis [2].

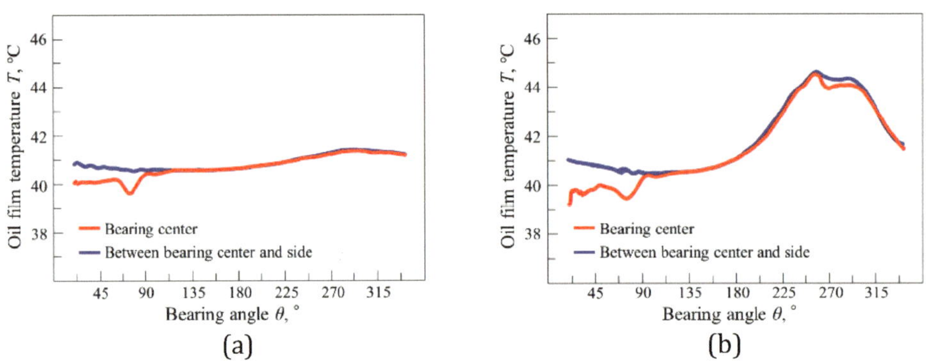

Figure 12.
Analytical results in temperature distributions [2]. (a) Transition Q = 3.7 cm³/s and (b) Starved Q = 0.5 cm³/s.

and (b) starved lubrication conditions. From **Figure 12(a)**, the temperature elevation in the bearing clearance against the supply oil temperature which is 40°C is not so high. In addition, there is a little difference from 22.5 to 90°. On the other hand, from the results for the starved lubrication condition, as shown in **Figure 12(b)**, the temperature distributions are extremely different from those of the oil transition conditions. The temperature distributions of the starved lubrication condition increase with an increasing bearing angle after 135°, and the highest temperature area is found around the bearing angle of 250°. This is because the minimum clearance around the bearing angle of 250° decreases and the share friction resistance of oil film rises in the starved lubrication conditions. The temperature difference of the bearing center is found from 22.5 to 90° same as the transition region; however, the amount of difference is larger than that of transition region. Moreover, in the low-temperature region, the value of temperature is lower than the supply oil temperature of 40°C.

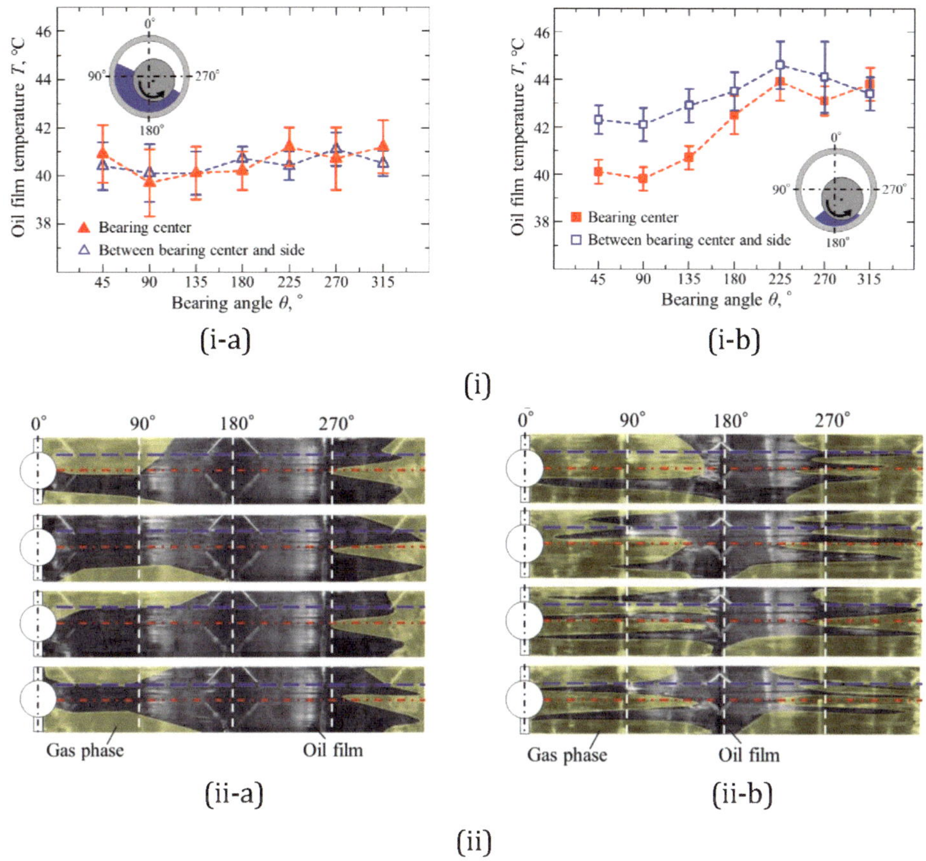

Figure 13.
Results of temperature distributions and gaseous-phase visualizations [2]. (i) Experimental results in temperature distributions ((a) Transition Q = 3.7 cm³/s; (b) Starved Q = 0.5 cm³/s). (ii) Experimental visualization results of gaseous-phase region ((a) Transition Q = 3.7 cm³/s; (b) Starved Q = 0.5 cm³/s).

Figure 13 depicts the experimental visualization results of (i) temperature distributions and (ii) results of gaseous-phase visualization. In **Figure 13(ii-a, b)**, there are typically four results which are indicated due to some fluctuation of gaseous-phase existing under actual rotations.

From the results in **Figure 13(i)**, it is found that the same tendencies of temperature distributions are shown between calculation and experiment. As shown in **Figure 13(i)**, the temperatures are almost constant against the bearing angle in the case of transition condition. Hence, the same tendencies of temperature distributions are found between the calculation and the experiment. On the other hand, under the starved lubrication condition, the temperatures are higher against the bearing angle, and the difference of temperature between two bearing positions is found. Here, the temperature at the center of bearing is lower than that of between the center and side despite being closer to outside the edge. Generally, the temperature of the bearing is larger than that of the side. Therefore, the reason could be the cooling effect from the visualization results shown in **Figure 13(ii)**.

As shown in **Figure 13(ii-a)**, the gaseous phase rarely existed at the center of the bearing at bearing angles between 0 and 135°. However, the temperatures are almost constant. This is because the friction resistance is not so high due to the relatively large amount of oil-film thickness existing in this case.

On the other hand, under starved condition shown in **Figure 13(ii-b)**, the gaseous phase exits at the center of the bearing. The gas is considered the inflow air from the oil-filler port. Hence, it is believed that the oil film at the center position was cooled down by an inflow air. Comparing with it, the position of between center and side, oil phase exists. Therefore, the cooling effect is lower than that of bearing center position.

From the abovementioned results, the authors considered that the internal flow of the oil-filler port influenced the temperature characteristics of starved lubrication conditions. Therefore, we focused on the calculation results in the oil-filler port under starved lubrication conditions.

4.4.2 Results in oil-filler port

Figure 14 shows the calculation results of the oil-filler port. **Figure 14(a)** indicates the results of the internal oil-filler port at the center of the bearing width of starved lubrication condition. **Figure 14(i-a)** indicates the volume fraction, while **Figure 14(i-b)** indicates the temperature. Moreover, **Figure 14(ii)** indicates the analytical results in the front view of the oil-filler port. From **Figure 14(ii)**, the temperature is high near the shaft, and the volume fraction decreases in that area. From the above, it is considered that the gas phase in this region is a circulating flow. In contrast, the gas phase exists in the wide area in the inside of the oil-filler port, while the temperature of the gaseous phase is smaller than the temperature of supply oil. Moreover, it is found that the temperature of oil is commensurate with the temperature of supply oil or less than. Furthermore, the temperature of the gaseous phase from the bearing clearance is about 40°C at the center of the oil-filler port. In **Figure 14(ii)**, the gaseous phase exists in the wide area in the inside of the oil-filler port as with **Figure 14(i)**, while it exists also in the oil-supply groove. The temperature of gaseous phase around the side end of the oil-supply groove is the same as the ambient atmosphere temperature set on the analysis; thus, it is found that outside air counterflows the oil-supply groove. From these results, in the case of the starved lubrication condition, it is considered that the outside air flows from

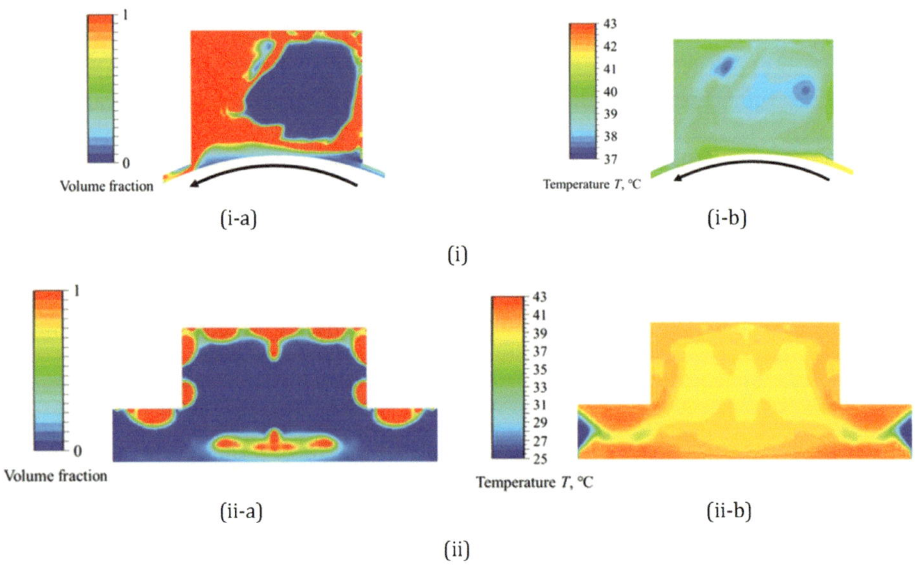

Figure 14.
Calculation results in the oil-filler port under starved lubrication conditions [2]. (i) Results in the center surface from the view of the side ((a) Volume fraction; (b) Temperature). (ii) Results in the front view ((a) Volume fraction; (b) Temperature).

the oil-supply groove and the side end of bearing cooled the supplied oil and the circulating flow; thus, the temperature of the center of bearing is controlled.

5. Conclusion

In this chapter, using a two-phase flow CFD analysis, the calculation of gaseous-phase areas in journal bearings under flooded and starved lubrication conditions was conducted, and the surface tension effect on multiphase flow CFD analysis of journal bearing especially generating the gaseous-phase area was studied.

As a result of comparing the calculation results and the experimental results, the VOF calculation considering the surface tension and vapor pressure was observed to be in good agreement under both lubrication conditions.

Furthermore, under starved lubrication, the calculation results of the interface of the oil film and gaseous phase during oil-film rupture agree rather well with experimental visualization result if they consider both vapor pressure and surface tension. While using these results, the effect of surface tension was discussed from the viewpoint of the Weber number, and it is concluded that the Weber number is strongly lower than one by using the supply oil speed as the representative speed and strongly influenced.

Moreover, thermal CFD analysis of a two-phase flow was conducted under two conditions of supply oil, and they were compared with the experiment. As a result, it is believed that in the case of the starved lubrication conditions, the air flowing outside of the oil-supply groove created a circulating flow; thus, the temperature in the bearing is controlled.

It is concluded that the two-phase VOF CFD analysis considering the vapor pressure and surface tension is applicable in reproducing the gaseous phase on the journal bearing.

Conflict of interest

The authors declare no conflict of interest.

Author details

Masayuki Ochiai*, Fuma Sakai and Hiromu Hashimoto
Department of Mechanical Engineering, Tokai University, Kanagawa, Japan

*Address all correspondence to: ochiaim@keyaki.cc.u-tokai.ac.jp

IntechOpen

© 2020 The Author(s). Licensee IntechOpen. This chapter is distributed under the terms of the Creative Commons Attribution License (http://creativecommons.org/licenses/by/3.0), which permits unrestricted use, distribution, and reproduction in any medium, provided the original work is properly cited.

References

[1] Ochiai M, Sakai F, Hashimoto H. Reproducibility of gaseous phase area on journal bearing utilizing multi-phase flow CFD analysis under flooded and starved lubrication conditions. Lubricants. 2019;7:74

[2] Sakai F, Ochiai M, Hashimoto H. Two-phase flow CFD analysis of temperature effects on oil supplied to small-bore journal bearing with oil supply groove. Tribology Online. 2018; **13**(5):232-240

[3] Hashimoto H, Ochiai M. Stabilization method for small-bore journal bearing utilizing starved lubrication. ASME Journal of Tribology. 2010;**132**:1-7

[4] Naruse Y, Ochiai M. Experimental study of safety supply flow rate on a small bore cylindrical journal sliding bearing. Journal of Advanced Science. 2012;**24**:24-28

[5] Heshmat H, Pinkus O. Performance of starved journal bearings with oil ring lubrication. ASME Journal of Tribology. 1985;**107**:23-31

[6] Gümbel L, Reibung und Schmierung im Maschinenbau. In: Everling E, editor. Berlin: Verlag von M. Krayn. 1925. VII + 240 S. Mit 37 Abb. Preis brosch. 12 M. ZAMM – Zeitschrift für Angewandte Mathematik und Mechanik [Internet]. 1927;**7**(5):415-415. Wiley; Available from: http://dx.doi.org/10.1002/zamm.19270070523

[7] Swift WH. The stability of lubricating films in journal bearings. In: Minutes of the Proceedings of the Institution of Civil Engineers, Vol. 233. London, UK: Thomas Telford-ICE Virtual Library; 1932. pp. 267-288

[8] Coyne JC, Elrod HG. Condition for the rupture of a lubricating film part i: Theoretical model. ASME Journal of Lubrication Technology. 1970;**92**:451-456

[9] Coyne JC, Elrod HG. Condition for the rupture of a lubricating film. Part II: New boundary conditions for Reynolds equation. ASME Journal of Lubrication Technology. 1971;**93**:156-167

[10] Ikeuchi K, Mori H. Hydrodynamic lubrication in seals with cavitation: 1st report, effect of cavity pressure on lubricating film. Bulletin of JSME. 1982; **25**:1002-1007 (in Japanese)

[11] Ikeuchi K, Mori H. An analysis of the lubricating films in journal bearings —Effects of oil supply condition on the static performance. Lubrication. 1982; **27**:533-540 (in Japanese)

[12] Boncompine R, Fillon M, Frene J. Analysis of thermal effects in hydrodynamics bearing. ASME Journal of Tribology. 1986;**108**:219-224

[13] Hatakenaka K, Tanaka M, Sizuki K. Thermo-hydrodynamic performance of journal bearings with partial reverse flow and finger-type cavitation being considered. Journal of Japanese Society of Tribologists. 2000;**45**:628-635 (in Japanese)

[14] Hashimoto H, Ochiai M. Experimental study on the stabilization of small-bore journal bearings by controlling starved lubrication and bearing orientation angle. ASME Journal of Tribology. 2009;**131**:011705

[15] Sakai F, Ochiai M, Hashimoto H. Stability characteristics and CFD analysis of two-phase flow of oil film journal bearing having two oil filler holes. Transactions of the JSME. 2017; **83**:16-00457 (in Japanese)

[16] Schnerr GH, Sauer J. Physical and numerical modeling of unsteady cavitation dynamics. In: Proceedings of the 4th International Conference on Multiphase Flow; New Orleans, LA, USA; 2001

[17] Ohta M, Sakai M, Shimada N, Honma S. Numerical Simulation of Multiphase Flow. Tokyo, Japan: Maruzen-Publishing; 2015 (in Japanese)

[18] Zhai LM, Luo YY, Wang ZW. Study about the influence of cavitation on the dynamic characteristics for the sliding bearing. IOP Conference Series: Materials Science and Engineering. 2015;**72**:1-10

[19] Dhande YD, Pande WD. Multiphase flow analysis of hydrodynamic journal bearing using CFD coupled fluid structure interaction considering cavitation. Journal of King Saud University - Engineering Sciences. 2016; **30**(4):345-354

[20] Egbers C, Gorenz P, Schmidt M, Wolf C. 3-D CFD simulation of the lubrication film in a journal bearing. Tribology International. 2008;**76**:40-47

[21] Olsson E, Kreiss G, Zahedi S. A conservative level set method for two phase flow II. Journal of Computational Physics. 2007;**225**:785-807

[22] Hirt CW, Nichols BD. Volume of fluid (VOF) method for the dynamics of free boundaries. Journal of Computational Physics. 1981;**39**:201-225

[23] Tryggvasona G, Bunner B, Esmaeelic A, Juricd D, Al-Rawahi D, Tauber W, et al. A front-tracking method for the computations of multiphase flow. Journal of Computational Physics. 2001;**169**: 708-759

[24] Bogdan RK, Abdollah AA. A Lattice-Boltzmann approach to fluid film lubrication. ASME Journal of Tribology. 2010;**132**(2):1-7

[25] Brackbill J, Douglas U, Koth B, Zemach C. A continuum method for modeling surface tension. Journal of Computational Physics. 1992;**100**: 335-335

[26] Tanaka M. Journal bearing performance under starved lubrication. Tribology International. 2000;**33**(3): 259-264

[27] Taura H, Kaneko S. Static characteristics of journal bearings under starved lubricating conditions. Transactions of the JSME. 2011;**77**: 3511-3521

Section 4

CFD Application in Thermal Engineering

Chapter 11

CFD Application for Gas Turbine Combustion Simulations

Valeriu Vilag, Jeni Vilag, Razvan Carlanescu, Andreea Mangra and Florin Florean

Abstract

The current chapter presents the use of computational fluid dynamics (CFD) for simulating the combustion process taking place in gas turbines. The chapter is based on examples and results from a series of applications developed as part of the research performed by the authors in national and European projects. There are envisaged topics like flame stability, pollutant emission prediction, and alternative fuels in the context of aviation and industrial gas turbines, growing demands for lower fuel consumption, lower emissions, and overall sustainability of such energetic machines. Details on the available numerical models and computational tools are given along with the expectation for further developing CFD techniques in the field. The chapter includes also some comparison between theoretical, numerical, and experimental results.

Keywords: combustion, gas turbine, numerical simulations, models, alternative fuels, experiments

1. Introduction

Gas turbines are energetic machines based on Brayton thermodynamic cycle [1] (**Figure 1**) meaning, among others, temperature rise using combustion at quasi-constant pressure.

In **Figure 1**, evolution (1–2) represents the real compression evolution of the working fluid into the compressor, (2–3) represents combustion at constant pressure, (3–4) represents real expansion of the working fluid into the turbine, and (4–1) represents the cooling down of the working fluid at constant pressure, usually the atmospheric one. Evolutions (1–2is) and (3–4is) are the isentropic compression and isentropic expansion, respectively, and they are shown on the graphic in order to emphasize the difference between real and ideal (isentropic).

Unlike compressors and turbines in which only pure gas-dynamic processes develop, combustion involves also chemical reactions between air and fuel, resulting in flue gases driving the turbine. Since combustion is known from ages, the overall efficiency of it is very high [2] and enhanced by the high pressure provided by the compressor. Still, many current studies are directed on this subject willing to reduce pollutant emissions or to accommodate alternative fuels such as biogas or both.

In general, the combustion process is organized into the gas turbine as shown in **Figure 2** [3].

In **Figure 2** the red arrows represent the burning working fluid (realizing close to stoichiometric mixture ratio with the injected fuel and very high temperature in the flame presented in orange), and the blue arrows represent the fluid which cools down the burning one down to the required temperature of the thermodynamic cycle (which is limited by the materials used to realize the combustion chamber and the turbine).

The arrangement of the combustion chamber is the subject of many research projects, some of the latest being lean burn program [4] where the reduction of NOx through the reduction of combustion temperature by multiple combustion is sought.

In **Figure 3**, depicted is a type of combustion which may lead to lower NOx production: the red zone (R) is rich in fuels, meaning more fuel than the calculated

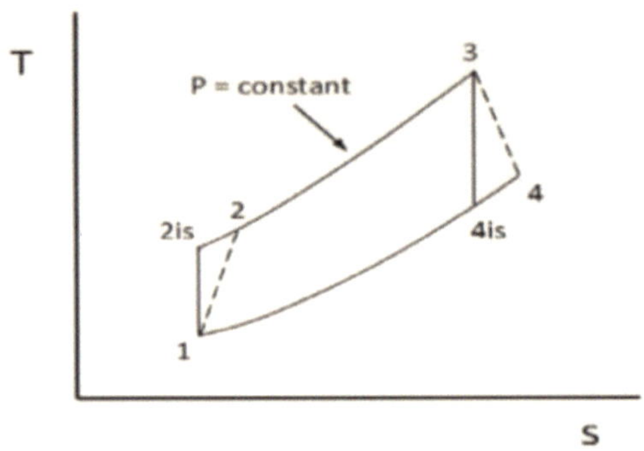

Figure 1.
Brayton cycle in temperature vs. entropy coordinates.

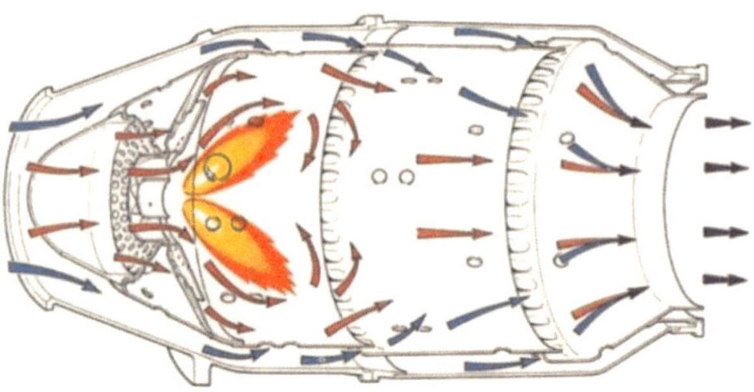

Figure 2.
Combustion chamber.

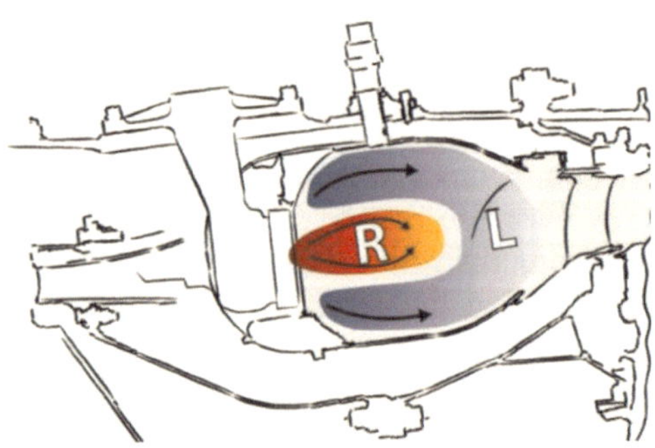

Figure 3.
Clean Sky research project [4].

from stoichiometric reaction, while the blue zone (L) is lean in fuel, meaning less fuel than the calculated from stoichiometric reaction. The Clean Sky research project aims at demonstrating that by realizing such an arrangement of the combustion process, the maximum flame temperature can be lowered reducing thus the dissociation reactions that create NOx, at the same final temperature reaching the turbine.

The thermodynamics of combustion is relatively simple and is based on the heat of reaction as shown in Eq. (1):

$$\text{Reactants}|_{\text{Tinitial}} \rightarrow \begin{cases} \text{Products}|_{\text{Tinitial}} + \text{HEAT} \\ \text{or} \\ \text{Products}|_{\text{Tcombustion}} \end{cases} \tag{1}$$

Still, the use of this equation gives good results only for the temperature at the end of the combustion process and not so accurate results on the resulting composition. Thus, the chemical reaction must be carefully studied and implemented in the study concerning the combustion chamber of a gas turbine.

Computational fluid dynamics (CFD) has been intensively used in the aerospace domain mainly for predicting the performances of the studied object which can be the entire aircraft or some particular part of it. For the gas turbines, all the main components can be studied using CFD: in the case of the compressor, the aerodynamic part is the most important seeking for high efficiency of transforming the available mechanical work into total pressure of the air; in the case of the turbine, the things are reversed, the study being focused on reducing the losses of transforming the potential energy of the fluid in the form of pressure and temperature, into mechanical work; in the case of the combustion chamber, aerodynamics plays a big role in the injection of fuel and mixing, but it must be coupled with chemical reactions and heat release within the transforming fluid.

2. Combustion and chemistry in CFD

The combustion process is the result of a strong exothermic chemical reaction as a result of energy exchanges that occur due to intermolecular collisions.

Generally, at ambient temperature, a chemical reaction occurs very slowly, because, although collisions at molecular level occur, they do not generate a sufficient amount of energy to trigger a chemical transformation.

According to the chemical kinetic theory, only "active" collisions, collisions involving molecules having an initial energy greater than or equal to an energy called activation energy, lead to a chemical reaction. This energy is needed to destroy or weaken existing intermolecular connections. The activation energy is given by Arrhenius equation [5, 6]:

$$k = \text{A} \cdot T^{\beta} \cdot \exp\left(\frac{-\text{E}_\text{a}}{\text{R} \cdot T}\right) \tag{2}$$

where k is the reaction rate coefficient, A is the pre-exponential factor, E_a is the activation energy, R is the universal gas constant, β is the temperature exponent, and T is the temperature.

The combustion process can be expressed in a simplified way through the global reaction mechanism, Eq. (3):

$$\text{Fuel} + \text{Oxidizer} \rightarrow CO_2 + H_2O + \text{Heat} \quad (3)$$

In CFD modeling more complex reaction mechanisms (e.g., mechanisms which take into consideration the formation of NO) are also available. Using a more complex reaction mechanism is more time-consuming and requires higher computational power. Thus, depending on the purpose of the conducted numerical simulation, a simpler or more complex reaction mechanism should be chosen.

In CFD modeling several combustion models are available [7–9].

2.1 Eddy dissipation model (EDM)

The eddy dissipation model is based on the concept that chemical reaction is fast relative to the transport processes in the flow. When reactants mix at the molecular level, they instantaneously form products. The model assumes that the reaction rate may be related directly to the time required to mix reactants at the molecular level.

By default, for the eddy dissipation model, it is sufficient that fuel and oxidant be available in the control volume for combustion to occur.

Because of the assumption of complete combustion, the eddy dissipation model may overpredict temperature under certain conditions (e.g., for hydrocarbon fuels in regions with fuel-rich mixture).

The eddy dissipation model was developed for use in a wide range of turbulent reacting flows covering premixed and diffusion flames. Because of its simplicity and robust performance in predicting turbulent reacting flows, this model has been widely applied in the prediction of industrial flames.

2.2 The finite-rate chemistry (FRC) model

The finite-rate chemistry model allows the computation of reaction rates described by the molecular interaction between the components in the fluid. It can be combined with the eddy dissipation model for flames where chemical reaction rates might be slow compared with the reactant mixing rates. The finite-rate chemistry model is best applied to situations where the chemical time scale is rate-limiting. This model can be used in conjunction with both laminar and turbulent flow.

2.3 The flamelet model

The flamelet model can provide information on minor species and radicals, such as CO and OH, and accounts for turbulent fluctuations in temperature and local extinction at high scalar dissipation rates, for the cost of solving only two transport equations. The model is only applicable for two-feed systems (fuel and oxidizer) and requires a chemistry library as input. The model can be used only for non-premixed systems.

The flamelet concept for non-premixed combustion describes the interaction of chemistry with turbulence in the limit of fast reactions (large Damköhler number). The combustion is assumed to occur in thin sheets with inner structure called flamelets. The turbulent flame itself is treated as an ensemble of laminar flamelets which are embedded into the flow field.

The main advantage of the flamelet model is that even though detailed information of molecular transport processes and elementary kinetic reactions are included, the numerical resolution of small length and time scales is not necessary. This avoids the well-known problems of solving highly nonlinear kinetics in fluctuating flow fields and makes the method very robust. Only two scalar equations have to be solved independent of the number of chemical species involved in the simulation.

Information of laminar model flames are pre-calculated and stored in a library to reduce computational time (PDF table). On the other hand, the model is still restricted by assumptions like fast chemistry or the neglecting of different Lewis numbers of the chemical species.

The coupling of laminar chemistry with the fluctuating turbulent flow field is done by a statistical method. The PDF used can in principle be calculated at every point in the flow field by solving a PDF transport equation.

The most often mentioned advantage of this method is that the nonlinear chemical source term needs no modeling. Even though the method avoids some modeling which is necessary if using moment closure, it still requires modeling of some of the most important terms, in particular, the fluctuating pressure gradient term and the molecular diffusion term. If combustion occurs in thin layers as assumed here, the molecular diffusion term is closely coupled to the reaction term, and the problem of modeling the chemical source term is then shifted towards modeling the diffusion term.

2.4 Burning velocity model (BVM) and extended coherent flame model (ECFM)

The burning velocity model (BVM) and the extended coherent flame model (ECFM) model the propagation of a premixed or partially premixed flame by solving a scalar transport equation for the reaction progress. The BVM uses an algebraic correlation for modeling the turbulent burning velocity (propagation speed of the flame in turbulent flow). When using the ECFM, the turbulent burning velocity is closed by solving an additional transport equation for the flame surface density.

The BVM is a combined model using:

- A model for the progress of the global reaction: burning velocity model (BVM), also called turbulent flame closure (TFC).
- A model for the composition of the reacted and nonreacted fractions of the fluid: laminar flamelet with PDF.

The ECFM is a combined model employing:

- A model for the progress of the global reaction: extended coherent flame model (ECFM), which is a member of the class of flame surface density models.
- Aa model for the composition of the reacted and nonreacted fractions of the fluid: laminar flamelet with PDF.

2.5 The monotone integrated LES (MILES)

The model solves the unfiltered Navier-Stokes equations for a global chemical reaction mechanism. The method uses no sub-grid closure models but employs the inherent numerical scheme dissipation to account for the energy transferred to the sub-grid scales.

2.6 The linear eddy mixing (LEM) model

LEM is a stochastic approach aimed at stimulating the turbulent mixing, molecular diffusion, and the chemical reaction in a one-dimensional domain embedded in the LES cells of the computational domain (LEMLES). LEM is the only known combustion model that does not use the scale separation hypothesis and is, therefore, valid even in regimes where the hypothesis fails. Also, the model is highly compatible with the large eddy simulation (LES) technique and very flexible in terms of the chemical reaction mechanism used to describe the chemical reactions. Nevertheless, the approach has some limitations. Most importantly, LEMLES is relatively much more expensive than conventional LES models, such as EBULES. However, it is highly scalable, so the overall computation time can be decreased by increasing the number of processors. Laminar molecular diffusion across LES cells is not included, but this limitation is significant only in laminar regions, whereas LEMLES is designed for high Reynolds number turbulent flow applications. Also, the viscous work is neglected in the sub-grid temperature equation but is explicitly included in the LES energy equation, which is used to ensure total energy conservation. Finally, the flame curvature effect is not explicitly present in the sub-grid.

3. CFD simulations for gas turbine combustion chamber and comparison to experimental results

3.1 Helicopter engine on alternative fuels

The desire to use top aviation technology in ground applications has led, through the years, to the transformation of a series of aviation gas turbines into drivers as part of industrial power plants. Some of these transformations have been made by the gas turbine producers, others even by beneficiaries or research institutions, such as COMOTI Romanian R&D Institute for Gas Turbines.

This section is focused on the behavior of a turboshaft with a structural construction allowing the modification of the entire fuel system, from the feeding lines, to the injection ramp and the actual injectors, as well as the relatively easy replacement of the aggregates.

From the theoretical point of view, several gaseous fuels have been studied as alternatives for the initial one, kerosene, such as methane and biogas with different chemical compositions (**Figure 4**). Obtaining chemical equilibrium for the

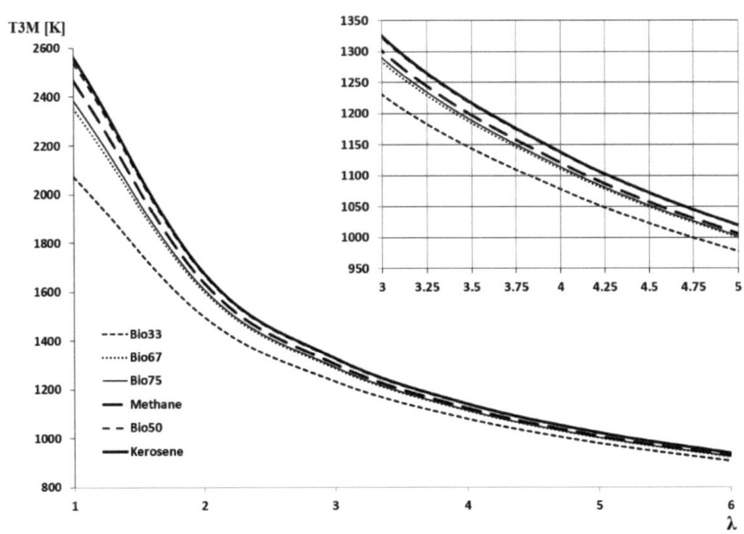

Figure 4.
The variation of the combustion temperature (T3M) with the air excess (λ) for different fuels (with focus on the usual zone for gas turbines).

combustion of these fuels has allowed to determine the parameters to be used as input data in numerical simulations of the combustion process in the gas turbine's combustor.

The numerical simulations, starting from data provided by either the producer, theoretical computations or experimental, include four cases, for the two mentioned alternative fuels, at two different operating regimes of the gas turbine: nominal and idle. The working fluids are defined as ideal gases: air as bicomponent mixture with 21% oxygen and 79% nitrogen; methane from the software library and biogas, as reacting mixture with 50% methane and 50% carbon dioxide. The cases are summarized in **Table 1**.

The numerical grid and boundary conditions are shown in **Figure 5**.

The eddy dissipation model, within ANSYS CFX [11], controls the formation of the reaction products, while the NO formation is controlled by two reaction schemes, WD1 and WDS. The advantage of the WDS scheme is that it also contains the CO creation model, through water-gas shift mechanism, allowing for higher accuracy, a fact also confirmed by the comparison with the experimental results, while the disadvantage consists in the necessity for higher computational resources and up to 50% more computing time. Some images with the temperature distribution in the combustion chamber are displayed in **Figures 6** and 7.

Using the 17 double thermocouples mounted on the engine, **Figure 8** containing a comparison between numerical and experimental results was obtained. It can be seen that the numerical results predict that two areas of maximum temperature exist at the end of the combustion chamber, and it was confirmed by the experiments on the entire engine.

Case/ parameter		Reference pressure	Air mass flow rate	O_2 mass fraction	Fuel mass flow rate	CH_4 mass fraction	CO_2 mass fraction
		[Pa]	[kg/s]		[kg/s]		
C1	Methane, nominal	760,000	0.825	0.233	0.010683	1	0
C2	Methane, idle	232,500	0.314	0.233	0.003212	1	0
C3	Biogas, nominal	760,000	0.825	0.233	0.0419	0.267	0.733
C4	Biogas, idle	232,500	0.314	0.233	0.0118	0.267	0.733

Table 1.
Input data for numerical cases.

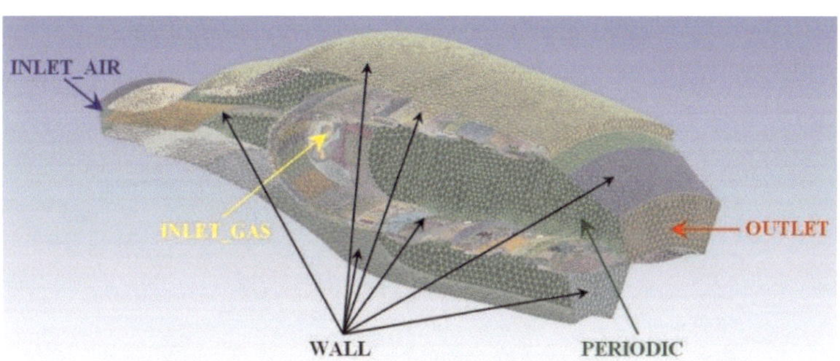

Figure 5.
Computational grid with defined regions for the boundary conditions [10].

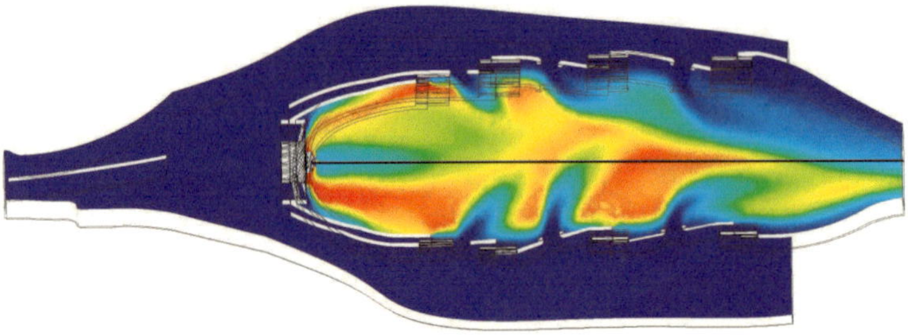

Figure 6.
General aspect of the temperature field in the combustor.

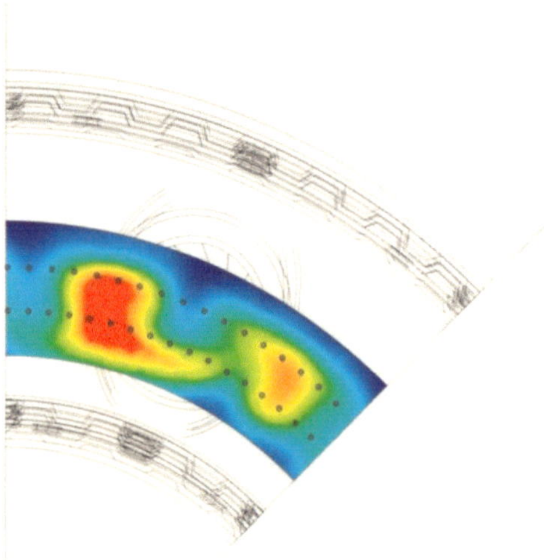

Figure 7.
Temperature fields at the end of the combustor.

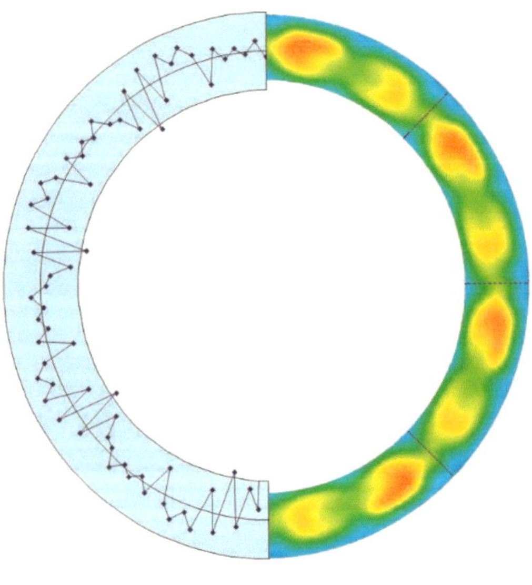

Figure 8.
Hybrid experimental diagram for biogas combustion [10].

3.2 Micro gas turbine for power generation

The numerical simulation of the gas-thermodynamic processes inside a Garrett micro gas turbine has been conducted using the commercial software ANSYS CFX. This work is part of the [12] PhD. thesis. The purpose of these numerical simulations was to validate the combination of numerical models used to simulate the turbulence process, combustion process, and liquid fuel atomization process and compare the numerical results with the functioning data of the used micro gas turbine engine.

The Garrett micro gas turbine is composed of an intake device, a single-stage centrifugal compressor, a tubular-type combustion chamber, a single-stage radial turbine, and an exhaust device (**Figure 9**).

The numerical simulation was performed only on the combustion chamber assembly. An unstructured-type computational grid, having 3.576.588 tetrahedral-type elements and 592.465 nodes, has been generated using ICEM CFD. A density was created near the injector to better capture the field near the fuel inlet (**Figure 10**).

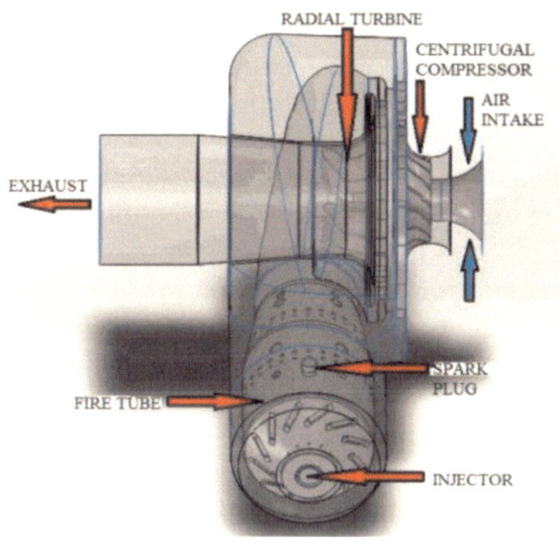

Figure 9.
Garrett micro gas turbine geometry.

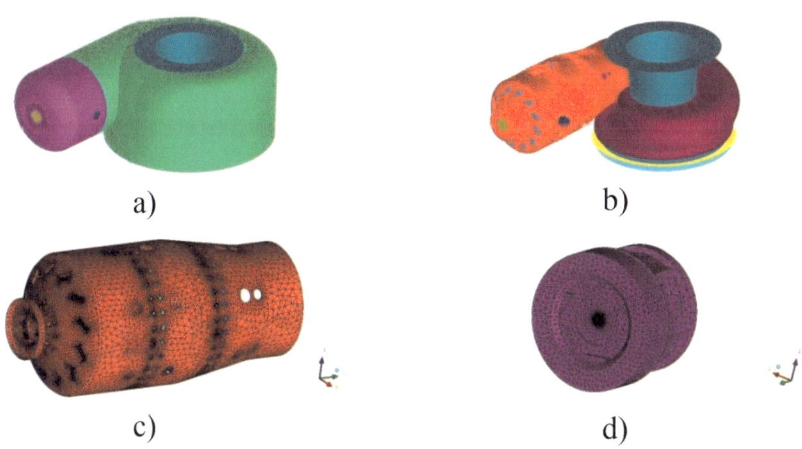

Figure 10.
Computational grid: (a) the exterior of the combustion chamber assembly, (b) the interior of the combustion chamber assembly, (c) the fire tube, (d) the injector.

The nominal functioning regime (20 KW load) has been considered for the simulation; thus, the fuel mass flow rate has been set at 0.0075 kg/s, and the air mass flow has been set at 0.522 kg/s. The air excess is of 4.7. Generally, for a gas turbine engine, the excess air should be between 3 and 5 [1]. The initial air temperature was set at 420 K. The used liquid fuel, Jet A, has been considered to enter the computational domain in the form of droplets. The mean diameter of the droplets has been assumed to be 30 μm, a value chosen based on the liquid droplet distribution diagram. The initial fuel temperature has been set at 300 K. The fuel spray cone angle has been set at 70°, based on the data presented in the micro gas turbine's maintenance manual [13].

A Reynolds averaged Navier-Stokes (RANS)-type turbulence model has been chosen, namely, the k-ε model, which is a numerically stable and robust model and very popular in the realization of technical applications numerical simulations [14–17].

The chosen combustion model has been the EDM model, based on a two-step kerosene-air reaction mechanism, imported from the ANSYS library. A simple reaction mechanism has been chosen because the purpose of these numerical simulations was to see if the used numerical models give a good approximation of the turbo engine functioning as a whole. The pollutant emission level has not been of interest at this stage. Using a more complex reaction mechanism would have been more time-consuming and would have required more powerful computational resources. The EDM combustion model has been chosen because of its simplicity and robust performance in predicting turbulent reacting flows. Because of these, the model is very often used in the realization of technical application numerical simulations [18–21].

The fuel droplet atomization and evaporation processes have been simulated using the cascade atomization and breakup (CAB) model, respectively, and the liquid evaporation model, both models imported from ANSYS library.

The reference pressure has been set at 101,325 Pa.

In **Figure 11** the pressure field through the micro gas turbine is presented. The pressure levels are relative to the reference pressure.

The pressure inside the fire tube is quasi-constant, as it can be seen in **Figure 3**, thus conforming the hypothesis that the combustion process inside a gas turbine combustion chamber takes place at constant pressure. The average air absolute total pressure at compressor exit-combustion chamber entrance is 275.466 Pa, thus obtaining an overall compression ratio of 2.7:1 which is close to the reported overall compression ratio of 3:1. The obtained pressure loss through the combustion chamber assembly is of 15%.

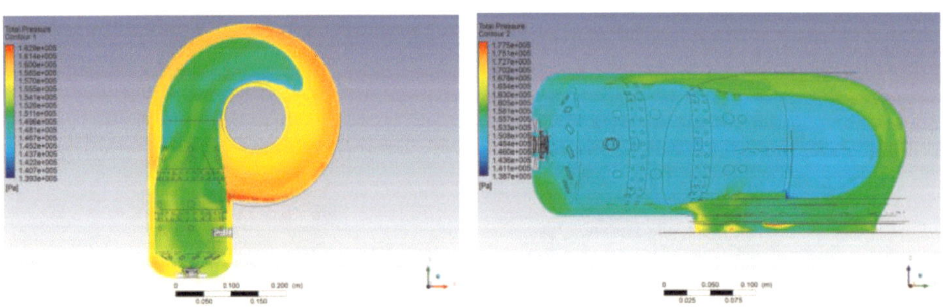

a) b)

Figure 11.
The total pressure field inside the combustion chamber assembly: (a) plan XY and (b) plan YZ.

In **Figure 12** the total temperature field through the micro gas turbine is presented.

From **Figure 12(a)** it can be observed that high-temperature zone is found only inside the fire tube and does not extend into the volute that redirects the exhaust gases to the turbine stator. The average total temperature at combustion chamber exit-turbine entrance is 992 K.

In **Figure 13** the fuel spray cone and the fuel droplet diameter distribution are presented.

From **Figure 13** it can be observed that the fuel is completely evaporated in the primary zone of the fire tube, before reaching the walls. This confirms that the numerical models chosen to simulate the spraying and vaporization processes of the liquid fuel are appropriate for the given application.

In **Figure 14** Jet A vapor mass fraction field is presented.

The results presented in **Figures 13** and **14** are in good correlation. Jet A vapors obtained from the vaporization of the liquid fuel are located in the primary zone of the fire tube. They are completely consumed inside the fire tube, as it should happen in the case of a properly functioning turbo engine. The average Jet A vapor mass fraction at combustion chamber assembly exit is of 7×10^{-7}.

In **Figures 15** and **16**, the CO mass fraction field and the CO_2 mass fraction field, inside the combustion chamber, are presented, respectively.

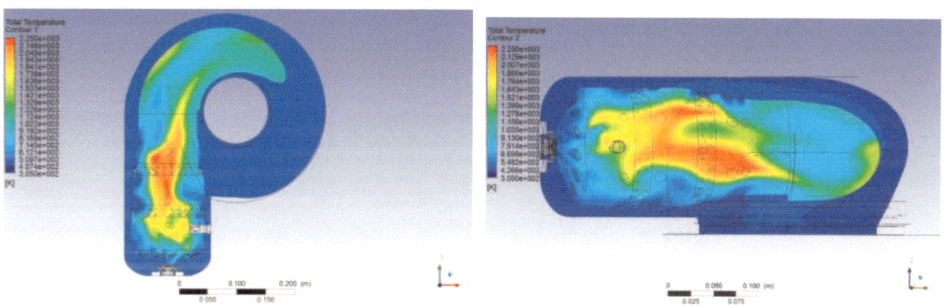

a) b)

Figure 12.
The total temperature field inside the combustion chamber assembly: (a) plan XY and (b) plan YZ.

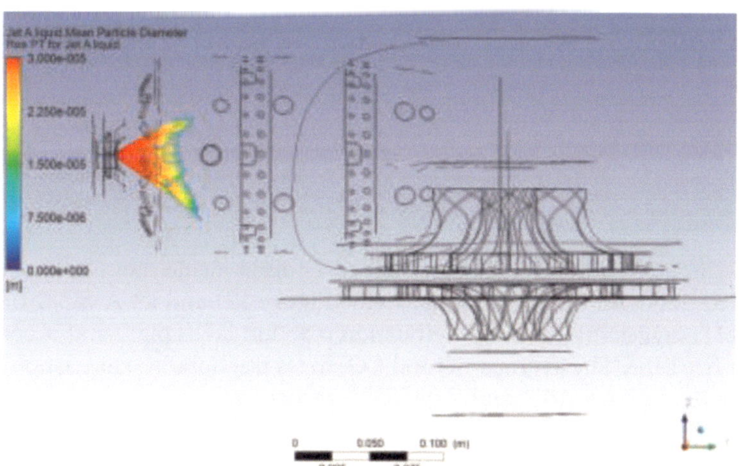

Figure 13.
Fuel spray cone and droplet diameter distribution.

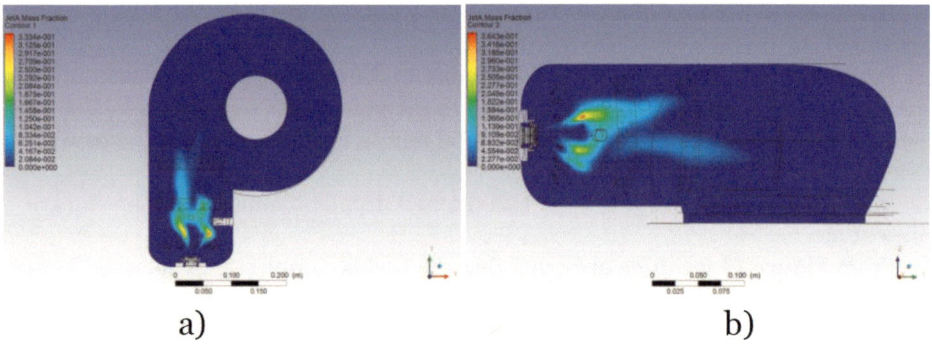

Figure 14.
Jet A vapor mass fraction field: (a) plan XY and (b) plan YZ.

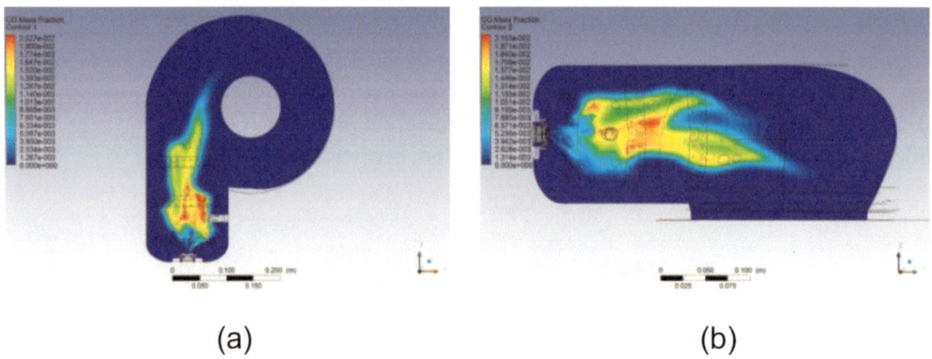

Figure 15.
The CO mass fraction field: (a) plan XY and (b) plan YZ.

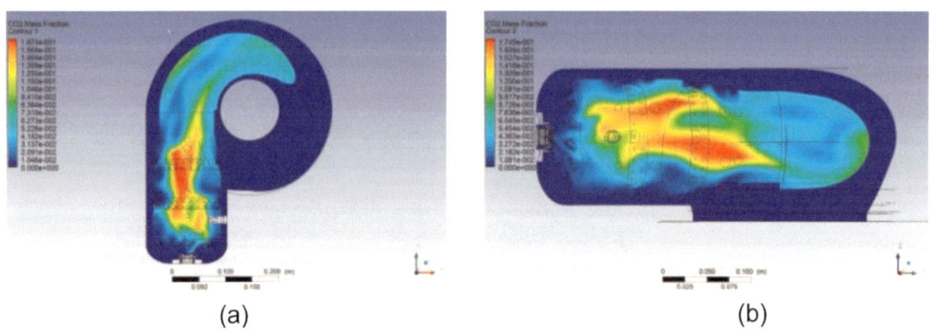

Figure 16.
The CO_2 mass fraction field: (a) plan XY and (b) plan YZ.

The highest CO and CO_2 concentrations are found inside the fire tube. This is in good correlation with the temperature field (**Figure 12**) and Jet A vapor field (**Figure 14**), suggesting that the combustion reaction takes place and is completed inside the fire tube. The average CO and CO_2 mass fractions at combustion chamber assembly exit are 7.4×10^{-6} and 0.0465532, respectively.

Based on the obtained results, the temperature and pressure fields, the fuel vapor, and CO and CO_2 mass fraction fields, it has been concluded that the numerical models used for the numerical simulation of the gas-thermodynamic processes inside the combustion chamber are appropriate for the given application, the results being consistent with actual functioning data of the Garrett micro gas turbine.

3.3 Afterburning system

The afterburning system is a component that is added to aviation engines (gas turbine engines), usually military ones, in order to maximize the thrust force of the planes. But it also has industrial applications like cogeneration. Cogeneration is a modern solution which allows simultaneous production of electricity and heat. Due to cogeneration units, not only it is possible to lower the costs associated with heating and electricity producing, but it can also generate it in a way that is efficient and environmentally friendly. The fact that the combustion process in the gas turbine consumes only a small part of the oxygen from the intake air flow makes possible the application of a supplementary firing (afterburning) for increasing the steam flow rate of the heat recovery steam generator. A new patented afterburner installation was proposed, for use in cogenerative applications (**Figure 17**) [22].

This study focuses on Stage I of the afterburner (**Figure 18**) for which a special experimental installation was designed. Here experimental measurements and numerical results of mean velocity and temperature are presented. The velocity measurements are carried out using particle image velocimetry (PIV), and the temperature measurements are performed using Rayleigh spectroscopy. Supplementary, flame front position measurements are presented, obtained with the planar laser-induced fluorescence (PLIF) technique [23]. The experimental setup, closely reproduced by the numerical simulations, consists of a post-combustion system, designed and manufactured at COMOTI and installed behind a Garrett 30–67 gas turbine engine serving as a gas generator. The flame is stabilized by means of a V-shaped flame holder, placed in the gas generator exhaust flow. Methane is injected into the flow upstream of the flame holder and ignited downstream of it, at a location where premixed conditions are reached.

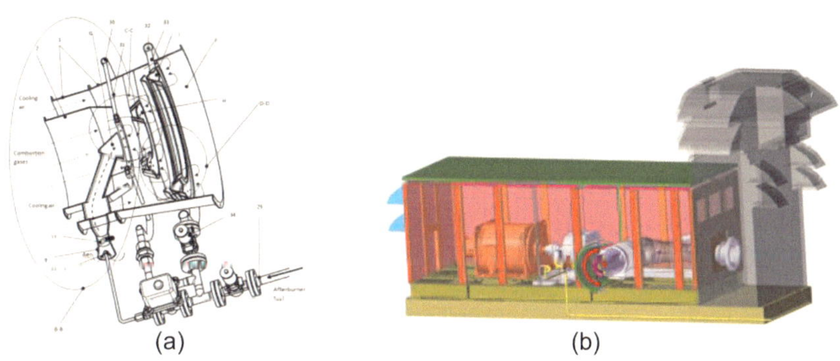

Figure 17.
Afterburning system for cogeneration: (a) partial 3D viewing and (b) assembly of the installation.

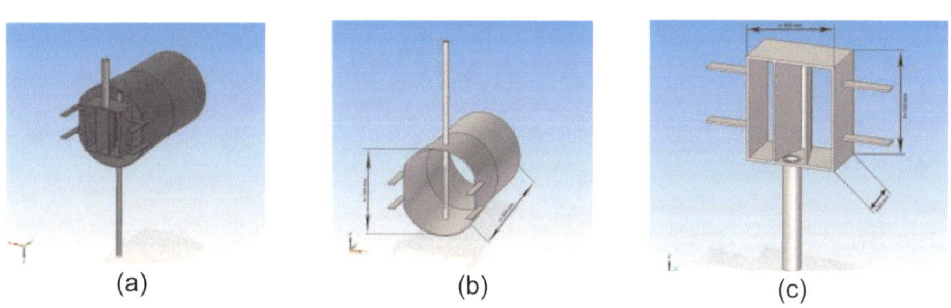

Figure 18.
Experimental setup design: (a) afterburner; (b) casing; and (c) flame stabilizer.

3.3.1 Experimental setup

3.3.1.1 The afterburning system

The afterburning system, shown in **Figure 18**, has the following overall dimensions: length = 304 mm, height = 228 mm, and width = 168 mm. The afterburning system is composed of a casing and flame stabilizer assembly. The casing has 240 mm in height and 304 mm in length. The casing also includes a gas fueling pipe with the following dimensions: diameter = 10 mm and height = 470 mm. The gas pipe has 20 equally spaced holes of 2 mm in diameter. The flame stabilizer assembly includes the actual, "V"-shaped flame stabilizer. The assembly also includes the ignition pipe, of the following dimensions: diameter = 16 mm and height = 115 mm. The post-combustion system presented above can raise the exhaust gas temperature up to a temperature of maximum 1800 K.

3.3.1.2 The gas turbine engine Garrett 30-67

The gas turbine engine Garrett 30-67 [24] was fitted with a pipe that allows PIV flow seeding and transfers the seeded exhaust gas to the afterburning system (**Figure 19**).

3.3.1.3 The PIV measurement system

The experimental program presented here aimed at determining the instantaneous three-dimensional velocity field in the exhaust gas downstream of the post-combustion system. For the measurement a medium intensity laser beam was used, emitted by a Nd:YAG double-pulsed laser (Litron Lasers, wavelength of 532 nm and a maximum output power of 1200 mj), simultaneously with the triggering of two fast charge-coupled device (CCD) cameras that record the images thusly formed. The laser beam is passed through a light sheet optic device that converts the beam to a light sheet in the experimental zone. The time interval between two laser impulses was of 10 μs [25].

3.3.1.4 Rayleigh spectroscopy

Rayleigh scattering (RS) is a nonresonant elastic effect in contrast to the commonly used laser-induced fluorescence. RS is instantaneous and therefore completely independent of the molecules' environment [26].

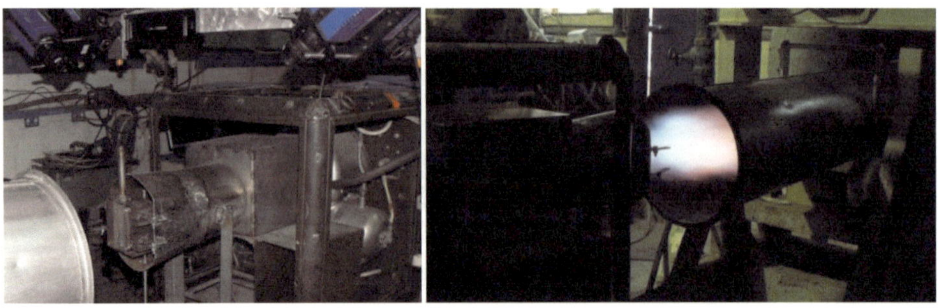

Figure 19.
Experimental setup.

3.3.1.5 The PLIF measurement system

The planar laser-induced fluorescence method is presented here. When laser radiation is tuned to specific wavelengths, it will excite certain species (molecules) within a flame to a higher energy level. Fluorescence occurs when this excited state decays and emits radiation of a longer wavelength than the incident laser radiation. In the atmospheric pressure flame created by the afterburner, quenching is negligible, and the fluorescence signal is proportional to the OH concentration. For the OH measurements presented here, the coumarin 153 dye was used. Laser light has a fundamental frequency of 1064 nm. The fluorescent light photons are captured by an intensified charge-coupled device (ICCD) camera equipped with a filter that lets through only the fluorescent light wavelength [23].

3.3.1.6 CFD software and numerical simulation setup

In order to evaluate the effect of the combustion model validity on the accuracy of the numerical simulation, the reference numerical simulation will use an extended EDM [27] combustion model, implemented in a fully three-dimensional numerical simulation conducted using the commercial software ANSYS CFX. In this study, the shear stress transport (SST) model has been used. The computational domain includes the post-combustion system described in the previous section and extends 350 mm downstream of the bluff body stabilizer. In the transversal direction, the extension measures 300 mm, centered on the post-combustion symmetry axis, and in the spanwise direction, it reaches the edges of the post-combustion chamber.

The following results will be shown. Combustion temperature has a significant effect on NO_X. NO emissions increase, but N_2O emissions decrease, with increasing temperature. Velocity—the speed at which premixed laminar and turbulent flames propagate—is a fundamental parameter in many combustion applications, such as engines and gas turbines. Flame speeds influence knocking events in spark-ignited engines and play an important role in their performance and emissions. OH concentration shows flame front shape and stability. **Figures 4** and **5** present, respectively, the mean temperature and velocity components along the symmetry axis of the afterburner. The length of the recirculation region that is created in the flow by the presence of the bluff body stabilizer is of about 90 mm, and the maximum absolute value of the negative velocity reaches about 25 m/s. The far-field free stream velocity is about 35 m/s.

In **Figures 20–23**, various comparisons between numerical and experimental results are presented.

The mean velocity field of the flow inside the previously defined computational domain and its axial and transversal components are presented.

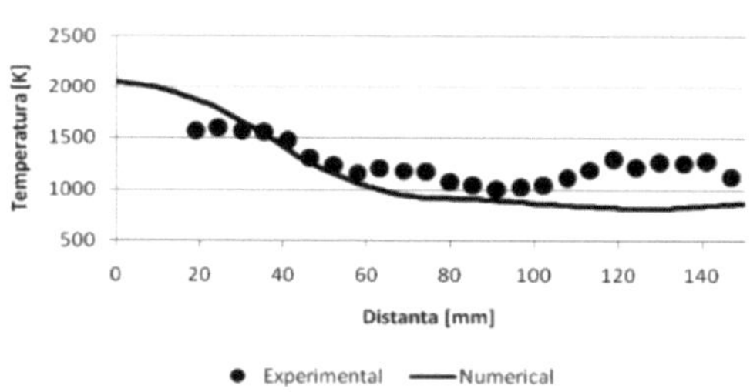

Figure 20.
Temperature profile along the centerline.

Below the variation of the mean OH concentration along the axis of symmetry is presented. The reaction mechanism used in numerically simulation did not allow the concentration of hydroxyl (OH) to be captured so that it is presented experimentally and not numerically validated. It must be noted from the beginning that the OH radical is a very fast radical, which is created and destroyed rapidly in the combustion process. For this reason, its presence can be detected in the flame front only, being a very precise indication on its position. As seen in **Figure 24**, the position of the mean flame front coincides to the recirculation region that forms downstream of the flame stabilizer. The turbulent flame brush, clearly visible in **Figure 24**, determines a significant increase of the mean flame front, as compared to its instantaneous thickness. The turbulent flame brush is an effect of the turbulent intermittency, which causes, through the effect of the turbulent fluctuation of the flame, a given point in space in the flame front region to be part of the time inside the flame front and part of the time outside it. Therefore, the averaging

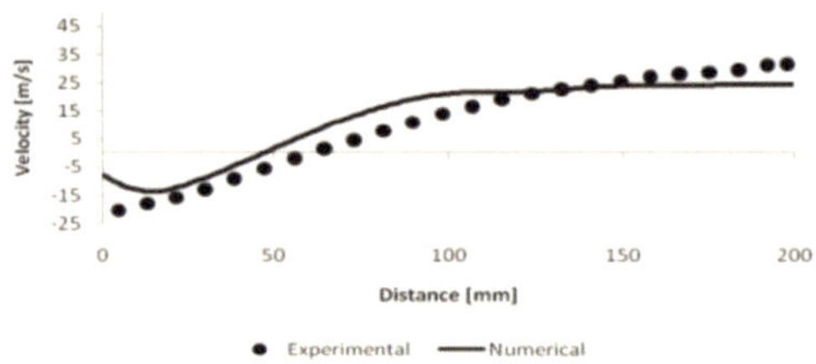

Figure 21.
Axial velocity profile along the centerline.

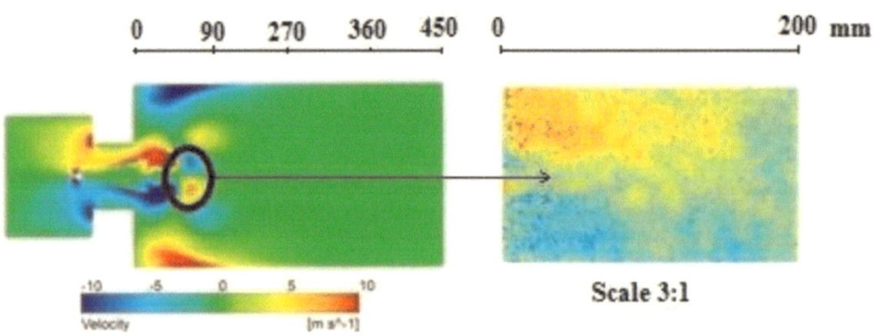

Figure 22.
Left: numerical mean axial velocity field. Right: PIV experimental mean axial velocity field (the same velocity color scale).

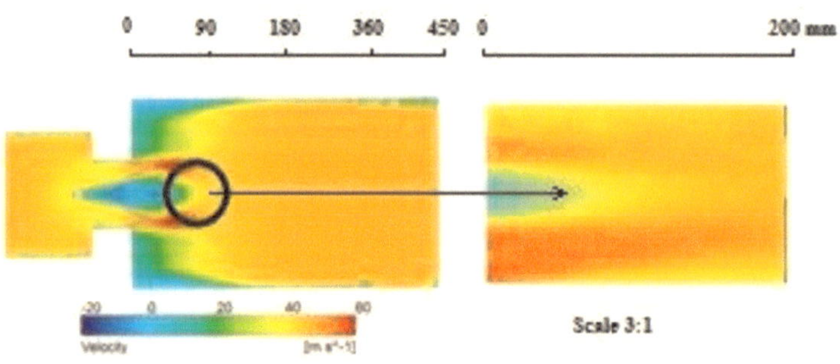

Figure 23.
Left: numerical mean transversal velocity field. Right: experimental mean transversal velocity field (the same velocity color scale).

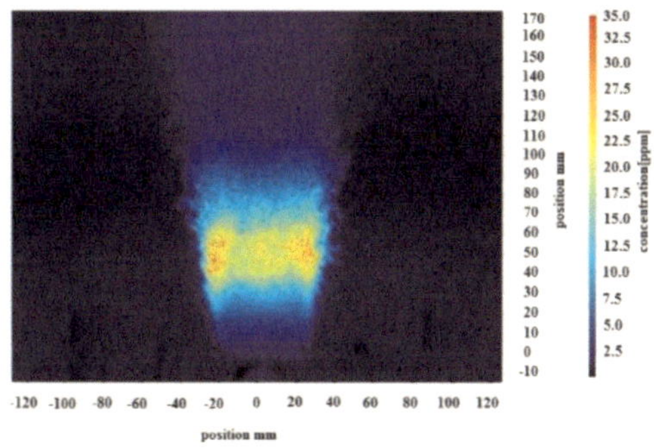

Figure 24.
Mean OH axial concentration.

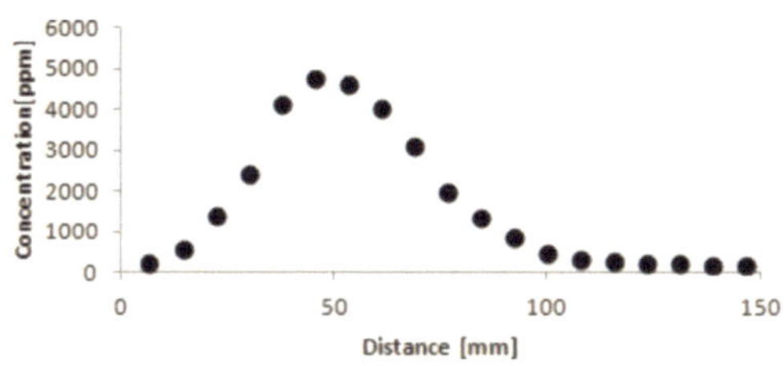

Figure 25.
The variation in the mean OH concentration along the axis of symmetry.

process leads to a region much thicker than the very thin flame front characteristic to a laminar flame, where the mean fields have characteristics corresponding partially to the flame front, partially to the preheating region, and partially to the oxidation zone [28]. In the axial direction in **Figure 25**, the maximum OH concentration, of about 5000 ppm, is reached at about 50 mm from the flame stabilizer trailing edge, and the turbulent flame brush extends between 0 and 100 mm with respect to the same axial coordinate origin.

3.4 Hydrogen use in gas turbines

Hydrogen is studied as a possible fuel in gas turbines due to its high calorific value and promising results in the field of environmental protection. More, hydrogen become actual again, since new ways for producing and transporting it developed lately. One interesting idea is to produce the hydrogen on site by electrolysis, using wind power or solar energy, and to transport it using the existing natural gas distribution network. Combining hydrogen with natural gas strongly influences the combustion parameters, due to the different properties of the mixture. Using the existing equipment would face new problems, like the modification of the flame front, the risk of flashback, and higher temperatures.

By numerical simulation on the combustion chamber of a small gas turbine, for 100% CH_4 and 100% H_2, a clear difference can be observed in **Figure 26**, indicating probable working problems and a possible installation's component damage.

Thus, the idea of searching for a new solution was born, by changing the type of injector and part of the geometry of the combustion chamber.

In the process of designing of the new injector solution, numerical simulations were used, testing and comparing different types, in order to obtain an optimized variant to be produced and experimentally tested later on. For example, in **Figure 27**, two types of geometries for swirl injectors were compared in respect to the flow characteristics and temperature and velocity fields.

The second type of injector shows better volume flame repartition and more intense recirculation zones, benefic characteristics for flame stability, and pollutant emission formation/inhibition. Also the higher swirl number resulted for type 2 is promising, influencing the inlet turbulence, with the mention that other authors of [30] suggest that it is not the case that a rise in the swirl number leads always to an increase in combustion efficiency, but there is an optimum angle for swirling vanes at which the combustion efficiency, temperature, and radiation heat transfer of the flame stand at their maximum. In the same time, the higher local temperatures can

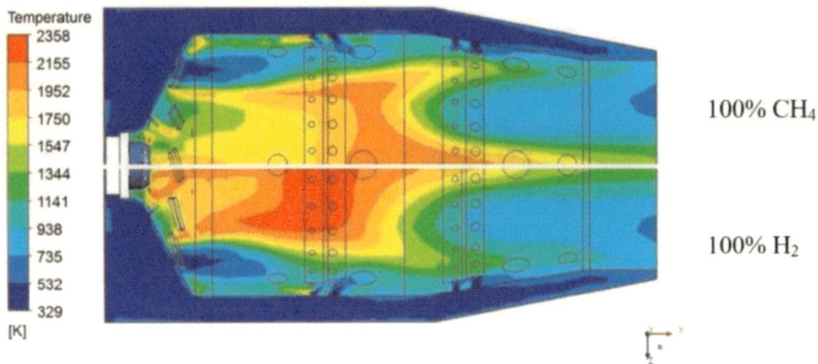

Figure 26.
Temperature field distribution for CH_4 and H_2 combustion.

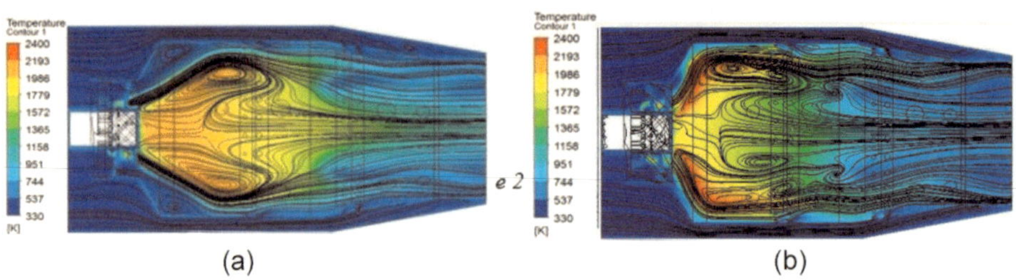

Figure 27.
CFD analysis for two different types of swirled injectors [29]. (a) Type 1 and (b) Type 2.

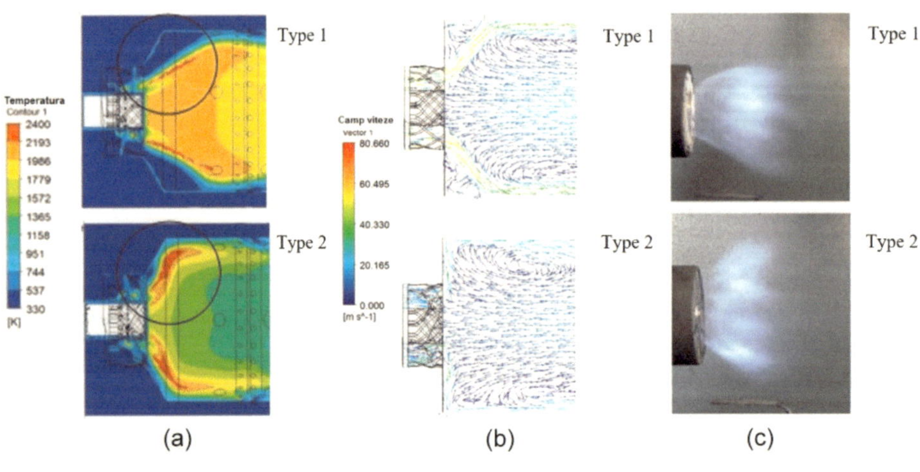

Figure 28.
Comparison of CFD results and visualization during the experiment. (a) Temperature field (combustion simulation); (b) velocity fields (cold simulation); and (c) experimental images.

lead to thermal NO_x formation, a disadvantage that can be controlled by flame cooling technics.

The CFD results show a clear picture of the differences between the two types, with very similar results to the later conducted experiments (**Figure 28**), helping in choosing the right geometry and leading to the optimized version to be manufactured and tested (**Figure 29**) [31].

The design of the new swirl injector was patented, considering some innovative ideas. For example, the convergent shape of the nozzle avoids the uneven velocities between the base section and tip section at the exit of the channel and leads to higher velocity in the exit section than the burning velocity of the fuel, eliminating the danger of flashback phenomenon.

The specific numerical simulation methods also helped for studying the main subject of the work, mentioned above, focusing on the different aspects of combustion of the mixtures CH_4-H_2, with various volumetric proportions (**Figure 30**). The results show relevant data, with very similar results as in the experiments,

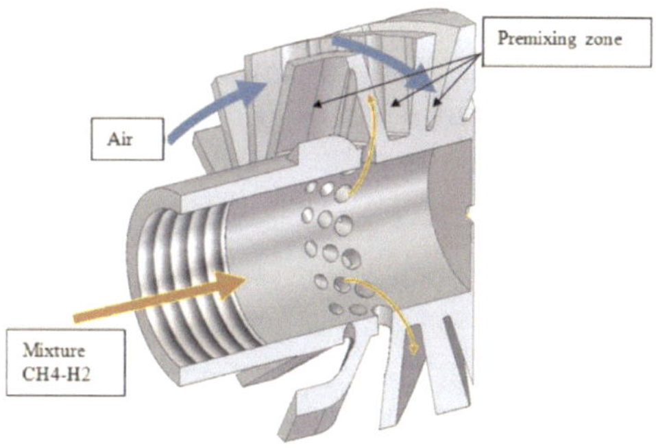

Figure 29.
The swirl injector and working principle [29].

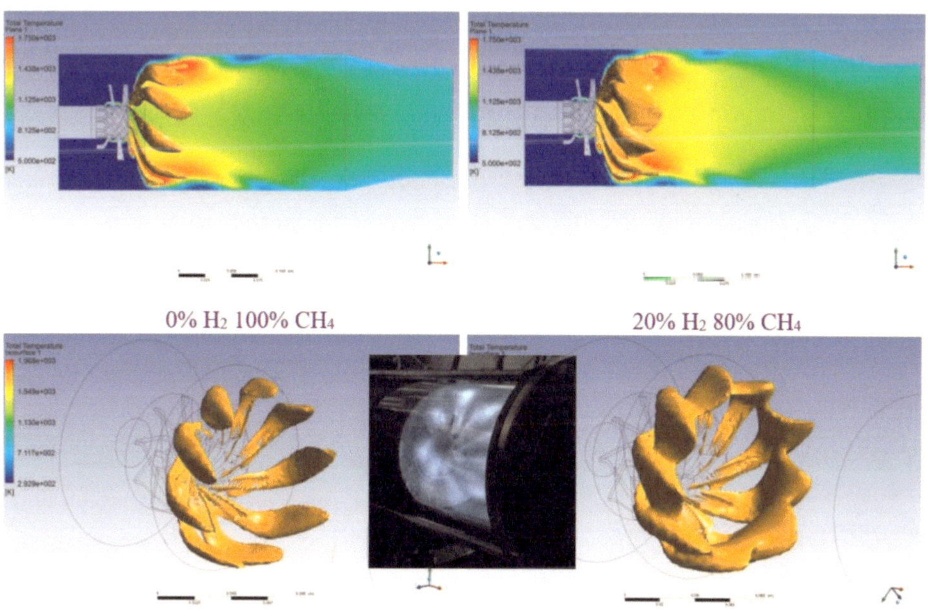

Figure 30.
Isometric section for 0% H_2 and 20% H_2 (airflow 0.04 kg/s, excess air 3.5).

validating the chosen CFD methods and input values, as part of a PhD. thesis [32] and part of a Romanian Research Authority-funded project [33].

For all these RANS simulations, k-ε model turbulence model was used. As for combustion model, the flamelet probability density function (FPDF) model was chosen, because of the CFX available kinetic reaction library, which provided a fast and easy way to mix the two gaseous fuels. This mathematical model has some known drawbacks too, like slightly higher temperature estimations and the absence of NO_x calculation, but this case did not require high precision; the only purpose was just getting a correct image of the combustion process for different cases, in order to optimize the solution and to have a clear preview to the experimental phase.

For this specific case, the combustion of CH_4-H_2 mixtures, there is also an interesting important issue that has no solution for the moment: the existing mathematical models do not take into consideration the very different reaction times of the two fuels in this mixture and thus cannot capture and validate the hypothesis that the hydrogen from the mixture burns faster and consumes at a higher rate from the oxygen required to burn the entire mixture, thus resulting in incomplete combustion of methane.

Considering all these aspects, depending on the studied case and on the purpose of the research, different CFD methods should be chosen, considering the resources, the allocated time, and the requested detail level of the results.

4. Conclusions

CFD represents a powerful tool that can be used even for studying the complex process taking place into the combustion chamber of a gas turbine. There exist many simple models giving relatively good results since of main interest in this field is the combustion temperature which is a measure of the entire gas turbine performance. Still, if someone needs to predict the pollutant emissions, complex models including reaction mechanisms must be employed.

The given examples show that good correlation can be obtained between CFD simulations and experiments at gas turbine assembly level, no matter the size of the gas turbine nor the fuel. An afterburning application was also analyzed since higher velocities are expected, but the results are still conclusive.

Author details

Valeriu Vilag*, Jeni Vilag, Razvan Carlanescu, Andreea Mangra and Florin Florean
COMOTI Romanian Research and Development Institute for Gas Turbines, Bucharest, Romania

*Address all correspondence to: valeriu.vilag@comoti.ro

IntechOpen

© 2019 The Author(s). Licensee IntechOpen. This chapter is distributed under the terms of the Creative Commons Attribution License (http://creativecommons.org/licenses/by/3.0), which permits unrestricted use, distribution, and reproduction in any medium, provided the original work is properly cited.

References

[1] Boyce M. Gas Turbine Engineering Handbook 4th Edition, Hardcover ISBN: 9780123838421, eBook ISBN: 9780123838438, Imprint: Butterworth-Heinemann. 2011

[2] Jeni VP. Contributions regarding the utilization of alternative fuels in reusable aviation gas turbines [PhD. thesis]. Bucharest: Politehnica University of Bucharest; 2013

[3] The jet engine, Rolls Royce 1996

[4] Clean Sky. Available from: www.cleansky.eu

[5] Lefebvre AH, Ballal DR. Gas Turbine Combustion. Alternative Fuels and Emissions. 3rd ed. New York: CRC Press; 2010

[6] Pimsner V, Vasilescu CA, Radulescu GA. Energetica Turbomotoarelor cu Ardere Interna. Bucureşti: Editura Academiei Republicii Populare România; 1964

[7] Peters N. Turbulent Combustion. Cambridge, UK: Cambridge University Press; 2000

[8] Peters N. Product-Manual for ANSYS CFX v16. ANSYS INC, Released 13.0. 2015

[9] Porumbel I. Large Eddy Simulation of Bluff Body Stabilized Premixed and Partially Premixed Combustion. Atlanta: Georgia Institute of Technology; 2006

[10] Popescu J, Vilag V, Petcu R, Silivestru V, Stanciu V. Researches Concerning Kerosene-to-Landfill Gas Conversion for an Aero-derivative Gas Turbine, ASME Turbo Expo 2010: Power for Land, Sea and Air; 14–18 Iunie 2010; Glasgow, UK. NEW YORK, USA: ASME. ISBN: 978-0-7918-3872-3

[11] ANSYS CFX Help

[12] Petcu AC. Research regarding the use of camelina vegetable oil as fuel [PhD. thesis]. Bucharest: Politehnica University of Bucharest; 2016

[13] Petcu AC. Technical Manual Shaft Power Gas Turbine Engine model GTP 30-67

[14] Dumitrescu O, Gherman BG, Porumbel I. Importance of a second entrance in a test cell. INCAS Bulletin. 2018;**10**(1):63-72

[15] Carlanescu R, Prisecaru T, Petcu AC, Porumbel I. Numerical analysis of the effect of the fuel composition on the flame characteristics in hydrogen methane diffusion flames. In: International Conference on Jets, Wakes and Separated Flows; Stockholm. 2015

[16] Petcu AC, Sandu C, Berbente C. Numerical simulations of jet-A combustion in a gas turbine combustion chamber. International Journal of Engineering and Innovation Technology. 2013;**3**(2):487-491

[17] Carlanescu R, Prisecaru T, Mangra A, Kuncser R, Florean F, Enache M. The analysis of the combustion of premixed methane-hydrogen mixtures stabilised by an inovative swirl injector. In: 10th Mediterranean Combustion Symposium; Naples, Italy. 2017

[18] Gherman B, Malael I, Florean FG, Porumbel I. Experimental combustion chamber simulation at transient regimes. In: E3S Web of Conferences 85. 2019

[19] Florean FG, Gherman BG, Porumbel I, Carlanescu R, Dumitrascu G. Experimental measurements and numerical simulations in bluff body stabilized flames. In: 1st International Conference

Newchallenges in Aerospace Sciences NCAS 2013; November 7–8, 2013; Bucharest, Romania

[20] German BG, Florean FG, Carlanescu C, Porumbel I. On the influence of the combustion model on the result of turbulent flames numerical simulations. In: ASME Turbo Expo 2012. Combustion, Fuels and Emissions, Copenhagen, Denmark. Vol. 2. 2012

[21] Petcu AC, Gherman B, Florean FG, Sandu C, Porumbel I. Numerical simulations of round turbulent jet flames. In: Starik AM, Frolov SM, editors. Nonequilibrium Processes in Plasma, Combustion and Atmosphere. Moscow: Torus Press; 2012. ISBN 978-5-94588-121-1

[22] Patent number 128845A0, Florean FG, Petcu AC, Carlanescu R, Porumbel I, Sandu C, Carlanescu C. Multistage afterburner installation in a self-ventilated turboengine skyd. Official Bulletin of Industrial Property no. 7/2016

[23] LaVision GmbH. Product-Manual for DaVis 7.2 "LIF in Gaseous Fluids", Anna-Vandenhoeck-Ring 19, D-37081 Göttingen. 2009

[24] Florean F. Research on the afterburner systems using gas fuels [PhD thesis] S.L.: "Gheorghe Asachi" Technical University of Iasi, Faculty of Mechanics; 2013

[25] LaVision GmbH. "Product-Manual for PIV", Anna-Vandenhoeck-ring 19, D-37081 Göttingen. 2009

[26] LaVision GmbH. "Product-Manual for DaVis 7.2, Rayleigh Therometry", LaVision GmbH, Anna-Vandenhoeck-Ring 19, D-37081 Göttingen. 2009

[27] Florean FG, Popescu JA, Porumbel I, Carlanescu C, Dumitrascu G. Experimental measurements and numerical simulations in isothermal turbulent flows, GT2012-69377. In: Proceedings of the ASME TURBO EXPO 2012. Vol. 1: Aircraft Engine; Ceramics; Coal, Biomass and Alternative Fuels; Controls, Diagnostics and Instrumentation. 2012. pp. 283-292. DOI: 10.1115/gt2012-69377

[28] Florean FG, Porumbel I, Carlanescu C, Dumitrascu G. LIF experiments in a turbulent reactive flow using an afterburner. In: 4th CEAS Air and Space Conference–Innovative Europe, Linkoping, Sweden, September 16–20, 2013, Proceedings. pp. 878-884

[29] Carlanescu R, Prisecaru T, Prisecaru M, Soriga I. Swirl injector for premixed combustion of hydrogen–methane mixtures. s.l.: Journal of Energy Resources Technology. 2018;**140**(7). Paper No: JERT-16-1506. DOI: 10.1115/1.4039267

[30] Pourhoseini SH, Asadi R. An experimental study of optimum angle of air swirler vanes in liquid fuel burners. ASME Journal of Energy Resources Technology, 2016;**139**(3). Paper No: JERT-16-1147. DOI: 10.1115/1.4035023

[31] Carlanescu R, Prisecaru T, Kuncser R, Pop EA. The optimization of a swirl injector for combustion of hydrogen fuel mixtures. U.P.B. Scientific Bulletin, Section D: Mechanical Engineering. 2018. ISSN 1223-7027

[32] Carlanescu R. Numerical and experimental research on the effects of hydrogen mixing with natural gas in the combustion chamber of a gas turbine [PhD thesis]. Bucharest: Politehnica University of Bucharest; 2019

[33] COMOTI. Romanian Research and Development Institute for Gas Turbines, "Combustion Chamber with Hydrogen-Natural Gas Mixture Fuel", Romanian Government founded research program, name HIDROCOMB, Contract UEFISCDI no. 76/2014